THE *SPECULUM ASTRONOMIAE* AND ITS ENIGMA

ORIENT AUSTER

Saturnus louffet XXX ior
Jupiter louffet XII ior
Mars louffet II ior
Sol louffet ein ior und VI stunden
Venus CCC und LXV tag und VI stunden
Luna louffet CCCLXIIII tag und VIII stunden
Ignis louffet onne zal
Aer louffet onne zal
Aqua louffet ane zal
Terra stat stil

BOREAS OCCIDENT

Fenix doctorum, vas fundens dogma sacrorum,
Dittus Albertus preclarus in orbe repertus, maior
Platone, vix inferior Salomone,
Corona phylosophorum, artium magister, dux theologorum.

Dominus Albertus magnus
Diser gottes knecht der grosz albrecht
über alle meister wol geleret
Albrecht der grosz In den got sin wiszheit beslosz
der kunsten ist er geeret.
felix dogma
doctor sacrorum
vas
vundens

Photo: Salzburg, Universitätsbibliothek, ms. M III 36, f.243v (15th century). With permission.

Cfr. Jungereithmayr A., Feldner J., Pascher H. P., *Die deutschen Handschriften des Mittelalters der Universitätsbibliothek Salzburg*, Wien 1988, pp. 211-213.

BOSTON STUDIES IN THE PHILOSOPHY OF SCIENCE

VOLUME 135

PAOLA ZAMBELLI

THE *SPECULUM ASTRONOMIAE* AND ITS ENIGMA

Astrology, Theology and Science in Albertus Magnus and his Contemporaries

SPRINGER-SCIENCE+BUSINESS MEDIA, B.V.

Library of Congress Cataloging-in-Publication Data

Zambelli, Paola.
The Spectrum astronomiae and its enigma : astrology, theology, and science in Albertus Magnus and his contemporaries / by Paola Zambelli.
p. cm. -- (Boston studies in the philosophy of science ; v. 135)
Includes translation of: Spectrum astronomiae.
Includes bibliographical references and index.
ISBN 978-90-481-4098-5 ISBN 978-94-017-3467-7 (eBook)
DOI 10.1007/978-94-017-3467-7
1. Spectrum astronomiae. 2. Albertus, Magnus, Saint, 1193?-1280. 3. Astrology--History. 4. Occultism and science--History. I. Albertus, Magnus, Saint, 1193?-1280. II. Spectrum astronomiae. English. 1992. III. Title. IV. Series.
Q174.B67 vol. 135
[BF1680]
001'.01 s--dc20
[133.5] 91-24566

ISBN 978-90-481-4098-5

Printed on acid-free paper

TABLE OF CONTENTS

EDITORIAL PREFACE

To the historian of science, the cognitive debates and criticisms over astrology tell a wonderful story. To the philosopher, this 'fossil science' is not only fair game for critical fun but serious as well. An old story, we may say, but questions from the far past may rise again. To understand the formation of the early modern world, we should see, as one strand, how close were the arguments about natural theology and its companion, law-like astrology, to those a few centuries later over laws of nature with their mathematical forms and empirical security. One of the most revealing sources, the fascinating and sophisticated *Speculum astronomiae*, is the focus of this insightful and complex study by Paola Zambelli.

Complex it is because Professor Zambelli is faithful to the text itself; faithful to the debates of medievalist historians of old and nowadays over the authorship and the significance of the work, and even concerning its place in the understanding of Albertus Magnus, that "greatest contributor to the science of observation between Aristotle and Vesalius"; faithful to the subtle interplay of determinism, necessity, possibility, of what is predicted to be, to be possible, to be flexible, and to be freely chosen; and finally faithful to methodological implications of this case study in the history of science for the general historiography of scientific explanations.

To defend astrology against prejudiced and supernaturalist anti-scientism, and to keep an open mind toward the newly challenging but already ancient astrology of careful observations, plausible hypotheses, and the rational presupposition that the universe is well-ordered, such was the purpose of Albert's *Speculum*. It was his guide for the inquiring reader of the 13th century. Did he resolve the puzzle over creative causation, the chain from First Ultimate Cause to nearest immediate cause? from divine (via natural) necessity to either the contingent or the miraculous? or the turn from black or holy divination to open, public and reasoned observation of regularities? Did he persuade readers to respect the Arabic astrologers? And, for us, does Paola Zambelli firmly establish for her readers the deep, essential, *and progressive* role of astrology in the European medieval setting? At any rate, Albert and Zambelli will demonstrate in this book that they are undogmatic but decidedly supported by the evidence they adduce as she cites Albert, he is going "to oppose, not to assert", to offer critical argument and not in a voice of mystery and authority, a lesson for us all, epistemological, metaphysical, and otherwise.

We have then an historical work on the development of a precious phase in the philosophy of the scientific understanding of nature. Recently, Tullio Gregory set forth a magisterial survey of the confrontation of theology and astrology in high medieval culture (*Bull. Soc. francaise de Philosophie 84* 103–1290). He presented the issue sharply: "God's foreseeing vs. freedom, fate vs. contingency, universal order vs. evil and disorder". Was sacred explanation, indeed was divinity itself, subject to "the laws of universal heavenly causation"? Was this the genuine challenge of astrology?

I am so grateful to Paola Zambelli for her work of lucid, genial, and impeccable scholarship, so beautifully presented in this book.

ROBERT S. COHEN

PREFACE

The greater part of this study was written more than twelve years ago, and it was designed to serve as an introduction to the critical edition of the *Speculum astronomiae* undertaken in collaboration with Caroti, Pereira and Zamponi. The work was the result of a seminar held under my direction a few years earlier. When the edition of the *Speculum* and the commentary were in press, I decided to delay the publication of my introduction, due to new material which came to my attention. I felt the need to investigate thoroughly the *Summa de astris* by Gerard of Feltre, a text of great relevance to the understanding of the *Speculum*: I wrote a paper on this *Summa* for the conference held at Paris for the 1980 centenary of Albertus. Then I decided to delay publication further, in order to read the studies to be published on the same occasion. As far as the last point is concerned, I confess that my expectations have not been fulfilled. The various congresses, miscellaneous publications and special issues of journals devoted to Albert have paid no attention to the *Speculum astronomiae*, not even to deny its authenticity. On the contrary, there has been great interest in Albertus's mathematical concerns, even though the editor of his *Commentary on Euclid's Elements* acknowledged that "Albertus is not famous for his special interest in mathematics".

That fact alone might simply have provided an excuse for my indolence: my expectations were largely frustrated when new, unexpected insights were provided by the recent symposium of the diffusion of Islamic science in the Latin Middle Ages organized by the Accademia dei Lincei. From very different points of view, David Pingree and Richard Lemay have taken up and discussed the subject-matter of this study of mine.

Another important novelty had been represented in 1975 by the inclusion in Albertus's *Opera omnia* of a controversial work (the *De fato* previously attributed to Thomas Aquinas) now edited by the Rector of the Albertus Magnus Institut responsible for the *editio coloniensis* of Albert, Paul Simon. This inclusion may appear as paradoxical, in view of the defense of astrology put forward in this dense and complex philosophical text: it should be remembered that Albertus's authorship of the *Speculum* has been questioned on the grounds of the defense of astrology, which characterized the anonymous work as well as the *De fato*. The critical apparatus to this as well as the various other works edited at the Albertus Magnus

Institut, does contain several references to the *Speculum astronomiae*. The need to refer to the *Speculum*, however, has not produced the need to reconsider the case of this "pseudo-Albertinian" work. The *Speculum* continues to feature among Albertus's spurious and dubious works, and has suffered a kind of "*damnatio nominis*" during the recent celebrations. Yet, in the past the *Speculum* has been the subject of a historiographical debate which lasted longer, and was more inflamed than similar cases (see for example the rejection of the *Liber de retardatione accidentium senectutis* and the *Epistula de secretis operibus naturae et artis et de nullitate magiae* from the canon of Roger Bacon's works). The authenticity of the *Epistola* (a pseudo-epigraphic work composed with great accuracy from Bacon's passages and addressed to Guillaume d'Auvergne, one of Bacon's teachers), for instance, had already been questioned a century ago by Charles, Bacon's biographer, and more recently by Thorndike. Their doubts have not been taken up nor discussed by Crombie and North, who in their entry for the *Dictionary of Scientific Biography* do not hesitate to include the *Epistola* in the list of Bacon's definite works. This could not have happened with the *Speculum*, as witnessed by the lively debate that lasted from 1910 up to 1955. To some extent, the cool – if not hostile – reception of our 1977 edition could be seen as a prosecution of the debate, though the two papers of mine, which in 1974 and in 1982 anticipated some of the arguments of the present study favoring Albertus's authorship, have never been commented upon.

The long-lasting exchange of arguments and counter-arguments concerning the authorship of the *Speculum* has not come to an end, as the arguments favoring the attribution to Albertus that Pingree and Lemay put forward in 1987 clearly show. The very length of the debate offers a prime element of interest to this case-study: it is my intention to devote the first part of this work to a thorough reconstruction of this debate. In doing this, I am far from motivated by trivial anticlericalism. From the point of view of the history of historiography, it is very interesting to compare the standpoint defended by a liberal scholar such as Thorndike with the one defended by authoritative representatives of the neo-Thomist school such as Pierre Mandonnet, and, more recently, Bernhard Geyer and J. A. Weisheipl. The concern for the history of historiography is a typical feature of Italian history of philosophy and science. The combined use of primary sources and of the relevant historiographical interpretations enriches the analysis of the texts under examination. The *Speculum astronomiae* is the more interesting in view of the fact that it has been the subject of a tense debate, which started, not by chance, at a time characterized by scientistic and positivistic prejudices, according to which it was unthinkable that the great medieval masters concerned themselves with occultist beliefs.

The study of Albertus's known authentic works, as well as of texts by contemporaries such as Bonaventura da Bagnorea, Roger Bacon, and Thomas Aquinas, makes it clear beyond any doubt that the astrological viewpoint characterized thirteenth-century conceptions of the natural and historical world. The principle of the influence of celestial movements on natural processes was unanimously upheld by medieval theologians, philosophers and scientists. It also constituted the foundation of all the natural sciences, from cosmology to medicine, two disciplines which at the time were put on the same level as alchemy and natural magic. The discussion of, and the attempts to establish the limits of astral influence, occurred when one wished to study the effect of such influence on individual free will and on the events of sacred and universal history. These fields of investigation were undoubtedly dangerous from the point of view of religious orthodoxy and susceptible to bringing philosophical apories. There were very few people, however, even among the most severe critics of astrology active during the thirteenth century and later, who denied all "inclination" of individual temperament towards certain given passions induced by the configuration of birth horoscopes, or the usefulness of great astral conjunctions for the periodization of universal history and the prediction of the tragic events of empires or religions and major natural disasters. People tended to differ on the extent and power of the influence they all admitted. The reading of Albertus's definitely genuine works was undertaken mainly to provide a relevant comparison with the doctrines put forward in the *Speculum*. I have therefore selected a series of texts of great interest for the understanding of the ideas of the great Patron of science, the man responsible for the most important contribution to the sciences of observation during the period between Aristotle and Vesalius. Albertus Magnus was a scholar more eager to collect facts than to organize them in a rigorous system. This attitude also characterized his approach to astrological doctrines, which he listed and examined in an objective way. Even though the *Speculum astronomiae* never became part of university curricula as the *Sphaera* by Sacrobosco did, it enjoyed a very wide circulation, as proved by the fifty or more manuscripts which have survived. The work not only provided a complete and accurate bibliographical guide to astrology, but also offered a clear and exhaustive definition of the main components of the discipline, and in the concluding section surveyed some of the main objections raised against astrology.

In my view, the *Speculum* was written to answer the criticisms against the astrological doctrines which had become prevalent after the translation of the *libri naturales*, including philosophical as well as astrological treatises written by Aristotle and Greek or Arabic authors. The very fact that the work listed so many astrological texts at a time of great importance for the scholastic culture, makes it even more interesting. The *Speculum* enjoyed wide circulation and achieved high authority as a classic systematic account of the discipline and a guide to those wishing to study it. Lastly, the *Speculum* represented the point of reference for those authors who embarked upon a discussion of the foundations and legitimacy of astrology, from Petrus of Abano to Oresme, from Pierre d'Ailly to Gerson and the two Picos.

The chief goal of this study, the establishment of the authorship of this short thirteenth-century text, is well defined and limited. I do however hope that I have succeeded in showing the compatibility of the doctrines put forward in the *Speculum* with the ones defended by Albertus Magnus, and by other important authors active in the thirteenth century. I also hope to have made it clear that the system of astrological ideas was integral to the medieval worldview.

A book which has been in the making over such a long period of time has necessarily incurred so many debts of gratitude that it is impossible to list them one by one. In the first place, I am very grateful to all the librarians who have been unsparing of patience and skills both to myself and to my initially youthful students. Mention must be made of all the Florentine libraries, which, when we began, were just drying out after the flood (from which they have still not entirely recovered), and especially of the Biblioteca Nazionale, whose Director was then the immensely missed Emanuele Casamassima. Later a professor of Paleography in my university Faculty, Casamassima was always extraordinarily willing to share his considerable paleographical skills, and two of my collaborators, Caroti and Zamponi, became his students. At that time the reference rooms at the Nazionale were the kingdom of Ivaldo Baglioni and his colleague Omero Bardazzi, both of whom I would like to thank, together with all their successors. From the library of my Facoltà di Lettere e Filosofia, mention must also be made of the sadly missed Renzo Ferretti, together with the entire present staff, especially those at the circulation desk and in charge of the inter-library loan, the latter so essential to a work of this kind. While initially the Domus

Galilaeana in Pisa and the Istituto Nazionale del Rinascimento in Florence provided the microfilm of all the manuscripts of the *Speculum astronomiae*, more recently the Istituo e Museo di Storia della Scienza in Florence has been very useful to us both for its library and for electronic equipment. Twenty years ago, I had acquired the fruitful habit of brief but regular stays in London in order to use the British Library and to do research at the Warburg Institute (University of London), where I had the honor of being introduced to the privilege of working "after hours" already under the directorship of Sir Ernst Gombrich. In more recent years, after a fellowship for the academic year 1983-84 at the Wissenschaftskolleg zu Berlin, I have prefered several times to take advantage of the exceptional competence and helpfulness which Gesine Bottomley, its librarian, and all her colleagues have always shown me.

Among many long-distance consultations, mainly concerning manuscripts, I remember in particular two cases of outstanding courtesy and competence: Doctor Eva Irblich of the Österreichische Nationalbibliothek Wien and a friend of mine, Albinia de la Mare, former Manuscript Librarian at the Bodleian Library and now Professor for Paleography at King's College, London.

To the deeply regretted Marie-Thérèse d'Alverny I am very grateful not only for information of this type, but above all for the generosity with which on the eve of the 1980 Centenary she agreed to read the manuscript of my almost completed book. Other colleagues have also had the patience to read it and have provided criticisms and suggestions which, however, does not make them in any way responsible for the final result, for which I hold myself entirely responsible. They are professors Kurt Flasch, Tullio Gregory, David Pingree, Stefano Caroti and Barbara Faes de Mottoni: for their critical reading and valuable advice I am very grateful. It also nice to remember that Prof. Eugenio Garin and the deeply regretted Prof. Raoul Manselli presented our 1977 edition to a friendly audience in the Biblioteca Laurenziana in Florence.

Professor Pingree and Doctor Charles S.F. Burnett also agreed to collaborate with me and the very competent Doctor Kristen Lippincott on the English translation of the text of the *Speculum astronomiae*, which has been painstakingly and repeatedly revised. Lippincott suggested the frontispiece photo , found for her personal art history research. The translation of my own essay should have been a simpler task, but such was not the case. It was begun and almost completed by Professor Pietro Corsi when, due to commitments more germane to his field as an historian of 19th and 20th century science, he was courteously and efficiently replaced by Doctor Ann Vivarelli, who translated sections of Part II, Part III and several other pages. Since the first translator used a computer it was necessary, in order to maintain this advantage in his absence, to confront various problems. They were resolved by two very kind typists, Mrs. Erika Gilser Caroti and Annarosa Muller Ciappelli, and this preparation for printing was funded by a contribution 0883312534 000 (cap. 106027; mand. 24244) from the Italian CNR (Comitato Nazionale delle Ricerche). But these problems were solved primarily because of the unexpected contribution of a computer genius, Prof. Franco Andreucci.

I am very grateful to Prof. Vincenzo Cappelletti for permission to reproduce in the appendix the text and commentary given together with Caroti, Pereira and Zamponi in the edition of the *Speculum* published in Pisa, Domus Galilaeana, 1977, as well as my article printed in its journal *Physis*, 1974 and now used for Part III. I also thank the most reverend Fathers of the *Recherches de théologie ancienne et médievale*, for the same permission concerning my article of 1982, now to be found in chapters II.2 and II.3. I thank Columbia U.P. for permission to quote from L. Thorndike *History of Magic and Experimental Science*.

I will always be most grateful to all the individuals and institutions mentioned here by name or merely alluded to. However, in closing I would like once more to thank my former students and collaborators since the 1970s, not only for innumerable telephone consultations, xeroxes and checking of page references, but also for discussing more significant questions.

The suggestion to publish this book in English for a larger audience was made to me by Prof. Robert S. Cohen, when we were both fellows at the Wissenschaftskolleg zu Berlin in 1983-84 (itself a very special year for living in Germany, as it was then generously involved in the anti-nuclear-pacifist movement – which helas seems today so far in the past!). He asked me to submit the work for publication in the *Boston Studies in the Philosophy of Science* and I will always be very grateful for this invitation, which gave me renewed energy to bring my manuscript up to date. I wish, therefore, to dedicate this book to Bob and to all the other good friends I made in Berlin.

PAOLA ZAMBELLI
Dipartimento di Filosofia
Universitá di Firenze

31 January 1990

PART ONE

A Historiographical Case-study

CHAPTER ONE

MANDONNET, THE *SPECULUM ASTRONOMIAE* AND THE CONDEMNATION OF 1277

Lynn Thorndike could not understand why "Dominicans seem so anxious to prove that the *Speculum astronomiae* was not by Albert the Great, or to saddle it on Roger Bacon, a Franciscan. It is a very valuable treatise, shows remarkable bibliographical information, and would be a credit to either Albert or Roger".[1] It is true that recently one scholar seemed proud not "to have paid much attention to it [...] because it is a rather trivial compilation interesting only inasmuch as it tells us what was being read at the time it was written".[2] This judgment by an "internalist" historian of medieval science, however, is not confirmed by the fully sympathetic reception the *Speculum astronomiae* enjoyed from the end of the thirteenth century up to and throughout the sixteenth, nor by the persistent historiographical discussion on the authorship and the importance of this work. The *Speculum* is in itself a precious bibliographical guide. It surveys all the information on Greek and Arab astronomy and astrology recently acquired by the Medieval Latin world. The greatest number of translations listed in the *Speculum* are in fact the product of the century which preceded the writing of this work, which I propose to place in the 1260s. The *Speculum* is even more important as an accurate and clear methodological introduction to the various parts and the fundamental problems of astrology, at a time when the discipline, being still a recent acquisition, was at the center of scientific, philosophical, and theological concerns. At the same time, no one denied that astrology had greatly superseded high Medieval divination, as a simple comparison with a Latin text of the twelfth century – for instance, the *Libellus de efficacia artis astrologiae* by Eudes de Champagne – would easily prove.[3]

At the time of Albert the Great, astrology was a commonly accepted discipline, classified in various ways as one of the sciences, often considered in the "quaestiones" of the schools of the Arts, and even in those of Theology, and not only in order to condemn it.

Like all his contemporaries, Albert was, therefore, faced with the reality of this discipline and its theoretical problems. He approached it from the perspective gained by reading the newly translated texts of Greek and Arab astrologers. Several of Albert's genuine texts dealt with astrology in depth, and Albert accepted its principles on many issues, though he did agree with the usual reservations that free will could not be compelled by astral influence, and did repeat Albumasar's definition that "astrological prediction lies somewhere between what is necessary and what is possible", since the influence was mediated and at times hindered by what was called by Aristotle "the inequality of matter" (*materiae inaequalitas*).

In the list of the numerous works attributed to the pen of Albert the Great, the *Speculum astronomiae*, the bibliographical as well as methodological introduction to the texts and the fundamental problems of astrology, enjoyed an enormous success from the end of the thirteenth century through the Renaissance, but it represents a true puzzle. The authoritative *Répertoire des maîtres en théologie au XIIIème siècle*, where Polémon Glorieux has listed the results of the latest scholarly research, mentioned the work we are considering several times. In the article devoted to the author to whom the tradition attributed the work –i.e. Albert– Glorieux maintains that the *Speculum* was "probably by the Chancellor Philippe". Philippe de Thory (or Thoiry), elevated to a canonry in 1270, was Chancellor of the University of Paris between 1280 and 1284. The only written text attributed to Philippe is a memoir in which he justifies his actions committed during his chancellorship. His authorship of four sheets of "quaestiones disputatae" is still a matter for debate. The hypothesis that Philippe could, in fact, have been the author of the *Speculum*, an idea originally put forward by Pierre Mandonnet and repeated with new arguments by Geyer, can be supported by citations contained in a few, scattered manuscripts. Yet the literary obscurity and mediocrity of Philippe is in sharp contrast with the scientific culture and speculative depth shown by the author of the *Speculum astronomiae*. Glorieux himself, in the section devoted to Philippe de Thory, listed the *Speculum* among the dubious works, adding that it might "also [be] attributed to Albert the Great or Roger Bacon". The last name had been put forward in 1910 by Mandonnet, who created a great sensation by dramatically revising the traditional attribution to Albert. In the article on Roger Bacon, however, Glorieux was forced by the data at his disposal to claim that "the *Speculum* is probably by Albert".[4] So Glorieux completed his vicious circle.

Mandonnet opened the question of the attribution of the *Speculum astronomiae* in an important study published in the *Revue néoscolastique de philosophie*. He concluded without hesitation, however, that the treatise showed the hand of Roger Bacon. Indeed, it exemplified yet again and dramatically Bacon's inconsiderate opposition to authority: in this case, to the famous 1277 condemnation of astrology by Tempier. As Mandonnet put it, "Roger Bacon, who cherished an exaggerated faith in the divinatory sciences, and had written with enthusiasm about them, must have felt particularly called upon by the action of the Bishop of Paris. Inconsiderate as he was, he wrote the *Speculum* [...]. Notwithstanding the moderation of the tone –commendable for Bacon– and the appellative of "friends" addressed to the promoters of the condemnation, the *Speculum* was nevertheless a very serious critical venture, as the work of a private individual who dared to oppose the effects of the episcopal condemnation [...]. The wrong position Bacon soon found himself in, as well as the sanction which immediately followed the publication of the *Speculum*, bring us to suppose that the Bishop of Paris had something to do with the serious subsequent events which overrun the Franciscan writer", namely, his imprisonment.[5]

In order to reach his peremptory conclusions, Mandonnet was unable to find documentary support in the manuscript tradition of the *Speculum* (which never mentions Bacon's name) or in the works and the biographical data referring to the Franciscan friar (which are still today open to a great variety of chronological interpretations). Yet, it would be unfair to propose some Dominican political perversity in order to explain Mandonnet's attribution, such as, for example, suggesting that he might have liked to make the rival order responsible for a work clearly condemned by the positivist spirit of his time. It should be noted that Mandonnet, whose conclusions seem closer to an historical novel than to scholarly research to present-day readers, succeeded in creating a rich and complex argument. His hypotheses induced many worthy students, Alexander Birkenmajer amongst others,[6] to accept his conclusions without further examination ("unquestioningly", as Thorndike put it), though it should be mentioned that Charles H. Haskins, for instance, was hesitant to endorse the theses put forward by his colleague from Louvain.

It is important therefore to stop and examine Mandonnet's arguments in some detail, in order to reconstruct the debate from its very roots. The first issue dealt with by Mandonnet was "the date and the occasion of the composition of the *Speculum astronomiae*". He then examined the question of the attribution of the work, and embarked upon a comparative study between the theses of the *Speculum* and those defended by Bacon and Albert. Mandonnet concluded that the astrological theses deployed in the treatise were extremely rare, and to be paralleled only in Bacon's works. He concluded, therefore, that the Franciscan friar was the only possible author of the booklet.

Mandonnet's identification of the cause and the date of the *Speculum* was probably determined by his predominant historical interest in the condemnation of 1277. This bias led to his interpretation of the "charges" against astrology to which the author of the treatise refers. In fact, the prologue to the work (duly quoted by Mandonnet) declares that the occasion which prompted the author to write the *Speculum* was the current confusion between astrological and necromantic works. The latter, in the opinion of the author of the *Speculum*, "which lack the essential of science, [...] are enemies of the true wisdom (that is, Our Lord Jesus Christ), [...] are rightly suspect by the lovers of the Catholic faith".[7]

The dangerous contacts between black magic (reported to be capable of evoking spirits, and not only the so-called planetary and zodiacal demons) and judicial astrology (which was in any case the foundation of natural magic, and included further problematic elements for Christian orthodoxy) were frequent and topical matter for discussion during the Middle Ages. As a result, both occult doctrines –that is, black magic and judicial astrology– or each of them separately, were often the subject of criticism.[8] As the author of the *Speculum* points out, he was answering one such campaign of criticism: "it has pleased some great men to accuse some other books, which are perhaps innocent".[9] The author was faced by denunciation or polemic and he felt it was urgent to distinguish between acceptable and illicit books, and to list them, "since many of the previously mentioned books by pretending to be concerned with astrology disguise necromancy".[10]

This declaration of intent was in itself a novelty. The long history of formal condemnations against the naturalistic books of Aristotle, for instance, never offered the case of a proposal to introduce correctives and distinctions as the one put forward by the author of the *Speculum*. As had already been the case with respect to the teaching based on Aristotle's texts (*Physica*, *De anima*, and *Metaphysica*), time provided a remedy to the 1277 condemnation of the 219 articles, though they were not quickly forgotten.[11] Edward Grant has argued that "frequent citations of, and implicit allusion to numerous articles of the Condemnation of 1277 should convince us that it was taken seriously throughout the fourteenth century".[12] It seems unlikely that a scholar such as the author of the *Speculum* could have chosen to try a frontal attack and to express his opposition in the harmless terms of a bibliographical disquisition. As late as 1295, eighteen years after the condemnation, when Geoffroid de Fontaines remarked in one of his *Quodlibeta* that some of the 219 articles were self-contradictory, and asked for a revision of the proceedings, he showed considerable courage, to use the expression recently employed by the same historian.[13] It seems that Lull was premature, at least, when three years later, in 1298, he claimed that this condemnation had already been forgotten, suggesting, as a follower of the "via antiqua", that the condemnation should be reissued.[14] The famous, general condemnation of 1277 had been prepared by the one of 1270 and by various "consultations" i.e. public advisories requested by the Master General from some eminent Dominican. But the rapid oblivion at least of the articles condemning Saint Thomas Aquinas, and the obvious influence of some of the condemned ideas in debates raging at the end of the thirteenth century, clearly shows that it did not fare better than the repeated and always forgotten decrees against the teaching of Aristotle. If any action was taken to stop a condemnation, this could only happen before the official proclamation, during the preparatory period. It was then possible to put forward accusations and defenses, censorships and apologies. That this was the case is clearly shown by the *Errores philosophorum* by Giles of Rome, published just before the condemnation of 1277, which, in the chapters devoted to Aristotle, Avicenna and Al Kindi, includes several astrological propositions.[15]

It should also be noted that the author of the *Speculum*, whose polemical commitment becomes more explicit and open as the treatise develops, never suggests any modification or withdrawal of the episcopal decree. Moreover, there is nothing in the work which could possibly be used in support of such a drastic request. On the contrary, the author appears to argue openly against those who might wish to reach a condemnation of astrology. He is convinced that the science constituted an essential contribution to epistemology: "No human science attains this ordering of the universe as perfectly as the science of the judgement of the stars does".[16] The author is openly polemical against those who would have liked to suppress astrological texts, but felt that regardless of present suppression they would soon have been taken up again:

> And these are the books, which if they are removed from the sight of men wanting [to study them], a great and truly noble part of philosophy will be buried at least for a certain time, that is, until it would rise again due to a sounder attitude; for, as Thebit, the son of Chora, says: "there is no light in geometry when astronomy has been removed". And the readers of the aforementioned books already know that not even a single word is found in them that might be or might seem to be against the honour of the catholic faith; nor, perhaps, is it fair that those who have never touched these [books], should presume to judge them.[17]

It is rather unusual – to say the least – to find a text so full of recriminations and sharp allusions to the ignorance of the judging authorities to have been published after the promulgation of an episcopal condemnation. Moreover, this feature suggests that the pretext for the conception, writing and publication of the *Speculum* was not the aftermath of the 1277 condemnation. In fact, confirmation of this opinion can be located in the text itself. In chapter XII, after having criticized a book on necromantic images "and others equally damned, which no one having a healthy mind would ever excuse", the author repeats the distinction between necromantic texts and scientific works on astronomy and astrology:

However, the occasion having been [provided] by them, as has been said, many of the aforementioned books, [some being] perhaps innocent, stand accused [and] even though their accusors may be our friends, we must, nevertheless, honour the truth, as the Philosopher says. I swear, however, that if I say anything that I wish to use in defense [of these books], I do not speak as in a determination [i. e.: conclusion], but instead [I speak] in opposition, offering exceptions [to present opinion, so as] to provoke the mind of those who are reaching a decision to pay [careful] attention [to the criteria they are using] for their conclusion[s].[18]

Once again, this important passage – which exactly reproduced the sentence of the incipit of the treatise, and which opens the theoretical part of the *Speculum* after the copious and accurate bibliographical section – clearly refers to accusations and never mentions condemnations. The author addresses the "accusers" as his peers, and indeed as friends with whom the philosopher had the duty to discuss without impediments. As he notes, Aristotle had done so with Socrates and Plato, never putting friendship above truth. But this approach could never have been followed by a magister or a friar – either Franciscan or Dominican – discussing these matters with his bishop or his chancellor who, in his dogmatic pronouncement, was the interpreter of a superior Truth. The author of the *Speculum* does not feel less dignified than his opponents, nor less credible. He insists on the dialectical, undogmatic character of his theses. He does not settle the question, but feels rather that the discussion is still open. He is going "to oppose, not to assert", as he repeats in chapter XIV.

It is unquestionable that the condemnation of 1277 addressed astrological beliefs with a depth and breadth far superior to what even Mandonnet noticed. Nevertheless, these sections of the condemnation simply reflected the large circulation of astrological ideas and, as far as the interpretation of the *Speculum astronomiae* is concerned, cannot be taken as the provocation which the author of the treatise was answering. Indeed, starting from the twelfth century and the early translations of the great Arabic astrological treatises, several documents voicing similar views can be found.[19]

These older astrological documents were probably responsible for the most drastic and subversive of the theses condemned in 1277: "If the sky stopped, the flame would not rise from the wick, since there would be no God". Within this conception, which excluded the possibility of chance (art. 21: "there is nothing that happens by accident, when all causes are considered"), and even the gratuitous nature and the freedom of God's will, since He was seen as compelled to use the celestial bodies (art. 38: "God could not make first matter except through celestial bodies"; art. 59: "God is the necessary cause of the motion of the superior bodies and of the conjunction or division happening in the stars") the 1277 edict condemned a view of the Heavens which was alien, and indeed opposed to the ideas expressed by Albert the Great:

> That the celestial bodies are moved by an intrinsic principle, which is the soul; and that they move through the agency of a soul and a virtus appetitiva, like animals do; and indeed like an appetens animal moves, so does the sky.[20]

CHAPTER TWO

FURTHER CONDEMNATIONS, DEBATES AND "CONSULTATIONES"

In 1271, when Albert together with two other eminent Dominicans, Thomas Aquinas and Robert Kilwardby, was consulted by the General of his Order, he was the most outspoken of the three in his denunciation of this view.[1] In the articles discussed in 1271, the Aristotelian "intelligences" had been identified with the angels of the Scriptures. In 1277, the soul of the heavens was condemned together with the Aristotelian intelligences, though the latter beings were not specifically identified with angels. This ruling reflected not only a certain terminological shift – which was perhaps due to the rhapsodic nature and origin of many of the condemned articles – but revealed an unusual and interesting application of the doctrine of the intelligences in solving one of the major contradictions of the astrological doctrine: namely, the question of free will and what "influence" or effects the stars might have on it, as in the 1277 condemnation (art. 161). Indeed, the astrologers condemned in 1277 did not show the sort of caution demonstrated by Albert and Roger Bacon (not to mention Ptolemy, Albumasar, and several other astrological authorities), by excluding free will from the range of astrological determinations. They claimed "that will and intellect are not moved and put in action by themselves, but they are moved by an eternal cause, that is, celestial bodies".[2] This meant that because of the variety of *loci* and of *signa*, will and intellect are determined by "the necessity of events" (art.142). It followed that "in men [there were] different conditions not only concerning spiritual gifts but also earthly goods" (art. 143). "Health, illness, life and death" (art. 206) were dependent on the heavens, since the will or the healing power of the doctor also depended on it (art. 132). We know that this type of event[3] was regarded by critical supporters of astrology as merely natural; but in addition, man's free will was sometimes also subjected to the stars: "Our will is subjected to the power of the celestial bodies" (art. 162).[4]

The distinction between the corporeal level, which was admittedly subjected to the natural and determining influence of the stars, and the spiritual level of man, which was usually considered as free, was used here not only to diversify, but, more subtly and paradoxically, to support the thesis of determination on both levels: "The intelligence moving the heavens has an influence on the rational soul, in the same way in which the body of heaven has an influence on the body of man" (art. 74).[5]

The astrologers condemned by Étienne Tempier were practitioners of "interrogationes" as well as of birth horoscopes. The former were used to put questions as to the "intentions of men and the change of intentions, [...] the events occurring to travelers and pilgrims, the capture or the restitution of prisoners" (art. 167).[6] The latter horoscopes were decidedly opposed to the principle of free will:

Because of the order of superior and inferior causes, in the hour of man's generation, in his body and, as a consequence, in his soul, which follows the body, is inherent a disposition which inclines him towards certain actions or events.[7]

In fact, the distinction between these two fundamental branches of astrology, birth horoscope and "interrogationes", had been a recurrent topic for several centuries, and had been the subject of a series of special treatises.[8] As was the case with other doctrines condemned in 1277, the thesis discussed here was a rather common one: what might, perhaps, confer a certain originality to it, thereby allowing for a possible identification with other documents, could be the formulation offered by Tempier. Yet, his formulation had nothing in common with the one in the *Speculum*. On the contrary, this text is much nearer to the objection expressed by Tempier against article 207, dealing with birth horoscopes: "This is wrong, unless it is said referring to natural events and as a disposition".

The author of the *Speculum astronomiae* insists on denying that astral influences were strictly deterministic; he defines them as mere dispositions or inclinations. A far greater sphere of action was given to natural events, and there was no prevision attempted for non-natural categories of phenomena. The same approach was taken by Albert in *De quindecim problematibus*, a text certainly attributable to Albert and edited by Mandonnet, which represents the author's answer to the questions addressed to him in 1273-1274 by Giles of Lessines. The first thirteen questions corresponded to the theses condemned by Tempier in 1270, which were again proscribed in his great condemnation seven years later. It is interesting to note that contrary to his arguments developed for the dating of the *Speculum*, this time Mandonnet claimed that Albert's silence concerning the two condemnations clearly dated the text to a period prior to 1270 (whereas Van Steenberghen and Pelster have preferred a slightly later date).[9] In any event, two of the thirteen questions discussed by Albert reveal the contemporary tenor of the astrological debate, and confirm the coherence of his approach towards the issue. In the *De quindecim problematibus*, he is clearly well acquainted with Greek and Arabic astrological sources (the *Quadripartitum* by Ptolemy, with Hali's commentary; the pseudo-Ptolemaic Centiloquium; Alpetragius; Abubacer and Hermes. He places particular emphasis on the fundamental text "which is called *Alarbe* in Arabic, and *Quadripartitum* in Latin". It is interesting to note that the same expression is used to describe this work in the *Speculum* and in one of its sources, as will be shown below. The *Quadripartitum* offered Albert ammunition to refute the third thesis, "that the will of man desires and chooses by necessity", "or, if they signify, it is determined by destiny and the constellations". Albert makes further reference to the *Quadripartitum*, when he writes

that destiny, deriving from the constellation, cannot impose necessity due to three reasons. One of these is that the influence is not exercised directly, but through a medium, the inequality of which can act as an impediment; the second reason is that the influence acts on newborns through accidents and not in itself; for it acts through first qualities, which in themselves do not receive the virtues of the stars; the third reason is that it acts where it does through the diversity and power of newborns' matter, which matter cannot receive heaven's virtues in a uniform way and as though they are in the heavens.[10]

The fourth article of *De quindecim problematibus* examines the more general thesis "that whatever happens in the inferior regions is subjected to the necessity of the celestial bodies". In discussing this thesis, Albert relies on the *De generatione et corruptione*, and lists further causes for change in the bodies of the inferior regions:

Though the motion of the sun and the planets on the ecliptic circle is the cause of generation for inferior things, and the retrograde movement on the same circle is the cause of their corruption, and the periods of generation and corruption are equal, nevertheless inferior things do not attain an equality and order for their period due to the inequality and the disorder of their matter. Who can doubt therefore that the intention [plan, resolution] of man is less unequal and disorderly than that of nature? [Human] intention is far less subject to necessity than is nature.[11]

Albert was convinced that the celestial influence was an important cause of change in sublunar bodies, yet it was not the only sufficient one:

Superior beings do not impose necessity on inferior ones. No one among astrologers ever said so. If this were the case, free will would be lost, choice would be impossible, and there would be nothing contingent in the widest sense of the term: a very absurd tenet indeed".[12]

Albert had no doubt that absolute determinism (which was about to be condemned by Bishop Tempier) was absurd and heretical. He was also aware, nonetheless that so radical a thesis had never been defended by astrologers, not even with respect to natural phenomena, since it was essential to take into account the unequal disposition of matter. Absolute determinism could never be asserted in the case of man, his rational soul or his will. Albert's conception of the human will was based on a view of the microcosm and its partitions which has several points in common with article 74 of the condemnation of 1277 quoted above:

According to philosophers, the soul of man is the image of the world; for this reason, in that part where the image of the first intelligence and of the first cause is, it is impossible that the soul be subject to celestial motions. But in that part where it is in the organs, though it is moved by sideral sparkings, yet it does not attain the necessity and order of superior beings; and so not even in that part is it subject to necessity or in the control of superior beings.[13]

As we have seen, Albert was convinced that, in principle, only natural events were foreseeable astrologically, though not in a necessary way, but only in disposition. At times, however, the *Speculum* also discusses spiritual events, and embarks on the horoscopes of religions, a theory which is subsequently defined as "an elegant testimony of the true faith". The doctrine of horoscopes as applied to religions was commonly regarded as highly heretical. It probably escaped the attention of Tempier, who otherwise would not have missed the chance of condemning it. On the contrary, if one supposes that the *Speculum* was written after the condemnation, as Mandonnet argued, this inclusion of the horoscopes of religions would have contradicted the intention of defending astrology from these charges of heresy, as Mandonnet supposes of the *Speculum*. Indeed, why should the author include and defend a well-known heretical feature of astrology which had not been listed and condemned by Tempier? Why should he, in other words, provide new ammunition to the censors of the discipline? Instead he should have devoted his attention to a slightly different doctrine, of Pythagorean origin, which is related to the same cyclical ideas concerning the vicissitudes of religion and civilizations.[14] Indeed, this doctrine of the "great year" was firmly condemned in article 6:

When all celestial bodies go back to the same point, which occurs every 36.000 years, the same effects which occur now will be repeated.[15]

Though Albert disliked the Pythagorean doctrines, also as far as the question of the animation of the stars was concerned, he did not forget the issue of the "great year". He discusses it in the *Summa theologiae*, a text certainly composed after 1277, in which he examines the theme of man's free will and of the cyclical conflagrations of the universe as developed in the *De natura hominis* by Nemesius of Emesa (a work that was wrongly, but commonly attributed to Gregory of Nissa).[16]

As the Stoics say, when planets return, with respect to their latitude and longitude, to the same sign in which each planet was at the beginning, when the world was first created, in those cycles of time there will occur the burning and the corruption at what things there are, and the world will be restored again to the same state it had from the beginning, as will each of the stars *figens* in the previous cycle in longitude and latitude; hence a new world will be similarly brought into being. And once again there will be a Socrates and a Plato, and each man with his same friends and compatriots; he will persuade them of the same things, and he will have with them the same conversations; all towns and villages and estates will be similarly restored as they were before.[17]

Following this Pythagorean doctrine, only gods and demons were free from cyclical events; indeed, their power to foresee events was based on their continuous existence, and on their will to follow the repetition of the cycles:

[The Pythagoreans] also add that there are gods (be they corporeal or incorporeal, coelestial or terrestrial or infernal) who are not subject to this corruption which affects mortals, since once they have completed one cycle, that is, once they knew it perfectly, they also knew all those things which will happen in all the cycles which occur thereafter. For they also say that there are no further events in the future save those which have already happened before, but all things, down to the smallest detail, occur similarly and unchangeably in one cycle just as in every other. They also say that the time of one cycle is 36.000 years, and they call it the "great year". In this year, as Aristotle reports in the first book of the *Metaphysics*, the celestial gods swore they would have come back to the same beginning of the cycle, and that they would repeat a cycle similar to the previous one. And since [the Pythagoreans] thus located fate and fortune in the celestial gods, with sacrifices and prayers they paid hommage to fate and fortune instead of the gods.[18]

In fact, demons too were subject to destiny since they did not possess the dignity of gods. They reflected an anthropomorphic view of astrology alien to the Aristotelian mentality of Albert. It may have been for this reason that he decided not to discuss this matter at length in his works, nor in the *Speculum*.

One last remark will conclude the present examination of the *Speculum* and of the condemnation of 1277. In the prologue to his compilation, Tempier makes explicit reference to very few books, which were certainly insufficient in number to provide adequate references to all the propositions he condemns. Together with the *De amore* by Andrea Cappellano, he loosely mentions "books, rolls or notebooks on necromancy containing experiments of diviners, invocations of demons or spells dangerous for souls". It is far from certain that these books of diviners, invocations and magic formulas can be identified with the more specialized books on celestial or demonic images listed in chapter XI of the *Speculum*, as Mandonnet has suggested. In any case, the defense of these themes by the author of the *Speculum* is insufficient and not exhaustive. Moreover, it is strange that an author so keen to list the bibliography of his subject, would have missed the opportunity to quote the third precise reference made by Tempier, namely, to the *Estimaverunt Indi [...] Ratiocinare ergo super eum*, a geomantic treatise Hugo of Santalla translated from the Arabic.[19] This seems even more strange, considering that this text was very near to the interests and polemical intentions of the author of the *Speculum*, who in his general conclusions spoke with scorn of geomancy, as well as of other elementary forms of divination based on the four elements.

Mandonnet concluded the first point of his demonstration by claiming that the date of composition of the *Speculum* "immediately follows the decree of March 7, 1277". Indeed, he continued, the text was written "quite early in the year 1277". A confirmation of this assertion was the mention in the work of an Aristotelian thesis concerning celestial intelligences cited by Albumasar. The author of the *Speculum* claimed that the thesis could not be found in any known Aristotelian works, suggesting that it might be discussed in the books of the *Metaphysics* still awaiting translation. By comparing this citation to two similar passages in the *De unitate intellectus* and the *De anima* by Thomas Aquinas, Mandonnet argued that the reference to the Aristotelian doctrine confirms that the *Speculum* was written sometime after 1270 (the date of 'publication' of the first of the two books by Aquinas) since "the information our author relates is not original. It is derived from Saint Thomas". Now, if Thomas knew of the existence of the Aristotelian books yet to be translated "by his co-religionist William of Moerbeke, who following his request had undertaken the task of collating the existing translations of Aristotle with the original Greek texts, as well as of translating the books still in the original",[20] there is no reason to suppose that the author of the *Speculum* could not obtain the same information. Mandonnet was able to argue in this way only because he had already decided that the author was not Albert, but was Roger Bacon.

This issue –which is still open– was taken up again by Thorndike and Bernard Geyer. It is my opinion that the reference to the Aristotelian doctrine cannot be understood by comparing it solely to "the information first announced in the *De unitate intellectus*" by Thomas –which in any case was not a bibliographical bulletin– but should be seen in the context of the chronology of the previous translations of the *Metaphysics*, and the way in which these were used by Albert. As will be shown, this approach will alter considerably the dating of the *Speculum astronomiae*. The argument takes its lead from information provided by Geyer in his critical edition of Albert's *Metaphysics*. The editor tells us that the commentary Albert prepared in 1262-1263 followed the "translatio media", which he used constantly between 1250 and 1270. This translation was also employed by Thomas for his *Quaestiones de veritate* (1256-1259). It included book XIV (Nu), but not book XII (Lambda), which was only later introduced by William of Moerbeke in his "translatio nova".[21] Following the hypothesis that Albert was the author of the *Speculum*, the fact that the translations of both books were unknown suggests that Albert wrote the treatise before starting his commentary on the *Metaphysics*, that is, before 1262. It was indeed this problematic quotation from Aristotle that made Thorndike suggest an even earlier dating for the *Speculum*, proposing the year 1256 for the composition of the work, a date which reflected the current views on the chronology of the Albertine commentaries in his time.[22]

The second issue tackled by Mandonnet concerns the attribution of the work according to manuscript and literary evidence. It is interesting to note that Mandonnet was convinced that "the question rests, with any seriousness, only between Albert the Great and Roger Bacon". This is rather curious, since the name of Bacon, absent from the manuscripts, was introduced, according to Mandonnet, by later authorities, such as Giovanni Pico and Gabriel Naudé. The historian probably did not know that one manuscript put forward the name of Thomas Aquinas, an attribution certainly more ancient and plausible than the one of Bacon: an ignorance which might save him from the accusation of having fallen victim once again to his Dominican partisanship.[23] Mandonnet also listed the attribution to "Philippus cancellarius parisiensis" found in some codices and noted by Echard and Borgnet. This hypothesis was immediately discarded by Mandonnet, on the grounds that it was clearly impossible that the chancellor who took part in the condemnation procedures could be the author of a text intended to oppose the condemnation itself. According to Mandonnet, this attribution can be explained "by the very anonymity of the *Speculum*"; indeed, the proclamation of 1277 had established that suspicious books should be given "to us [Tempier] or to the chancellor". Mandonnet pointed out, however, that in 1277 the chancellor was Johannes Aureliensis, and not Philippe de Thory, who took office after Johannes, from 1280 to 1284. Nevertheless, he claimed that this "false attribution" supported his thesis concerning the chronology of the work: "some did not ignore the relationship between the *Speculum astronomiae* and the condemnation of 1277".

Mandonnet did not pursue an analysis of the manuscript tradition of the *Speculum*, though by 1910 some spade-work had already been done by Steinschneider, whose results were later improved with additions by Cumont and Thorndike. He did not know therefore that the predominant attribution in the manuscript tradition was to Albert. He listed some such attributions, not taken from the codices but from catalogs of Albert's work (Tabula Stams, Peter of Prussia, Nicholas of Dacia, Laurent Pignon), as well as the "critique of dubious value" supplied by Jean Gerson. He was also aware that the authoritative study on Albert by F. von Hertling "claimed that the rights of Albert [to the authorship of the work] had been wrongly disputed". Mandonnet had started his disquisition by noting that the *Speculum astronomiae* "has been almost universally attributed to Albert, and has found a place in the edition of his works"[24], but concluded that the authorities quoted in support of the attribution were "for the most part late ones". This is of course true for a only few of the attributions. The Tabula Scriptorum O.P., i.e. the so-called *Tabula Stams*, dates to 1310, and contains the oldest list of Albert's works.[25] Jean Gerson, as chancellor of the University of Paris, is also an authoritative and trustworthy witness. The data discussed by Mandonnet needs to be integrated with several manuscript confirmations, with information taken from further catalogs of Albert's works, and with the literary tradition.

Mandonnet himself was forced to admit that very few authorities supported the attribution of the *Speculum* to Roger Bacon. In order to emphasize their testimony he abandoned the criterion of antiquity, which he had previously used to negate the attribution to Albert, and appealed to competence: "it is remarkable that those who have studied the questions concerning astrology and magic in particular were the ones nearer to truth". In fact, Giovanni Pico, the expert quoted by Mandonnet, never attributed the *Speculum* to Bacon. In the first of the two passages from the *Disputationes adversus astrologiam iudiciariam* Mandonnet cunningly put together, Giovanni Pico correctly attributed the *De erroribus in studio theologiae* and the *Opus maius* to Bacon. In the second passage quoted by Mandonnet, the philosopher – coherently with his own polemic against astrology – does his best to deny that the *Speculum* had been written, endorsed, or authorized by Albert. But this second passage never mentions Bacon. When answering the objection of a defender of astrology, who had mentioned "Albert, excellent theologian, although a supporter of astrology", Pico declared that not everything to be read in the works of Albert was contrived or maintained as a personal opinion.

If by chance you are going to quote as an objection the work *de licitis et illicitis* [the *Speculum*], where he [Albert] indeed condemns magicians, but approves of astrological authors, I will retort that many believe this to be a work by Albert, but neither Albert himself, nor the book's inscription ever indicates this, since the author, whoever he was, deliberately and explicitly dissimulates his name. For what reason? Because in this book there are many things unworthy of a learned man and a Christian [...] Either Albert never wrote them, or, if he did write them, we must say with the Apostle 'I praise him for other things, but not for this'.[26]

Pico's theological criticism, dictated by his desire to deny the right of astrologers to use Albert as their most authoritative patron, did not introduce historical or philosophical arguments. Moreover, as Mandonnet himself acknowledged, in the conclusion of the second passage Pico does allow that Albert might have written the *Speculum*.

The second, and last testimony Mandonnet quoted in support of his attribution to Bacon, is the one offered by the seventeenth-century bibliographer Gabriel Naudé. It is clear however that he has been unaware of the steps linking Naudé to Giovanni Pico, namely via the *De rerum praenotione* by Gianfrancesco Pico and the *Atheismus triumphatus* by Campanella. He was also unaware of Bayle's critical remarks on this and other similar testimonies. And as far as Naudé himself was concerned, he did not care to find the passage which, according to Échard –the author of the information– "gave back the *Speculum astronomiae* to Roger Bacon".[27]

Mandonnet was aware that the body of data he had gathered was insufficient to support his attribution. He therefore embarked upon "a comparative study of the *Speculum astronomiae* and the doctrines professed by Albert and Bacon". The result of this study, in his view, "offers the maximum of evidence to conclude" that Bacon was the author of the work. Unequivocally, he claimed, the style of the *Speculum* is Bacon's own, though in this work, "written ten years after the *Opus maius*, it is more refined and less prolix". Against the Dominican biographer Échard, who saw "Albert's manner of writing" in the *Speculum*, Mandonnet pointed out that three expressions (very common indeed) in an astrological work – "radix scientiae, sapientia, libri nobiles" – were, in fact, to be found in Bacon.

Clearly more important was a comparison between the doctrinal contents of the works under discussion. Mandonnet claimed that to his knowledge Bacon "is the only ecclesiastical author who, in the second half of the thirteenth century, defended judiciary astrology and all the occult sciences which are more or less directly dependent on it".[28] Today, however, having benefited from the first two volumes of Thorndike's *History of Magic and Experimental Science* (published only thirteen years after Mandonnet's article) and from the many recent studies on medieval astrology and magic, Mandonnet's remark can only be judged by repeating what has been said by Thomas Litt, another Dominican historian: the historiography practiced in Mandonnet's time was totally unprepared to appreciate scientific and astrological issues.[29]

Mandonnet also pointed out that "Bacon was not an uncritical follower of astrology and of other suspicious sciences. The same restrictions and concessions characterized both the *Speculum astronomiae* and his authenticated works, the *Opus maius* in particular. But, these reservations notwithstanding – verbal rather than real – Bacon shows amazing faith in the practice of the divinatory sciences". Those who are familiar with the twentieth-century studies examining the works of several medieval authors interested in science and astrology, know well that distinctions and reservations, or beliefs and practices found in Bacon, are far from "amazing", but were, in fact, common to the point of constituting a topos.

If this is the case, the beliefs and qualifications present in the *Speculum* do not constitute a criterion for the attribution of the work to one or the other of the authors who may have shared the same attitudes. Bacon was not the only author who "rejected with indignation all that belonged to magic, the invocation of demons and idolatry"; who instead "accepted astrology in its true meaning".[30] As we have seen, in his *De quindecim problematibus*, Albert developed a substantially similar "critical" conception of astrology. There too, as in the *Speculum*, "the heavens and the stars, with their motions and the combination of their positions, rule over the inferior things, including singular or collective human affairs. The constellations under which men are born rule over their destiny, their health, their fortune". Even the example quoted by Mandonnet in a footnote, referring to the danger of bloodletting when the Moon was in the sign of the Twins, was rather commonplace in astrological literature. Even more common was the reservation introduced in the *Speculum* in order to save man's free will. According to Mandonnet, "It is true that this undermines his judiciary science; nevertheless, Bacon does not try to reconcile the two elements". After having discussed "ex professo" in chapter XIV, how to reconcile free will with the knowledge of contingent future (a point which, according to Mandonnet, showed acquaintance with discussions then beginning on this issue), the author of the *Speculum* fell back into the identification of divine foreknowledge with celestial "signs": the latter "do not offend free will, since they do not produce what they cause, but only signify events in obedience to divine orders". Here too we are faced with a topical solution to astrological disputes, going back, in this case, to a Plotinian theme. If it is true that this passage displays "a desire for orthodoxy", and an explanation which is philosophically open to criticism, it is less plausible to argue, as Mandonnet does, that it shows "the lack of consistency typical of Bacon's ideas when he deals with philosophical issues", since this incoherence and analogous compromises are also found in several other philosophers. That astrology "reveals what is going to happen in nature and in the social order: famine, floods, earthquakes, war and peace, the occurrence of new sects, of great prophets and of heretics, of horrid local or universal schisms, etc." was a belief common to many, and indeed it corresponded to the content of numberless prognostications. It is incorrect therefore to regard this conception of astrology as simply "the summary or essence of the doctrines put forward by Bacon in the two treatises of the *Opus maius* entitled the *Judicia astronomiae* and *de Astrologia*". That the organization of the *Speculum* is made according to the same subjects (that is, astronomy and judicial astrology) roughly corresponding to the headings of Bacon's treatises does not mean much, since both follow the classical distinctions established by Ptolemy and Albumasar. Finally, it should be noted that the discussion of this distinction betwen astronomy and astrology was by far more clearly argued in the *Speculum* and the *De fato* by Albert, than in Bacon's *Opus maius*.

The *De fato* has been re-attributed to Albert in a recent volume of his *Opera omnia*, on the basis of a reliable codex, and despite the poor manuscript tradition (which by a slight majority favored Thomas Aquinas, to whom it had been attributed in some editions). The *De fato* was written in 1256 at the papal court of Anagni, and offered a clear distinction between astronomy and astrology, similar to that found in the *De animalibus*, the *Super Dionysium de divinis nominibus*[31] and in the *Speculum* itself. One must admit, however, that this distinction is topical, and can also be found in Thomas:

As Ptolemy says, in astronomy we must distinguish two parts: the first one concerns the positions of the superior bodies, their measures, and their passions; this part of astronomy can be reached through demonstration. The other part concerns the effects of the stars on inferior things, effects which are differently received in these mutable objects. Thus, this second part can only be reached through conjecture. It is required therefore that the astronomer who deals with this second part be also, at least in part, a physicist, capable of framing conjectures from physical signs.[32]

One has, however, to admit that such a distinction is topical [33] (it can be found also in Thomas and many other authors). What is interesting in the *De fato* is the fact that this distinction opened the way for a critical examination of the epistemological status of each division of astronomy.

From conjectures, which derive from mutable data, comes a mental attitude endowed with less certainty than science or opinion. Since signs of this sort are in themselves common and mutable, because they do not possess a predictory value [significatum] true in every case or in most of them, it is impossible to deal with them through syllogism. In their essence some of these forecasts vary with various causes.... For this reason it is often the case that the astrologer may say things that are true, and nevertheless his prediction does not occur; his saying was true as far as the celestial disposition was concerned, but the disposition itself was hindered by the mutability of inferior things.[34]

In his comparison between Roger Bacon and the *Speculum*, Mandonnet touched upon a further point of crucial importance from a philosophical and theological point of view: the horoscopes of religions, and especially the one relating to the "nativity" of Christ. In the aftermath of the diffusion of ideas developed by Alkindi and Abû Ma'shar into the Latin West, this puzzling and worrying theme had been taken up by innumerable authors practicing a variety of literary genres: the Roman de la Rose, the last avatar of the Roman du Renard (Renard le contrefait), the *De mundi universitate* by Bernardus Silvester, the *Anticlaudianus* by Alain de Lille and the pseudo-Ovidian *De vetula* (probably written by Richard of Fournival), Herman of Carinthia, the *Liber Hermetis de sex principiis* (which was also based on Firmicus Maternus), the astrologer Guillaume de Reims, the theologian Caesarius von Heisterbach, Robert Grosseteste, Richard Fishacre[35], Roger Bacon and Albert. Following the diffusion of the *Introductorium* by Abû Ma'shar, the idea that the constellation of the Virgin announced the virginal conception of Christ was commonly accepted. For this reason, the fact that Roger Bacon had made a concrete reference to this thesis in his *Opus tertium* (after a passing allusion in his *Opus maius* and in the *Metaphysica*) is relatively unimportant when compared to the analogous passages in the *Speculum* (Chap. XII) and to the wide diffusion of this commonplace. Moreover, as Mandonnet himself has noted, the quotation in the *Speculum* follows the translation of Johannes Hispalensis. The one by Bacon was based on the translation by Herman of Carinthia, even though the English philosopher appears to have been familiar with both versions.[36]

We cannot read Mandonnet without smiling when he criticized a realistic remark by Bacon's biographer, E. Charles, who had suggested that "these doctrines, painful as they appear to us, were common during the thirteenth century; Albert was not immune to them; Saint Thomas expressed reservations, but did not deny the legitimacy of the science". On the contrary, Mandonnet expressed his conviction concerning the authorship of the *Speculum* with absolute certainty: "there seems not to be a single philosophical or ecclesiastical contemporary of Bacon, who maintained similar doctrines. All theologians were hostile to astrology and to the superstitious sciences. Even masters in liberal arts, then busy in following Aristotle, never would have professed doctrines so remote from the teaching of the Stagirite. Bacon, despite his continuous reference to Aristotle, rarely understood the thought of the Greek philosopher, since he accepted the lead of the apocryphal or dubious literature he was the only one in his time still to follow".[37] This evaluation of the relationship between Bacon, his contemporaries and astrology on one side, and Aristotle on the other, is undoubtedly anachronistic and incorrect. Mandonnet appeared to ignore that the Aristotelian exegesis current during the thirteenth century (and indeed from Avicenna to Campanella) had compiled the doctrines expounded in the *De caelo*, in the *De generatione* and pseudo-Aristotelian treatises[38] together with astrological themes. This practice was also common to Bacon, as well as to Albert and Thomas. All three deserve a place in the history of the fortune of pseudoepigraphic Aristotle writings. Albert too, as Mandonnet put it, took the very same "route upon which Bacon was marching alone, with confidence and pride".

CHAPTER THREE

MANDONNET'S HYPOTHESIS: ACQUIESCENCE AND DOUBTS

Despite the fact that reactions were mixed, Mandonnet's study gained a great audience. Birkenmajer and other scholars took for granted the hypothesis put forward by the Dominican historian, and abandoned further examination.[1] But the specialists in the field were soon divided on the issue. Among Franciscans, Father Raymond uncritically summed up Mandonnet's conclusions.[2] Later, Father Vandewalle re-examined the entire issue questioning Mandonnet's basic assumption concerning the dating of the *Speculum*. He, like Thorndike, disagreed that the condemnation of 1277 had anything to do with the *Speculum* (which he considered as having been written before 1277). He felt that the *Speculum* was answering "accusations", not a formal condemnation or "declaratio". Vandewalle concluded that Bacon could not have been the author of the treatise, on the grounds that the name of the English philosopher is never found in the manuscript tradition, and the testimony offered by Naudé is irrelevant, since it was based on a misconstruction of Pico's passage discussed above. Moreover, the style and the anonymity of the *Speculum* did not fit Bacon's personality: the tone of the discussion is so moderate, the doctrines put forward so relatively orthodox, that the *Speculum* could never have led the alleged author to prison. Finally, the doctrinal content of the work "is part of the common scholastic heritage of the time", and is characterized by peculiarities which lead to Albert.[3] Writing soon after Mandonnet, the Jesuit Franz Pangerl quoted authorities such as Gerson, Pierre d'Ailly, and Giovanni Pico, who, without trusting the spurious writings or believing the legend of Albert magus and alchemist, did nevertheless "find in his authentic writings, especially in the *Speculum astronomiae*, several points worth considering".[4]

Among the specialists in the field, Robert Steele, Bacon's editor, vehemently rejected the attribution on stylistic grounds.[5] In his important introduction to the edition of Albert's works, G.G. Meerssemann subjected the *Speculum* to a thorough internal analysis, and concluded that the work "perfectly fits the views Albert expressed in his other works".[6] The arguments deployed by Vandewalle and by Meerssemann convinced the author of one of the "Serta albertina" published in the journal *Angelicum* that the *Speculum* was indeed by Albert.[7] Four years after Mandonnet's article, the

Italian modernist Giovanni Semeria (writing under a pseudonym) published a thorough discussion of the issue in the *Rivista di filosofia neoscolastica.*[8] Semeria championed the attribution to "Chancellor Philippe", though he now identified him with Philippe de Grève, at whose instance Bacon composed the *Epistola de accidentibus senectutis*. We now know, however, that this more famous Chancellor Philippe, author of a *Summa de bono*, probably did not survive the year 1236.[9]

Semeria distinguished weak and strong points in the arguments put forward by Mandonnet. For example, he agreed that authors and quotations referred to by the *Speculum* were present also in Bacon's works, such as a passage from Hali and one from Abû Ma'shar. The definition the author of the *Speculum* offers of himself, as a "man zealous for the faith and philosophy" who started the work "at the command of God" ("nutu Deï"), in Semeria's opinion also fits Bacon. He was convinced that Bacon's style can also be seen in the ironies of the *Speculum*, though the work seemed unusually meek, particularly towards Aristotle. "Finally, the philosophical doctrines of the *Speculum* seem to be expressed more precisely than in the *Opus maius*, where Bacon similarly defended judiciary astrology from the point of view of orthodoxy".[10] Semeria appeared to be better equipped to appreciate the depth of some of the themes of the *Speculum*, such as the hint to the Antichrist surreptitiously introduced under the guise of the advice not to destroy necromantic books,[11] which could have been useful, perhaps, to recognize and oppose him. He underlined the symbolic, not deterministic value of the horoscopes of religions, and the distinction between inclination and determination of man's will.[12] Mandonnet had considered these ideas as indicative of a typically Baconian incoherence, whereas Semeria took the opposite view. He did not consider these views of the *Speculum* as incoherent, nor did he think that Bacon was inconsistent: he was thus ready to return to Mandonnet's argument, maintaining that, in view of its coherence, the *Speculum* might well have been the work of Bacon.[13]

Later the Dominican Bernhard Geyer took up the issue, and unequivocally expressed his viewpoint in the very title of his article: *Das* Speculum astronomiae *kein Werk des Albertus Magnus*. After having offered a perceptive summary of the treatise, Geyer remarked that

> the author of the *Speculum* attempts to approach the problem –then embarrassing– in a totally scientific way, of finding a justification of astrology, and of its reconciliability with the Catholic faith. In doing so, he goes to the limits, and defends whatever position is not clearly

opposed to the Christian doctrine. For centuries, therefore, he has provided the theoretical justification for the astrological delusion of the West. For that reason, the question of the authorship of the treatise has aroused the interest of friends and foes of this doctrine, and still does. To solve the problem, we have to remember that the text was issued anonymously. In view of the many daring theses it contained, the author had good reason to keep the secret. This said, it is clear that the question of attribution simply involves raising the curtain, if possible, and determining the true author.[14]

This remark, based on Pico –who probably read one of the few, but ancient and authoritative anonymous manuscripts,[15]– represented the starting point for Geyer, who not only considered the *Speculum* anonymous and "adespota", but almost "clandestine". In his edition of the Ueberweg textbook, Geyer accepted the attribution of the work to Chancellor Philippe.[16] In the later article on the *Speculum* he acknowledged that the information provided by the Oxonian manuscript which bears the Chancellor's name is not of much use since we know absolutely nothing of Philippe de Thory's scientific views.[17] Geyer did not, however, accept the attribution to Albert, nor what he considered the more plausible attribution to Roger Bacon. The Dominican Historian allowed that anonymity was not typical of Albert, since we know of no other anonymous text by him, and we know that he often engaged in sharp personal polemical exchanges.[18]

In his capacity as editor of the critical *Opera omnia* of Albert, Geyer was forced to concede that the manuscript tradition for the *Speculum* is virtually unanimous ("fast einhellig") in favor of the great Dominican. Regardless of the fact that the attribution to Bacon put forward by Mandonnet had been rejected by many (and with reason), Geyer maintained that the impressive parallels between the *Speculum* and the *Opus maius* still deserved attention, as they showed that the treatise was without doubt nearer to Bacon than to Albert.[19] For this reason he felt that one should reconsider the attribution even after the fundamental contribution by Thorndike. In fact, Albert was less interested in mathematics and astronomy than Bacon and than the author of the *Speculum*,[20] even though he cultivated strong methodological and original interests for the observational natural sciences. The full appreciation of Albert's standpoint is made notoriously difficult by his habit of quoting authorities without making it clear whether or not he was accepting their position. As far as magic was concerned, it was certainly his empiricist inclinations which made him relate the marvels his sources talked about without any critical comment. "For Albert, the *scientia naturalis*, which deals with natural phenomena that can be empirically described and explained in terms of causes, is complemented by the *scientia magica*, which deals with extra-terrestrial causes and with occult natural forces". As far as our immediate problem is concerned, that is, the question of the theoretical premises of astrology, Geyer pointed out that Albert "agreed with Bacon and the author of the *Speculum*. However, concerning the issue of the legitimacy of the practice of astrology, Albert argued in favor of a more limited sphere of action".[21]

Geyer based his conclusion –which will be discussed below– on three passages by Albert. The first, found in his *Commentary to Matthew*, is nothing more than one of the polemical expressions commonly uttered by supporters of astrology against those who practice the art in a cavalier and dishonest fashion. The passage is indeed intended to endorse and to justify explicitly the correct use of astrology:

> If someone should make an astrological prognostication concerning things which are within the order of natural causes, and if his forecast concerns these things only in as far as they belong to the natural order, and if it deals with them only in so far as the first order of nature (which is in the position and the circle of the stars) inclines towards them, he does not do wrong. Instead, he prevents many damages and usefully furthers adventages. However, he who, without considering all data, prognosticates on future events in a way different from the one we described, is a swindler and a rascal and must be rejected.[22]

Those who are familiar with medieval discussions on astrology will recognize immediately that the passage contains a common admonition against those "swindlers" who are discrediting their art. And that the distinction between serious and admissible astrology, and the one practiced by quacks, is part of a typical classification which draws a distinction between *màthesis* (the science which deals with abstract objects) and astrological *mathésis*, and lists various features of magic:

> The magus, enchanter, astrologer, wizard or necromancer, *ariolus*, *haruspex* and diviner differ from each other, since the magus is really nothing but 'magnus', because he, having knowledge and making conjectures on all things through the necessary [consequences] and effects of the natures, is sometimes able to indicate beforehand the marvellous things of nature.[23]

This passage where Albert employs topical terminology – which can also be found, with minor variations of meaning and judgment, in the works of Roger Bacon – is therefore greatly in favor of astrology and magic. It also shows a similar preoccupation with classification and legitimization to that which characterizes the *Speculum astronomiae*. The second passage quoted by Geyer was, in fact, part of a wider discussion of the legitimacy of magic prompted by a comment in *The Book of Sentences*: "That the magic arts owe their strength to the power and the knowledge of the devil" (1.II, dist.VII). Treading on the border between magic and astrology, Albert discusses the use of planetary images in order to obtain extraordinary events and sortileges. Article IX contains the theological condemnation of such "science of images". Albert's answer to the question: "Does the devil avail himself of constellations in his operations, or not? And also, is, or is not, the science of images the work of the demon?", was as follows: "It is clear that the science of magic happens by the work of the demons, and for this reason it is prohibited; if it were based on natural power, it would not be prohibited".[24] Such an explicit circular reasoning is not surprising in a theological work, where Albert faithfully deploys the method of scriptural and patristic authority. It was not by chance, perhaps, that Geyer chose all his examples from this kind of writing! Yet, in this same *Commentary to the Sentences*, Albert shows his familiarity with the literature on images, and quotes extensively from a *Liber de mansionibus lunae* which he attributes here to Aristotle, but which may perhaps coincide with the one quoted in the *Speculum* amongst the Hermetical books,[25] and a passage quoted from the *Quadripartitum* on the astrological basis for explaining monstrous births.[26] As usual, Albert's judgment of astrology in the text is on the whole positive, even though greater allowance is made here than in the *Speculum* for the demoniac character of images (condemned in the text upon which he was commenting) without completely excluding their astral nature:

> As we have shown above with the authority of Augustine, there is no doubt that the rising and the aspect of the stars has a great effect on the works of nature and of art, but not on our free will, as [John] of Damascus said. But the art of images is wicked, because it leads to idolatry through the divinity that is believed to be in the stars, and because no images have been found but those which are useless and wicked.[27]

These two passages cited by Geyer had already been examined by Thorndike, who pointed out that Albert often showed a certain interest for these practices that he had completely condemned in the *Commentaries on the Sentences*, whereas in the *De mineralibus* he reverted to a more thorough and favorable discussion of the issue.

The third passage was found by Geyer himself in the *Summa theologiae*, and concerned the horoscope of the virginal birth of Christ: "What Albumasar says is the worst of mistakes, and those who have quoted him to prove that philosophers had also predicted the delivery by the Virgin, should be condemned".[28] Geyer commented that this passage contained a polemical hint not only against Albumasar and Bacon –who, as several other Latin authors had done, quoted this Arabic source[29] – but also against the *Speculum astronomiae*, where the horoscope was described as "something far more elegant, namely a testament of faith and of eternal life, not acquired save by faith". The author of the *Speculum* had already asked whether "gentile" philosophers like Albumasar had acquired "merit" through their prophecies: "What was its [his book's] value, if there was written in it that the birth of Jesus Christ from the Virgin [...] was figured in heaven, from the beginning?",[30] and answered that eternal life can be gained only through faith. The discrepancy between the *Summa* and the *Speculum* on this point, however, can be explained, first, by calling attention to the different methods Albert employed and admitted in theological and in philosophical works, and, second, by noticing the late date of this work[31] (not to mention the doubts as to the authorship of the *Summa* voiced by Lottin).[32] If the critical examination of the *Summa theologiae*, even without definitely denying the authorship or ascertaining the compilatory nature of the text, fixes the date of the work to a time after the condemnation of 1277, this would in itself explain such a change of tone and the increased preoccupation with orthodoxy. This would explain also the peculiarities Lottin noticed in the work, asking himself "why Albert so often contradicted Saint Thomas Aquinas, and why did he recant his previous opinions, and adopt the Franciscan theses; yet, at the time of the condemnation of 1277, the old man went to Paris in order to defend the memory of his pupil, Thomas Aquinas"[33] – and, one might wish to add, to defend some of his own views as well.

It should be pointed out, however, that even in the *Summa theologiae* there are texts referring to the horoscope of Christ, which have inexplicably escaped Geyer's attention. I refer to the quaestio LXVIII in particular, which constitutes an exemplary case of parallelism and dissociation with respect to the *De fato*, since the latter text deals in philosophical terms with the same problems, as were discussed in the theological work of twenty years earlier. In his *Summa*, Albert did not follow the terms of Aristotelian philosophy, which were the ones he had been asked to dwell upon at the request of the Papal court at Anagni. In *De fato*, a work which, in my opinion, shares a common date and motive with the *Speculum*, Albert did not have to deal with dogmas and their violations; therefore, as a consequence, he did not consider the most difficult and heretical point of the deterministic doctrine he was thereby substantially endorsing. In the

Summa, he was forced to deal with the extremist doctrine, which probably enjoyed a worrying circulation.[34] He had to discuss "Whether [or not] Christ was subjected to destiny in his soul and body". Albert's answer is negative, but in order to reach it, the author has to expand upon the arguments of the culprits:

> Christ took up our ineliminable defects, as [John] of Damascus says. One special defect of ours is to be subjected to fate and fortune. And therefore Christ assumed this defect. So far all mobile entities possess a disposition which consists of fate. It is agreed that Christ was mobile relatively to his body. He was also mobile relatively to free will [...]. It therefore appears that he, relatively to body and soul, was subject to fate and fortune.[35]

Albert did not recoil from the blasphemous theses which were popular within the Faculties of Arts up until the time of Pietro d'Abano and Cecco d'Ascoli, Pomponazzi and Cardano. He examined and as a highly responsible theologian rejected them, pointing out that Christ had become a man through his own choice, and, therefore, "had not necessarily acquired our defects", especially excluding the effects of the astral determinism which he himself had established.[36] He used the same line of argument in order to refute the horoscope of the Virgin. The transition from the philosophical discussion of astrology, to the consideration of the theological "determination", revealed to Albert that some of the theses he had once supported on account of their intellectual elegance presented enormous difficulties. Aware of the obligations and the aims of a *Summa theologiae*, Albert decided to refute such tenets, without, however, acknowledging the contradiction with respect to what he had already written when dealing "philosophically" with the issue of astrology.

CHAPTER FOUR

THORNDIKE'S CONSISTENCY. HIS RESEARCHES ON THE *SPECULUM ASTRONOMIAE* FROM 1923 TO 1955

Lynn Thorndike first became interested in the *Speculum* when he was writing the second volume of his *History of Magic and Experimental Science*, and this interest accompanied him throughout his career as he devoted one of his last works to the *Further consideration of the* 'Speculum astronomiae'.[1] In the conclusion to the section of the *History* covering the period from late antiquity to the thirteenth century, Thorndike devoted three chapters to three great scholastic authorities: Albert, Thomas Aquinas and Roger Bacon. A fourth chapter, the LXIIth of the *History*, was centered specifically on the problem of the attribution of the *Speculum astronomiae*. The arrangement of the material was not a matter of chance, and the author was right not to exclude Aquinas. Indeed, Thorndike hoped to assemble the data required for a thorough discussion of the attribution proposed by Mandonnet.[2] After having refuted Mandonnet's "extraordinary contention that Albert did not believe in astrology",[3] Thorndike showed how this could have logically led to the absurd result of attributing to Roger Bacon a work like the *De mineralibus*, though it is undoubtedly by Albert.[4] The *History* did more than that. Indeed, Thorndike's enterprise was not limited to an assessment of the role of magic in medieval thought, but also offered a close examination of the science of observation throughout this period. This strategy allowed him to write an important chapter (LIX) on Albert's attitude towards science, a term rightly understood by Thorndike in all its various thirteenth-century connotations. Thorndike made good use of the best available scholarship on Albert. He argued that the thought of the medieval philosopher gradually developed along two parallel lines, which found expression in his theological and philosophical works, and he summarizes the remarks on Albert's psychology and metaphysics put forward by Schneider and Baeumker: "They note that in his *Commentaries on the Sentences* he [Albert] is still glued to the Augustinian tradition, while in his *Summa*, he is strongly influenced by Aristotle" and by "spurious Arabic ... additions".[5] Later, Martin Grabmann in his *Zur philosophischen und naturwissenschaftlichen Methode* agreed with Thorndike's reconstruction of the development of Albert's philosophical and scientific method. Grabmann, the famous specialist on scholastic method, pointed out that "Albert's attitude towards precise scientific research is revealed by his remarks on the method of the natural sciences, and in particular by his emphasis on the role of observation and experimentum in scientific knowledge".[6] The appreciation of the role of direct independent observation is made possible by Albert's well-known independence with respect to the *auctoritates*. He was remarkably able to distinguish between the various specialties and true competences of his "authorities". For instance, "on faith and ethics" he acknowledges the superiority of Augustine with respect to gentile philosophers, but he does not accept the former's views on "the nature of things" (where he accorded Aristotle the primacy) and on medicine, a field he thought better served by Galen and Hippocrates. These methodological comments, taken from the *Commentaries on the Sentences*, had already been quoted by Mandonnet.[7] Yet Albert's methodological remarks were not restricted in the theological works, where they naturally belonged, but were also present in the philosophical commentaries. Grabmann quoted at length from the latter, to show that "Albert tried his best to have a correct dialogue with Aristotle", whom he defined as "the prince of the peripatetic, the archdoctor of philosophy", though he never followed him uncritically or dogmatically.[8] In order to ground scientific observation within Aristotelian methodology, the first problem was to overcome the contradiction existing between the logic of universal scientific judgment and the concrete descriptions Aristotle had offered on contingent beings such as animals, etc. In one of the works he added to the commentaries on Aristotle, Albert embarked upon the explanation of the causes and properties of elements and planets, remarking:

> It is not sufficient to know in a universal way, but we desire to know each thing according to what is proper to its nature. This is indeed the best and most perfect kind of understanding.[9]

Albert's idea of science was opposed to the allegorical view of nature represented, for example, by the *Physiologus*,[10] but he was not happy with the simple registering of observations. "He was looking for the knowledge of causes, of laws and of natural processes", which ran parallel to the critical assessment offered by the scientific authorities. "Natural science is not simply to record what is said, but is the search of the causes of things following nature".[11]

According to Grabmann, in view of his critical and selective approach to science, Albert well deserved the appellation of "Doctor Expertus" contemporaries attributed to him. As Albert himself put it, "let us take up from the ancients all that is well said", but we must reject the old mistakes made by Plinius and Solinus, and the more recent ones by Michael Scot are to be rejected. He also added that "any conclusion which contradicts the senses is not to be trusted".[12] Moreover, sense-data do not derive from observation only, but also require experiment. As Albert noted in the *Ethics*,

we need a long time, so that the experiment can be verified, to avoid all possibility of failure [...]. It is necessary that the experiment should not be verified in one way only, but according to all circumstances, so that it [will be verified] with certainty and rightly as the principle of the work.[13]

The method for the acquisition of data, Albert theorizes, does not exclude contributions from earlier times –which he often and rather naively accepts– but he does argue that all sources should be carefully screened, as he himself had done in the *De vegetabilibus*:

Of those things we put forward, we have verified some ourselves through experiment, others we have instead related from the sayings of those whom we have found not easily to say anything which is not proven through experiment. In these matters, only experiment gives certainty, since in matters so particular it is impossible to make syllogisms.[14]

This important passage represents well the problem Albert was facing when he tried to distinguish between universal syllogisms and the domain of probable knowledge – a domain including botany, zoology, and mineralogy, as well as descriptive astronomy and judiciary astrology.[15] We are now in a position to appreciate why Albert is credited with having established a true scientific team, which he used in order to acquire information for his naturalistic works, as Thorndike has documented in several passages from Albert's works.[16] The emphasis on the need to subject information to criticism also explains why Albert, who was less coherent and clear than Thomas in his metaphysical and theological arguments, was however remarkably organized in the data and sources of his scientific works. This peculiarity supports arguments in favor of Albert's authorship of the *Speculum*, an exceptionally well-organized, encyclopedic article both from the theoretical and the bibliographical point of view.

Thorndike was convinced that "Roger Bacon had hitherto been studied too much in isolation",[17] and therefore had been considered as a privileged exception to the philosophical and scientific norm of the second half of the thirteenth century. On the contrary, Thorndike emphasized the importance and originality of Albert's thinking, which he considered superior to Thomas's in both the variety and articulation of his commentaries on Aristotle, and in his scientific interests and observations. "But the modern eulogies of the scientific attainments of Roger Bacon, supposed to be a thorn in the side of the medieval church and falsely regarded as its victim, and as the one lone scientific spirit of the middle ages, have been rather more absurd than the earlier praise of Albert, who was represented both as a strong pillar in the church and the backbone of medieval and Christian science".[18] Thorndike pointed out the fundamental methodological distinction between philosophy and theology introduced by Albert, and developed by Thomas Aquinas, as well as by the Averroists Siger and Boethius de Dacia. With Albert, this distinction was not limited to the level of a declaration of principle, but, as Thorndike has shown, was implemented in his writings, and in the pages devoted to the discussion of occult sciences. Thorndike was "impressed by the differing and almost inconsistent attitudes in different treatises by Albert, for instance in his attitude towards magic, which seem to hint that his opinions changed with the years, although it may be attributable, as in some other authors, to the fact that in different works he reflects the attitude of different authorities, or approaches different subjects with a different view-point, writing of theology as a theologian, but of Aristotle as a philosopher".[19] Contrary to what Mandonnet maintained, according to Thorndike we must acknowledge that "Albert's astrological views crop out in almost all his scientific writings",[20] but we must also notice his changing opinions, always characterized by a great interest in magic and natural wonders.[21] In his monographic chapter on Albert, Thorndike devoted two sections (placed between the analysis of the scientific work of the philosopher, and the concluding part on astrology) to the discussion of magic and "natural wonders". "So far as mere classification is concerned, Albert's references to magic in his scientific writings are in closer accord with his discussion of magic in the *Summa* and *Sentences*, where too he associated magic with the stars, with occult virtues, and its connection with demons is now almost entirely lacking". Thorndike also observed

how closely magic, or at least some parts of it, border upon natural science and astronomy. And yet we are also always being reminded that magic, although itself a 'science', is essentially different in methods and results from natural science or at least from what Albert calls 'physical science'. Overlapping both these fields, apparently, and yet rather distinct from both in Albert's thought, is the great subject of 'astronomy' which includes both the genuine natural science and the various vagaries of astrology. It is all like some map of a feudal area where certain fiefs owe varying degrees of fealty to, or are claimed by, several lords and where the frontiers are loose, fluctuating, and uncertain. Perhaps the rule of the stars can be made to account for almost everything in natural science or in magic, but Albert seems inclined to leave room for the independent action of divine power, the demons, and the human mind and will.[22]

There is no doubt that the ten years before 1277 witnessed the climax of astrological theories within medieval schools. It is not surprising that Thorndike himself, independently from Mandonnet, had been tempted initially to connect the *Speculum* with reactions to the second condemnation. He also considered, in passing, the possibility that the time Bacon might have served in prison was related to his endorsement of astrology after 1277, but made the comment only to re-emphasize the presence of internal elements in the *Speculum* itself which made it impossible to assign so late a date to the work.

Thorndike was convinced that the passages where the author of the *Speculum* complains that there is no available translation of two books of Aristotle's *Metaphysics* could not be interpreted as a quotation from Thomas, as Mandonnet had suggested, but that it reflected the situation at a time when the *Metaphysica mediae translationis* or the translation by William of Moerbeke had not been completed.[23] For this reason, he was convinced that Albert was the author of the treatise. He put forward the hypothesis that the work was written before 1256, the year in which – following the rejected chronology Mandonnet proposed – Albert had completed his paraphrasis of the *Metaphysics*.[24] In chapter LXIII of his *History*, Thorndike also examined the question of the Albertinian pseudoepigraphs, with particular reference to the *Experimenta Alberti*, *De mirabilibus mundi*, and the *De secretis mulierum*. Furthermore, he devoted a few notes[25] to still unexplored manuscripts on natural philosophy and astrology attributed to Albert. Later he also prompted Pearl Kibre to continue further research on the alchemical literature attributed to Albert.[26]

These investigations of the pseudo-Albertinian corpus reinforced Thorndike's conviction that the *Speculum* was indeed by Albert. Of more weight still, of course, were the style and the method of presentation, the manuscript tradition, the confirmation by medieval testimonies and the ancient lists of Albert's works. Also of considerable importance to the attribution of the work was the individuation in the *Speculum* of theses Albert had often supported, such as the rejection (in Albumasar and his source Aristotle) of the doctrine of the animation of stars –which were seen as "dumb and deaf instruments" in the hand of God– and the differentiation between three different types of astrological images, in order to save the legitimacy of at least one. This latter point brings forward the question of the relationship between astrology and magic, an issue the author of the *Speculum* was keen to avoid, and the prime argument for Thorndike's attribution. Thorndike correctly pointed out that the astrological perspective defended by the author of the *Speculum* was not unusual. Indeed, it was commonly shared by Christian scientists active in the twelfth and thirteenth centuries. As far as "astrological" images were concerned, the author of the *Speculum* sided with Albert and Roger Bacon, rather than with William of Auvergne and Thomas Aquinas. The latter two saw the doctrine as advocating the direct intervention of demons, and therefore condemned it, believing that no image could work through astral influence alone.[27] "In general the astrological position of the *Speculum* closely parallels the attitude of Albert and Roger Bacon, who in turn held almost identical views. If anything, the *Speculum* is somewhat less favorable to astrological doctrine than Albert. Whereas he in large measure accepted the casting of horoscopes, although saving free will, it emphasizes the conflict between free will and nativities." And it emphatically denies that the stars are animated, a point upon which he seemed rather hazy in his scientific treatises. There is now available an edition of al-Kindi's *De radiis sive theorica artium magicarum*, in a Latin translation of "uncertain date" which, however, employs "the language of twelfth- and thirteenth-century schools"[28]. We are struck by the similarity of the theses found in the *Speculum* as well as in Albert to al-Kindi's theories. Robert Grosseteste knew al-Kindi's *De radiis*, a work Roger Bacon quoted, and Thomas Aquinas and Giles of Rome criticized. Because of his theory concerning the necessary influence of stars, al-Kindi had been implicitly censured by Étienne Tempier both in 1270 and 1277.[29] Before these condemnations, however, Aquinas had written a few pages against this "theory of magical arts" to prove "Quod opera magorum non sunt solum ex impressione caelestium corporum".[30] In his Chapter 6, "De virtute morborum", al-Kindi on the one side limits the power of ceremonial words to the natural action of imagination, and on the other hand he supports the opinion of those "who don't believe that the nature of spirits is of a sort, which can be reached by human knowledge. For that motions and images, which happen in the air or in another element or in an elementatum, which are not normal occurrences by nature as it is commonly known, is not the result of the operation of the spirits, but only of the condition of celestial harmony which makes matter fit to receive such motion and such images by the actions of other corporeal things, such as prayers and words, and also some other things like herbs and gems, which move matter accordingly to a similar harmony".[31] Albert and the *Speculum* do not quote the *De radiis*. In a recent paper on "al-Kindi vu par Albert le Grand", it has been argued that Albert did quote other works by al-Kindi, but not the *De radiis*.[31] This silence can be explained in many ways. Either the author of this research did not consider al-Kindi's work as philosophical, or he did not have d'Alvernay-Hudry's edition of the *De radiis* at hand, or, possibly, Albert himself chose not to mention the title, or the author censured it. What is certain is that Albert's teaching concerning talismans and images perfectly corresponds to that of al-Kindi, an author many contemporaries criticized, Aquinas included.

There is no actual contradiction between the *Speculum* and other works of Albert on these points, and we have already seen in the case of his theological and Aristotelian works that Albert is likely to state the same thing somewhat differently according to the point-of-view from which he writes. The writer of the *Speculum* is obviously desirous to conciliate a theological opposition to or suspicion of 'astronomy' and therefore naturally inclines to be moderate and conservative in his advocacy of astrological doctrine.[33]

According to Bacon, virtually no one had dared to write about astrological images for fear of being accused of magic. For instance, he tells us that "scarcely anyone has dared" to speak of astronomical images in public, "for those who are acquainted with them are immediately called magicians, although really they are the wisest of men".[34] As is well known, Bacon launched a sharp and hammering polemic against magic. He "associates 'magic' and necromancy, not like Albert with astronomy, but with deception".[35] "In certain passages, however, Bacon suggests that magic is not utterly worthless and that some truth may be derived from it".[36] Bacon reconstructs the historical origin of the generalized condemnation against magic, tracing its origin to the denunciations of pagan practices by Augustine and other Fathers, who were probably reacting against early accusations of magic brought against the first Christians. The polemic did not differentiate much between different kinds of magic. "Bacon complains that this confusion still exists in his own time and that contemporary theologians, Gratian in his works on Canon law, and 'many saints' have condemned many useful and splendid sciences along with magic".[37] In the *Epistola de secretis operibus artis et naturae et de nullitate magiae* (a work of uncertain authorship), as well as in other certainly genuine works,[38] Bacon tries to introduce distinctions and qualifications, but "fails in his attempt to draw the line between science and magic, and shows, as William of Auvergne, Albertus Magnus, and others have already shown, how inextricably the two subjects were intertwined in his time".[39] For instance, "Bacon's 'astronomer' is really a magician and enchanter as well –one more of the many indications we have met that there is no dividing line between magic and astrology: divination is magic; astrology operates".[40] Thorndike also examines the thesis, endorsed by Mandonnet, that the 1278 condemnation of Bacon (mentioned only in the *Chronica XXIV Generalium*, written around 1374, and therefore "of very doubtful authority") resulted from the friar's inclination towards astrology and magic. Thorndike noted that "Bacon's views were not novel", as were those allegedly submitted to censorship. He "shared them with Albert and other contemporaries, and there seems to be no good reason why they should have got him into trouble. His expressed attitude towards 'magic' is so hostile that it seems unlikely that he would have been charged with it, when other clergymen like Albert and William of Auvergne spoke of it with less hostility and yet escaped unscathed".[41] Thorndike admitted that Bacon was "more favorable [towards astrology] than some of his contemporaries. With his views on astrological images and his attribution of religious sects to conjunctions of the planets theologians like Aquinas and William of Auvergne would refuse to agree, but Arabian astrology supported such doctrines, and the views of an approved Christian thinker like Albertus Magnus concerning astrology are almost identical with those of Bacon".[42] Thorndike concluded: "Again therefore there seems to be no reason why Bacon should have been singled out for condemnation".[43] "It also seems somewhat strange that Bacon should always be so condemnatory and contemptuous in his allusions to magic and magicians, when both William of Auvergne and Albertus Magnus allude to it as sometimes bordering upon science, in which case they do not regard it unfavorably. The suspicion occurs to one that Bacon perhaps protests a little too much, that he is condemning magic from a fear that he may be accused of it".[44] Thorndike is therefore led to conclude that the silence observed in the *Speculum* on the subject of magic, and on the distinction between magic and the other sciences, helps to confirm the thesis that Bacon was not its author.[45]

The whole of Thorndike's *History* constitutes a final refutation of Mandonnet's general thesis that Bacon, being the only scholastic author who believed in astrology, was the author of the *Speculum*. The second volume of Thorndike's work is devoted to the examination of the astrological doctrines put forward by Peter Abelard, Adelard of Bath, Hugo of St. Victor, William of Conches, Bernard Silvester, John of Salisbury, William of Auvergne, Vincent of Beauvais, Robert Grosseteste and Peter of Spain, just to limit the list to the more serious representatives of scholasticism without mentioning less "respectable" authors such as Hildegard of Bingen, Arnald of Villanova, and Lull. Sixty-four years after the publication of Thorndike's masterpiece (which contributed as much as the better known works by Duhem, Haskins, and Gilson, to modify our understanding of medieval science and cosmology) the list of scholastic authorities endorsing astrology could be made longer still. Following Gilbert Paré's reconstructions of court scholasticism, a movement which embodied the intellectual context of the *Roman de la Rose*, I would like to mention that even Bonaventura in his *Commentaries to the Sentences* had no hesitation in making allowances for astral influence:

The celestial luminaries [the Sun and the Moon] exercise an impression on elements and elementary bodies; an impression, I say, which is not univocal, but acts in different ways.[46]

Paré connected this concession made by Bonaventura, the General of the Franciscan order, with analogous ones made by Thomas Aquinas, and the particularly numerous ones by Albert. He also mentioned the *Errores philosophorum* by Giles of Rome, and Tempier's 1270 and 1277 condemnations.[47]

Paré's interpretation deserves further comment. If indeed Bonaventura acknowledged coelestial influences (as everyone did in his time), and if in other works he deployed the doctrine of lunar nodes ("caput et cauda draconis") in order to indicate metaphorically the two "eclipses" even contemplative men can suffer from,[48] it is nevertheless true that he was one of the harshest opponents of astrological necessitarianism. His views on the subject were expressed in his sermons to the students and teachers of the University of Paris, where he said that astrological necessitarianism was as serious a mistake as the thesis of the unity of the intellect and the eternity of the world. Of the three errors listed above,

The second one is about the necessity of fate, as it were about constellations: [according to this error] if a man has been born under such a constellation, he will be a thief by necessity, or [he shall be by necessity] good or bad. This annuls free will and merit and rewards: since, if every man does what he does by necessity, what is the value of the freedom of will? What will he deserve? –As a consequence, God will be [considered] the source of all evils. It is true, that a certain disposition is allowed to remain from the stars; but God alone is the ruler of the rational soul.[49]

In these *Collationes de donis Spiritus sancti* Bonaventura concludes by uttering "opprobrium" as Jeremiah had done. One year later, in his *Collationes de decem preceptis*, Bonaventura alludes in passing to the errors of "investigating the contingent future" and to the "magical art of [ceremonial] words and inscribed talismans".[50]

During the years 1267-1268 Bonaventura felt he had to attack the necessitarian doctrine at the basis of birth horoscopes and of interrogations concerning contingent futures, as well as those practices which came near to the magic of talismans and charms. Thus, even though Bonaventura allowed that "a given disposition is allowed to remain (*relinquitur*) from the stars", he must be seen as one of the critics and censors against whom the *Speculum* defended astrology, a discipline the author of the treatise interpreted in non-necessitarian terms.

PART TWO

Discussions on Astronomy at the Time of Albert

CHAPTER FIVE

ALBERT'S *'AUCTORITAS'*: CONTEMPORARIES AND COLLABORATORS

It is inappropriate to consider further the historical debate concerning the authorship of the *Speculum astronomiae*. The previous pages have attempted to show the progress made in the last fifty years by historians of science and philosophy. Catholic historians such as Marie-Thérèse d'Alverny and Thomas Litt have abandoned the anachronistic hagiographical preconceptions still present in Mandonnet. Very recently, the authors of two historical syntheses on alchemie and magic in the Middle Ages came to the conclusion that the *Speculum* is very probably an authentic work by Albert.[1]

It is now important to reconstruct the spectrum of Albert's reflections and readings concerning the "duae magnae sapientiae" on the basis of his authentic writings, in order to respond to the interpretative issues proposed by Mandonnet. To do so, we need to expand our analysis of primary sources, and to proceed more systematically than we have done until now, when we called upon medieval authors only within the limits of the critical examination of the hypotheses put forward by Mandonnet and Thorndike.

The numerous passages undoubtedly written by Albert on the subject of the two "wisdoms", listed by Thorndike in 1923, are in themselves sufficient to exclude Mandonnet's surprising preconception against the Dominican author. Leaving aside for the moment the question of whether or not Albert was the author of the *Speculum astronomiae*, there is still no reason to deny that he might have been. For example, there is the important testimony offered by the Franciscan Bonaventura d'Iseo, who flourished during the middle decades of the thirteenth century, and was the "intimate friend of Albert the German and of Thomas Aquinas".[2] This testimony was brought forward by Martin Grabmann and was again used by Richard Lemay in order to attribute the *Speculum* to Albert. Though it is not explicit or conclusive, the passage does nevertheless provide a strong argument in favor of the attribution to Albert:

Thanks to his celebrated sanctity, intelligence and wisdom, during the days of his life Brother Albert obtained from his Lord the Pope the grace to be able legitimately to learn, know, examine and verify all the arts deriving from the sciences of good and evil, approving truthful books and condemning the false and mistaken ones. He then worked extensively to complete the books Aristotle had started, and he himself completed new books concerning many arts deriving from sciences such as astrology, geomancy, necromancy, precious stones, and alchemical experiments.[3]

It is appropriate to stress here that Lemay saw in Bonaventura d'Iseo's words a clear allusion to the *Speculum*: a work designed to distinguish truthful from false books,[4] and clearly compiled with the aim of completing the series of Aristotle's naturalistic works. This last characteristic also appears, according to Grabmann and Pelster, in Albert's naturalistic writings.

The attention the author of the *Speculum* devotes to documentation and to exact bibliographical information makes it difficult to think of him as an old man busy with his many official duties as Albert was in 1277. The date of that famous condemnation of 1277, however, is no longer a point of reference for the dating of the *Speculum*, which alluded, as seen above, to criticisms and accusations, and not to a condemnation, and can easily have been occasioned by other polemics. The possibility of an earlier date for the work, on the other hand, helps to explain the passage in which the author laments the lack of the Aristotelian pages on intelligences,[5] probably contained "in the twelfth or the thirteenth book of the *Metaphysics*, which are not yet translated".[6] Even if this allusion precedes Albert's commentaries on the *Metaphysics*, it has also been noted that, in 1270 Thomas Aquinas was still complaining about the missing chapters of the *Metaphysics* almost in the same words.[7] By 1250 Albert had already taken advantage of the Greek-Latin anonymous or *media* translation of the *Metaphysics*, which was almost complete, lacking only the eleventh book (K), not two books as the *Speculum* laments.In various contemporary texts the "books missing from the Metaphysic" were often connected with the *Liber de causis* in the same way as in the passage quoted above from the *Speculum*. However, the difficulty was mainly in knowing the exact consistency of the plan of the *Metaphysics*. The "recognitio and translatio" to which William of Moerbeke submitted the previous translations began only shortly before 1265[8] and was completed before 1271, when it became generally diffused and took the place that the anonymous translation of books I – X, XII – XIV had held for twenty years. But for William of Moerbeke the task of this "recognitio" should have been as heavy as that of the translation: when he started there were three Greek-Latin and one Arab-Latin translation in use (corresponding to the texts chosen by Averroes for his Great Commentary and lacking books XI, XIII – XIV and sections of books I, II, as well as the proemium of bk XII). Every teacher was then using Averroes's commentary and his "montage" of Aristotelian texts. The first two known Greek-Latin translations have been preserved for us only for books I – IV, and it is not clear if the "vetustissima" by Jacob of Venice ever contained the whole text: in any case it was certainly not widely diffused. So it is easy to understand why so many hypotheses flourished about the contents and plan of the *Metaphysics* and its relationship with the doctrines of "intelligences" of the pseudo-Aristotelian *De causis*. To cite only one extreme case, a commentator writing immediately after Thomas Aquinas, but not yet using Moerbeke's translation summarized the hypotheses currently discussed, and wrote:

> We have not the translation of several books of the *Metaphysics*, notwithstanding the fact that in greek – as it has been said – they are up the number of twenty-two".[9]

In the intellectual circle at the papal court gathered in Viterbo (where Albert came back for the second time in 1264, exactly when William of Moerbeke was starting his *recognitio*) the expectations for the results to be obtained from the latter's work must have been great. The translation was completed by William before 1271: more than five years before it was possible that the plan of the *Metaphysics* could have appeared unclear to his friends, especially for what concerns the order of the last books. This confusion was more likely to happen to a professor like Albert, used to teaching on the "montage" of the Great Commentary by Averroes. However, precisely for this habit of teaching on the Aristotelian text and on Averroes's commentaries (Albert's work on the *Metaphysics* was being written just after 1262-63) his curiosity must have been great. This may also explain why this digression – which otherwise could be surprising – was introduced in chapter XII of the *Speculum*: it brings an echo of the rumors diffused and of the expectations about a new, complete and revised Latin text of the *Metaphysics*. These expectations are likely to have spread within the Viterbo circle around 1265.

The manuscript and literary tradition of the *Speculum*, however, does not go back to the mid-thirteenth century, but only to the last decades (with the case of the still unedited *Philosophia* attributed to Oliverius Brito) and testimonies on the authorship do not start until the end of the century (with the ms. Parisinus latinus 7335, which bears the title: *Speculum domini Alberti)*. Thus, if we prefer not to consider the treatise as a relatively early work by Albert (going back, according to Lemay, as early as the 1250s), we could also consider it as a work begun perhaps shortly after that time, but continued and updated with the addition of new material gleaned from his visits or written questions to various libraries, and then published at a later time by him or by one of his pupils and assistants. Recently, during the 1980 celebration, it has been pointed out that Albert, as well as Thomas, were typical representatives of the medieval teacher, "who never worked alone, but was always surrounded by his assistants (*socii*)". It was often the case that their research was the result of a team effort.[10]

In the *Speculum*, we find at least two hints that make us consider the possibility that the work might have been prepared and discussed by a group of scholars. Firstly, the collection of bibliographical data is often a work that lends itself to cooperation. Secondly, the *Speculum*, though a work written in the first person, bears the mark of the presence if not of a second hand, of at least some other collaborator. Indeed, the primary author ("a certain man zealous for faith and philosophy")[11] indicated in the proemium is accompanied by a different "vir", ready to provide the translation of astrological terms from Arabic or other languages the primary author does not know.[12] The first "vir" declares that he is going to get help from a certain orientalist, since he will be able to dissolve with this translation the cloud of mistaken necromantic suspicion arising from the use of an unknown language in the otherwise innocent astrological treatises. The practice of keeping terms transliterated from Arabic found an equally annoying analogue in the use of Greek and Arabic terms in the Aristotelian corpus, a complaint humanists of a later age still raise.

A further passage in the *Speculum* provides the source for yet another hypothesis, one I am putting forward with great caution, and one which mainly represents an invitation to discussion and research. In the opening pages of the *Speculum*, and therefore soon after the reference that mentions the "vir zelator fidei et philosophiae", we find a rather curious reference to another "quidam vir", someone contemporary to the writing of the work. This "vir" is described as the compiler of the *Almagestum parvum*, written "secundum Euclidis stilum" (following Euclid's style) in order to display theories put forward by Ptolemy and by Albategni, Ptolemy's summarizer.[13] The *Almagestun parvum* has recently been attributed to Campanus of Novara, and Paravicini Bagliani, the specialist on Campanus's biography, considers this attribution as "beyond doubt".[14] He also notes "the presence at the Papal court in the second half of the thirteenth century of students of optics, astronomy, astrology, geomancy, alchemy, etc., who belonged to the 'intellectual élite' of the time: scientists such as Campanus of Novara, John Peckham, Simone da Genova",[15] as well as Witelo, Johannes

Gervasius and William of Moerbeke. Albert became acquainted with this group of scholars during his two visits to Italy, the first one to Anagni in October 1256, which lasted until the following June, the second one to Viterbo and Orvieto, July 1261 to February 1263.[16] The editors of the *De fato* provide confirmation that this text dealing with the theoretical foundations of astrology represented the result of "a quaestio Albert discussed at Anagni",[17] where the papal court stayed in 1256.

It would be interesting to pursue the suggestion that the earliest draft of the *Speculum* originated during these visits to the papal court and was possibly the result of exchanges of ideas and data with so great an astronomer and orientalist as Campanus. The editors of the *De fato* have pointed out that this work lacks accurate quotations due to the lack of a library specialized in astrological treatises at Anagni. This lack of adequate library holdings was obviously a severe limitation for the bibliographical guide the *Speculum* was designed to be; this might have made it indispensable to get Campanus's help, as well as to proceed to integrations, including later ones, following the model provided by the catalog of a library rich in astrological texts as was the one put together by Richard de Fournival.

The relationship of that work with the *Biblionomia* and the library of Richard de Fournival led me until now to think it probable that the bibliographical chapters were completed in Paris. Among the authors contributing to the centenary volumes, there are various opinions concerning Albert's returning to Paris after the condemnation of 1277. Weisheipl denies it, whereas the curators of the great exhibition held in Köln accept it.[18] It is, however, certain that Albert was in Strasbourg in 1257 and 1268; from there, he might well have gone on to Paris. In any case, he might have seen a copy of the *Biblionomia* and he might have transcribed it during the years of his teaching in Paris before the struggle between seculars and Orders, when Gerard of Abbeville kept the manuscript from the priors; one of these hypotheses might be sufficient to explain the coincidences between the *Biblionomia* and the *Speculum* (XVI/ 24-28) discussed below.

But now a circumstance kindly suggested to me by Prof. Paravicini Bagliani brings an interesting new hypothesis: Richard spent some time at the papal court at the end of his life and he could have brought the *Biblionomia* with him (if not the manuscripts themselves). A copy of his catalog could have been kept in Rome and later used by this group of scientists and by Albertus himself. He could have read and used the *Biblionomia* at the papal court.

As yet, I do not possess documents allowing me to identify the contemporaries the author of the *Speculum* cryptically refers to. It is, however, important to call attention to these allusions, unusual in a medieval text, rarely given to quoting contemporary writers, with the exception of those who had achieved an *auctoritas* comparable to the one accorded to the Fathers and the great philosophers of the past. Thus, contemporaries often made reference to Albert, because, as Bacon despondently pointed out, they wrongly felt Albert had become an *auctoritas*.[19]

Even before 1259, Campanus too was considered a great mathematical authority, who had written a commentary to books I-IV of Euclid. Albert himself commented on the *Elementa*, "but in no way refers to the new edition of Euclid made by Campanus before 1259"; however, according to Tummers – who edited the text – Albert's comment was written before 1260, so that it is not impossible that only after completing his own commentary, during one of his stays in Rome Albert became interested in the Campanus book.[19]

If the unusual and certainly planned hints could refer to Campanus, author of the *Almagestum parvum* and an expert in Arabic astrology, it is even more likely that the two met at the papal court during the years 1256-1257, or more probably 1261-1263. As yet, we do not possess documents attesting to the encounter. However, in view of Albert's stay at the papal court, it is more probable that they met. Moreover, in his *Theorica planetarum* inscribed to Pope Urban IV (1261-1264), Campanus employed a terminology often similar to that found in the *Speculum*. In the catalog of Campanus's minor works we also find a text on medical astrology, a few quotations from which have been preserved in a late fifteenth-century work. Yet, these fragments coincide with the examples found in the *Speculum* concerning the danger of practicing surgery without taking into account the zodiacal melothesia.[20] This coincidence leads us to think that Campanus was acquainted with the author who around 1264 was drafting the *Speculum astronomiae* or with the group that was discussing it.

CHAPTER SIX

ASTROLOGY IN THE EARLY DOMINICAN SCHOOL AND GERARD OF FELTRE

Astrology was a subject of interest quite early on in the Dominican school, as can be seen from the *Quaestiones in Sphaeram* (1263-1266) by Bernard de la Trille. The *Quaestiones*, still unpublished, were studied by Thorndike, who pointed out the astrological digressions, constructed along lines similar to those reappearing later in Robertus Anglicus.[1] Amongst the pupils of Albert, the case represented by Thomas Aquinas is famous; and Theodoric of Frieberg wrote a *De animatione coelorum*. But they are both too well known to merit further discussion here.[2] We should, however, mention Giles of Lessines, an author who, like Theodoric, was tormented by the typically Albertinian problem of differentiating between the "intelligentiae" (similar to demons) and the celestial motors. Sometime after 1264 Giles wrote his *Tractatus de essentia motu et significatione cometarum*. The astrological works by thirteenth-century Dominicans[3] have been studied far less than their theological production. Yet, it will be useful to follow what Thorndike and Grabmann have highlighted in these works. Thorndike has published the complete edition of Giles's treatise on comets, a work generated by the observation of the comet of 1264. Giles had studied the astronomical event when in Paris, as opposed to his fellow Dominican Gerard of Feltre, who "with many others" had examined the comet from the Lombard province of the order. Thorndike also published Gerard's writings on the comet, correctly pointing out that the work of both friars was indebted to the Albertinian comment on the Meteora[4]. Indeed, they completed some of the quotations to which Albert had only alluded in his text, thus showing that they had probably been listeners, rather than readers, of the commentary by their master. Gerard also added some historical examples of comets that Giles had taken from Seneca's *Quaestiones naturales* into his text, thereby showing his debt to his Dominican brother.[5] He also mentioned Albert, calling him a fellow-Dominican, and comparing him to Ptolemy. Gerard wrote about 1264, during the generalship of John of Vercelli (1264-1283), who six years later would have promoted the investigation on the XLIII Problemata designed to answer the doubts and the discussions which had troubled the Lombard Dominicans. Gerard's *Summa de astris*, still unpublished, is preserved in three manuscript copies. Thorndike published the pages relating to comets, and Grabmann independently provided a short and general, but penetrating, survey of this text, and an edition of other sections of the work.

Both documents are useful with regard to the *Speculum astronomiae*, even though this work never mentions comets, since Albert, following Aristotle, did not consider them to be stars, but only "secondary stars" (*stellae secundae*), that is, secondary phenomena. Giles and Gerard must have had access to a rich astrological library. They also provide evidence of the presence of much interest in the topic, and of an articulated astrological theory among Dominicans active during the decade 1260-1270. Giles of Lessines was probably more independent from the Albertinian text than Gerard. He was certainly influenced by the second phase of Thomas's teaching (1269-1272), which he had followed in Paris. Giles was also well acquainted with the ideas of Robert Grosseteste, whose views he at times (as was the case with his views on comets) preferred to those of Albert. He may have followed the latter's astrological theories, but not without several qualifications and much less clarity. Due to the treatises *De unitate formae* and *De usuris*, Grabmann considered Giles as "one of the eminent figures of the earliest Thomistic school", despite the fact that "his contacts with Albert must have been the first and the most precocious". In Giles's scientific production (the *Tabula Stams* had already claimed that "he wrote several works on astrology") "one can see the stimulus and the influence of Albertus Magnus's scientific idiosyncrasies and personality".[6]

As far as method is concerned, as Grabmann has pointed out, the influence of Albert on Giles's typical distinction between the viewpoint of "the astrologer, the physicist or the doctor" and that of the theologian, is most clearly expressed in *De concordia temporum* (still unpublished), in *De crepuscolis* (a work the historian was able to find after Mandonnet pointed it out) and especially in the treatise on comets. A final dating for these works has not yet been established, though one tends to agree with Grabmann that all these texts, and the *De cometis* in particular, were written prior to Giles's Thomistic period. It should be noted, however, that in chapter VIII of *De significatione cometarum et causa et modo significandi* Giles expresses some reservations about Albert, his first master, despite the fact that he still continues to invoke him as "Albert, sometimes Bishop of Regensburg", even in his markedly Thomistic work on the unity of form. In his later work, Giles also appears aware of the debate between his two mentors concerning the animation of celestial motors.[7]

As we have often pointed out, the frankness and directness displayed by the author of the *Speculum* makes it difficult to believe that he was arguing against, or indeed defying a canonical condemnation. Whoever the author was – even if, hypothetically speaking, he was such an authoritative figure as Albert – it seems highly likely that the *Speculum* addresses someone not representing official canonical authority, or indeed, more plausibly, a younger and less prestigious fellow Dominican.

In 1267, a special comet also aroused the interest of John of Vercelli, Master General of the Dominican Order.[8] Following the private submission of several questionnaires to Thomas Aquinas by Italian Dominicans, in the Spring of 1271 the Master General himself consulted with the greatest masters of the Order. He sent Albert, Thomas and Robert Kilwardby forty-three questions partly concerned with the issue of astral influences, which were causing controversies among Dominicans, especially in Italy.[9]

One of the most complex products of these debates is the still unpublished *Summa de astris* by Gerard of Feltre, which dates to 1264. Grabmann has provided a short analysis of this treatise, stressing the frequent use of texts by Thomas and Albert[10] (especially Albert's *Meteora*), but failed to mention the *Speculum astronomiae*, though elsewhere he attributed it to Albert. In view of the comparable structure of the two works, the similar terminology, and at times the common bibliographical peculiarities, I am inclined to argue that there is a close relationship between the *Summa* and the *Speculum astronomiae*. Lacking a critical edition of the *Summa*, however, in this instance I have limited my work to a partial transcription and an analysis of this text, and have formed the impression that the author was investigating astrological issues from a theological point of view, and not without a considerable amount of severe censorship.

This does not mean that Gerard was insufficiently versed in astrology. On the contrary, his *Summa* (on whose second sheet is the line "now, that is the year of the Lord 1264", though the text was not, perhaps, completed within the year, considering its length) is extremely well documented and reflects a wide-ranging theoretical proficiency, an uncommon feature for the time. In the second prologue, "in which is set forth the reason and method for studying astrology and the difference between astronomy and astrology", one reads:

> This summa on the stars has been compiled from the very words of Ptolemy, Albumasar, Alfargani, Alcabitius, Omar, Zahel, Masshallah, who wrote authoritatively about the mastery of the stars.[10]

This Arab-like definition of astrology and listing of its "masters" can also be found in the *Speculum astronomiae*. Leaving aside the other auctores, both texts give priority to the analysis of works by Ptolemy and Albumasar. Gerard knew both authorities well and quotes their work with propriety, so much so that more than once we can note that in the *Summa de astris* and in the *Speculum* the same choice of quotations is given and the same texts are examined.[11]

Gerard states a need to investigate thoroughly the discipline ("if anyone unacquainted with the stars would undertake to write against the mathematicians, he will be the object of laughter"),[12] according to the approach chosen by Albert, whom he calls "a friar of my order" and "a great philosopher".[13] Of the two sources Gerard consults, as Grabmann has indicated, Albert's *Meteorologica* is filled with astrological quotations: whereas in his *Quaestiones disputatae de veritate* Thomas devotes two long articles to the problem of "whether divine providence affects inferior bodies [and human acts] by means of celestial bodies"[14] without ever referring to the astrological authorities cited by Albert.

Gerard and Albert agree on many fundamental points – so much so that we can presume that the pupil had adopted the theses expressed by the master in his *Meteorologica*, and in his other commentaries on Aristotle. If, for example, in his *De caelo*, Albert had argued against the thesis of the animation of the skies, Gerard does the same, albeit adding his special brand of polemical vehemence:

> The astrologers say, blaspheming, that all human actions and customs, both good and bad, indeed even the choices made by the rational soul, occur by necessity, according to the disposition of the heavenly bodies; and to prove this I quote their more famous authors, Albumasar [...] Ptolemy [..].[15]

In fact, what characterizes Gerard is the willful denunciatory and condemning tone of his *Summa*, which is particularly vivacious in several passages of the work: "we will make their stupidity manifest at greater length: and indeed they become insane in the process of their uttering";[16] "we are going to show, concerning those authors whom they call philosophers, were not at all philosophers, but were people to be despised as swindlers".[17] The above criticisms may be looked at as being strictly limited to the philosophical realm, whereas, in fact, the condemnation is primarily theological: "This obviously goes against the Holy Page";[18] "then these adversaries of the Christian faith have to be silenced";[19] "having irrevocably condemned these judges of the stars as infidels and blasphemers";[20] "once we have heard these blasphemies";[21] "clear is the unfaithfulness of those who produce such judgments".[22] Finally, concerning the theory of the great year, which has nothing to do with the Christian dogma of the Resurrection, Gerard defines astrology as "this heretical plague which by itself tears out many articles of the faith; hence such authors *should be eliminated from the community of the faithful*".[23]

For Gerard, the Greek and Arab origin of astrological books, instead of seeming like a treasure generously handed down to "the poor Latin world", is a cause for diffidence and denunciation:

Furthermore, how can we believe the enemies of the Christian faith, even when they have passed on to us astrological judgments on matters which are not against the Christian faith? I have read in their books about the rites of the pagans, even of the Saracens. But still in those authors who have written after the Incarnation of our Lord I find no mention commending the Christian religion. More and more, they date their years starting from Muhammed; Albumasar's *Introductio in scientia iudiciorum astrorum*, Messahalla's *De receptione*, not to speak of others, were not written in our language, but were translated from the Arabic into Latin by Johannes Hispalensis.[24]

Gerard ended his denunciation with true fierceness: "astrologers are not gods, but enemies of God".[25]

Gerard likened astrology to the "arts which are worthless or dangerous". In this, he followed Saint Augustine who, in his *De doctrina christiana* "where he deals with the arts of the magicians, of the haruspices [i.e., soothsayers], the augurs, the enchanters, he associates astrologers with them".[26] This grouping can explain a fundamental, though still puzzling feature in the organization of the *Speculum astronomiae*. Starting from the proem, the author takes particular care to distinguish between the science of astrology and his "perhaps harmless" (*fortassis innoxios*) sources from those occult books, which are "the enemies of true wisdom, that is of our Lord Jesus Christ".[27] The discussion of this distinction is taken up again in the two short final chapters devoted to those practices "which in truth do not deserve to be called sciences, but exotic jokes [*garamantiae*]".[28] The association of astrology (which enjoyed a far more respectable scientific status) with other divinatory arts is not justified in the *Speculum* by means of an explicit quotation from Saint Augustine: this silence leads us to think that the author's polemical target may have been closer at hand, namely Gerard, his fellow Dominican.

There are numerous points of agreement between Gerard and Albert. For example, both held that "not by necessity the soul follows the complexion of the body, but by will [...], and similarly human actions depend upon a voluntary cause, not upon the position of the stars".[29] However, the two authors did come to different conclusions about the problem of the compatibility of free will with the astrological system. Gerard held that astrology radically excluded free will. Albert, on the contrary (as was already evident in his well authenticated works) felt that the conciliation between the two tenets was possible by introducing a necessary, simple correction to the system. This had been done by the more responsible astrologers and Albert naturally agreed with them.

By and large, Gerard provided a deterministic interpretation of astrology, quite the opposite from what Albert had always done. In the first place, Gerard argues against the cautious formulas[30] of the astrologers: "by the order of God (*nutu Dei*)", "if God wished to change (*si Deus voluerit immutare*)", and the like, "under which words lies a poison meant to kill simple souls".[31] These were indeed the very same formulae upon which the author of the *Speculum* often relied. Gerard dismisses the interpretation of the stars as mere signs rather than causes of the future. On the one hand he quotes Augustine, and on the other Ptolemy, who "blasphemes against God saying that heavenly bodies force man to sin or to be good".[32]

Within the context of a polemic against certain definitions of the stars as "rulers of the dispositions" and "rulers of days and hours", Gerard states that "God created the stars to be ministers, not lords".[33] He therefore appears to accept the thesis (so widely used by Albert in all his texts) that the stars are instruments of the divine will, that is, secondary causes. Very soon thereafter, however, Gerard singles out the only type of causality attributable to the skies: "not as a material cause, nor as a formal or as a final one; hence the heavens should be an efficient cause, not a cause acting by election [...], but by nature, and thus an efficient cause operating naturally".[34] But even by presuming that the stars are only the secondary instruments of divine causality, moral determinism (which for Gerard is fundamentally intrinsic to astrology) makes this doctrine absurd and, above all, heretical:

> The first cause does not take its operation away from the secondary cause, but strengthens it, as is clear from what is said in the book *De causis*; therefore, if the stars make a man a murderer and a thief, so much more does the first cause, that is God, and that is a wicked thing to say.[35]

Albert had taught Gerard to discard the "Platonic" thesis of the animation of the skies, dealt with in depth in his *De caelo*. Gerard devotes his fifth distinction, "if heavenly bodies have a soul", to a discussion of this issue, and to a critique of Avicenna, Albumasar, and Zahel.

Elsewhere Gerard had condemned Avicenna for having upheld the possibility of a "natural prophecy";[36] here, on the other hand, he unhesitatingly declared him a heretic: "Avicenna maintains that, as our bodies are affected by heavenly bodies, so our wills are altered by the will of heavenly souls, and this is completely heretical".[37] Albumasar is also condemned: "He, who is in filth, is worthy of getting more filthy, and, with a blind mind, of falling from one error into another. And Albumasar was victim of this error, the same Albumasar whom the astral judges so much venerate".[38]

Albumasar traces the thesis of the soul of heavens to Aristotle. As far as Gerard was concerned, he was less interested than Albert in denying that the doctrine was genuinely Aristotelian, since his discussion did not hinge on natural and philosophical grounds, but on the theological one, over which Aristotle had no authority. Gerard relates in other passages "Albumasar says that planets are rational animated beings [...] though they do not possess free will".[39] It is striking that this same passage is quoted in the *Speculum*,[40] followed by another text by Albumasar on the non-voluntary motion of the sun, a text which Gerard also quotes word by word.

Albert nevertheless did argue against Albumasar, "because he writes that Aristotle said this, even though this cannot be found in any of the books of Aristotle that we have",[41] adding that the books of the *Metaphysics* dealing with the intelligences had not yet been translated. This is another point of contact between Albert and Gerard, as both had made use of the so-called middle translation (*Metaphysica mediae translationis*) and not yet of the Moerbekana.

A more theoretical encounter between the two can be found in the thesis that matter interposes a mediation to the influence of the stars, even at the level of natural effects, and thereby determines their contingency. Albert had always insisted on this. Gerard was at least convinced of the possibility of astrometeorology and other forms of "natural astrology" which concerned tides, animals, plants and minerals, which are dealt with separately in the last eight distinctions of the *Summa*. In *Pars tertia de reprobacione iudiciorum* he often refers to "the power of the magnet or of other stones, to astrological prognoses of doctors and to medicaments, to the diversity of air and complexions, and to other natural effects",[42] and even considers astrometeorology as true scientific knowledge: "in its causes, as a future cold wave can be known by zodiacal signs and by the arrangement of the stars".[43] The natural mediation for the temperament of man is quite different:

In the same way the complexion of some man can be traced not to the position of the stars, but rather to the nature of his parents, to his food, to the exercise he does, to the quality of the air around him, etc.[44]

Gerard introduces the distinction between "near causes" and remote ones, the latter being less efficacious. This distinction was taken from the works of Augustine, and would later be treasured by Giovanni Pico: "But if the stars were causes of health and sickness, they would be remote causes".[45] Gerard uses a syllogism to prove that the stars with their necessary motion cannot reveal the contingent futures.[46] He had already defined "the anticipations of physicians and astrologers" as "conjectural sciences":[47]

The causes [which are studied by those sciences] incline more towards one result than the other, and for this reason the events they are dealing with are contingent; these events are the object of astrological questions, for the most part [*ut in pluribus*]) determined, as is the case for the accidents proper to the inferior bodies; although natural causes are pre-disposed and pre-determined to a certain event, they can find an impediment; these effects cannot be known infallibly through their causes, but can only be conjectured with some certitude, as is the case with some natural events in the inferior world, such as rain, and so on [...]. Thus, if men were able to know all natural causes (which cannot happen in the present life) measuring the weight of some of the causes they would know that those things which appear contingent are necessary, such as rains, storms and other meteorological phenomena.[48] This could happen only if they knew all causes; indeed, the disposition of the inferior matter – which is subjected to contingency – is a concomitant cause, in addition to the movement of superior bodies; for that reason the effect is contingent.[49]

Later on, in the distinctio IV, "Utrum omnia de necessitate contingant, Gerard quotes a few famous pages from Albumasar in full (*Introductorium*, I,v, *caput de secta tertia*), precisely those which Lemay has discussed in connection with Aristotle. Once again, it is striking that these texts dealing with contingent actions which, once realized, become necessary, and with astrological prediction which is "between the necessary and the possible", will be taken up in the *Speculum astronomiae* with such an emphasis.[50]

I am therefore inclined to believe that it was the *Summa de astris* which provoked Albert to write his *Speculum* in order to uphold the astrologers' views, "even though those who bring accusations against them are our friends" (*licet accusatores eorum amici nostri sint*, XII/2-3). It is also notable that texts by Gerard, Albert, and the *Speculum* share numerous terminological and bibliographical features. How is it still possible, then, to subscribe the fundamental argument employed by Mandonnet to reject Albert's authorship of the *Speculum*, according to which the body of astrological information and beliefs present in the anonymous treatise could never find a place in the mind of a Dominican master?

CHAPTER SEVEN

ASTROLOGY IN ALBERT'S UNDISPUTED WORKS

The self-portrait of the author of the *Speculum astronomiae* presented in the proem to the work, has given rise to numerous manuscript variants as well as interpretations. I mean not only to underly the sentence, in which his research is said to have been very long and consequently the author cannot be supposed to be young ("iamdiu est, libros multos inspexi", XI/36); but mainly I am looking at the proem, where the wording of our edition, which has "a certain man zealous for faith and philosophy, putting each in its proper place", allows us to glimpse the hand of an author subscribing to the methodological distinction between the two methods ("duae viae"), i.e. the so-called "two truths" which characterized Albert's production. This distinction, which Nardi and Grabmann have noted as the fundamental criterion adopted by Albert in his theological and Aristotelian studies, also characterizes his examination of the theoretical problems of astrology.

Thus, when in his *De quatuor coaequaevis* Albert discusses the issue of "the soul of heaven" ("anima caeli"), he accepts the concept only in a metaphorical sense. The reading of a series of authorities ranging from Augustine and Gregorius to Avicenna and the hermetic *De sex principiis*, however, confirms the tenet that "stars are living beings", and establishes the concept of a "soul of the world" ("anima mundi"). At this point, Albert feels it appropriate to substantiate his Aristotelian conclusion with the usual methodological distinctions: "We said all this following those philosophers who do not contradict those Saints who deny that the heavens have a soul, apart from metaphor".[1]

Even more interesting with regard to our discussion is a passage appearing in the commentary on the *Metaphysics*, where Albert distinguishes two roads which might be followed in order to discuss the problem of the superior or heavenly influence on inferior bodies. In Aristotelian terms, one has to consider the influence that is transmitted through the motion of animated bodies; yet, one has also to take into account another influence – one which Plato and the theologians acknowledged– which cannot be measured, since it consists of a kind of irradiation or direct inspiration:

This is not said because, according to the opinions of the Peripatetics, there is no influence of the superior on the inferior world, except through the motion of a body endowed with soul.

However, this happens because the first animated body has an influence on and turns towards the most undetermined and most universal forms, which become increasingly determined with regard to matter, according to their differing descent towards one or another matter.[2]

The influence of the superior bodies is also defined according to the matter which is going to receive it ("materia recipiens"), following the important principle quoted above. Albert emphasizes it again in his commentary on the *De generatione et corruptione*. This principle would have momentous consequences as far as the practice of astrological prognostication was concerned: prognostication had to consider material factors, but was also granted the possibility of making indicative or possible predictions. This was not the case as far as the other kind of influence was concerned, and Albert limits his discussion to quoting the "theological" definition of it:

If there is some other radiation from the superior world on inferior things, as Plato said and as the theologians still say, this radiation cannot be investigated by reason, but, in order to study it, it is necessary to call upon other principles derived from the revelation of the spirit and religious faith; we must not talk of this irradiation in the terms of the peripatetic philosophy, since this knowledge cannot be put together with its principles.[3]

This influence acts "per actum coniuncti", that is, it transmits the action of the "intelligentia" and of the first motor "as far as it is possible that what is moved can receive the form of that motor". This cautionary note is typical of Albert's interpretation of astral influence: the limitations inherent in contingent things explain the lack of determination, and therefore the relative predictability of the stars's effect, especially when the prognostication does not take into account particular physical facts. This lack of determination does not imply a limitation of God's omnipotence since, according to Albert, the stars are not gods, but only "instruments of superior motions", secondary causes, or (as we find in the *Speculum*) dumb and deaf executors of God's will. In comparison with man's intellectual agent,

[it is] in this manner [that] the active intelligence which moves the orbit and the star or stars, transmits forms through the agency of the luminous stars, [as it is] by means of the light of the star that translates them into the matter which it moves; and by touching matter in this way, the intelligence leads it from potency to act. This is shown by the fact that astrologers, committed to these principles, which are the localization of stars, are able to prognosticate about the effects, induced in inferior bodies by the light of stars.[4]

In this way we can thus understand Plato's saying,

Since the God of gods told the superior celestial gods that he himself was the one responsible for the sower of generation, and that he would have transmitted them that sower to accomplish it.[5]

It is important to point out that Albert considered astrology a science well established on the basis of the experimental and practical improvements contributed by natural philosophers. Astrology thus approximated the sister occult art of alchemy or the discipline of medicine, both of which were regarded by Albert as interpreters of nature:

[Art] does not come from nature, but from an extrinsic principle, and occurs through a certain amount of violence, with the exception of the case in which the artificer is a minister of nature, as the doctor and the alchemist are sometimes."[6]

But astrological prognostication reached the level of science when it was able to measure the motion of the stars and of their orbits, and when it was based on astronomy, not "only on the conception of the intellect". The latter approach produced disorderly variations, which were inadequate to legitimatize a regular forecast. On the contrary, "the wise astrologers are able to make prognostications following those principles which regard the locations of the stars and concerning the effects that the stars, through their lights, produce on the inferior bodies".[7]

While judiciary astrology was seen as a philosophical-metaphysical discipline ("the Philosopher says that astrology is the second part of natural philosophy, and Ptolemy writes that he who practices judicial astrology, elections, and the observation of stars, will err if he is not [also] a natural philosopher"),[8] the passages from the commentary on the *Metaphysics* we have quoted above show that Albert considered astrology as descriptive astronomy –as a mathematical discipline measuring heavenly motions, and not concerned with their "principles".

The number and the plurality of traces and of the circles from which we can know the number of the motors, must be known through that philosophy which belongs to the number of mathematical sciences, and is particularly apt to the investigation of those motions: this [philosophy] is astrology, which studies these motions through three approaches, that is, through sight, reason and instruments.[9]

If, as Weisheipl has pointed out, Albert rejected "Plato's mistake", taken up by Grosseteste and Kilwardby, in considering "mathematics almost as the mother of all other sciences",[10] he did not deny the importance of this discipline. Several times he claims to have devoted his time, and also his pen, to mathematics.[11] In the program he announces at the beginning of the *Physica*, Albert postpones the treatment of "all mathematics" until an unspecified future date. In the commentary on the *Metaphysics*, he describes the field as already dealt with ("already clarified as far as possible in its natural and doctrinal features"), and as characterized by pedagogical steps, which, as they increase in abstraction, enable the student to approach metaphysics ("according to the fact that something divine is given to us, when natural knowledge accomplish our intellect, in the course of time and becoming perfect thanks to the doctrines concerning continuum"). Astronomy enjoyed a well defined and peculiar status among mathematical or doctrinal sciences. It possessed the rigor of all other abstract disciplines.

That [science] does in fact deal with the sensible incorruptible substance, that is, the heavens and their motions: the science builds a theoretical speculation from their basis. On the contrary, other mathematical disciplines do not deal with substances, as is the case with arithmetic, which is concerned with numbers, and geometry, which is concerned with immobile spaces, and music, which is concerned with the harmonic properties in melodic singing. Since astrology is the only discipline which considers the motion of heavenly orbs, we take from it those results which appear to be the wisest. [...] we will not investigate the truth of those things following our own opinion, but by relating what has been said by some of the most reliable mathematicians.[12]

This surveying mood, accompanied by critical observations, is a constant feature of Albert's writings. In the commentary on the *Sentences*, the *De quatuor coaequaevis* and the *De Caelo*, as well as in the passages of the commentary on the *Metaphysics*, all opinions are discussed and compared with each other and with elaborations on the one side by the Chaldeans, Eudoxus, Callippus and Ptolemy, and on the other by Aristotle and Alpetragius. The very same approach is used by the author of the *Speculum astronomiae* in his well organized and clear compilation of texts and problems in judiciary astrology.

The ranking of astronomy in between mathematical sciences and metaphysical knowledge reflects the definition of the discipline as "intermediary between the necessary and the possible", a definition which goes back to late Antiquity and probably Albert has borrowed from the *Centiloquium*, not only from Abu Ma'shar. In the *De fato*, heavenly influence is characterized by simplicity, derived from "the simplicity of the circulation of the common circle", and by the multiplicity of virtues, derived from the number of stars carried around by that circle: the heavenly influence

derives from many stars, many places, and spaces, and images, and rays, and conjunctions, and interferences, and from various angles produced by the intersection of the rays of the heavenly bodies, and from the production of rays towards the center in which only, according to Ptolemy, all the virtues of those bodies which are in the celestial circle are collected and concentrated. This is a middle form between the necessary and the possible; whatever belongs to the motions of celestial orbs, is, in fact, necessary, whereas whatever belongs to the matter of things which can be generated and corrupted is mutable and possible. This form produced by the celestial circle and related to things which can be generated and corrupted occupies a middle position between the two [i.e. the necessary and the possible].[13]

Albert refers here to the *Centiloquium*, where, in the "first word", it is said emphatically that "just as we acquire a dubious knowledge of reality from matter, and certainty by means of forms, so the prognostications which I am giving you are intermediary between the necessary and the possible".[14] Referring again to Ptolemy, Albert adds that the so called astral-destiny

though it derives from a necessary principle, it is nevertheless mutable and contingent. The reason for this is well explained by Ptolemy in the *Quadripartitum*, where he says that the virtues of stars are realized in the inferior things through what is different and accidental in them (*per aliud et per accidens*):

1. through what is different, because they are realized in inferior bodies through the sphere of active and passive things, that influence inferior things through their active and passive qualities;
2. through accident, because this form [the virtue of stars] emanates from a necessary and immutable cause, it happens in contingent and mutable things. Its mutability thus derives from two reasons, firstly from the quality of elements, through which it is transmitted to things which are generated, and secondly from the nature of things which are generated, on which it acts as upon a subject. This is actually fate.[15]

It was not only Ptolemy, therefore, but also Aristotle, who confirmed the lack of determination of heavenly influences. This was particularly true even concerning the wholly natural case of the length of individual life; its end did not always correspond to the data of each "nativity", but its prevision was in itself simpler than that of orders of phenomena which involved spiritual faculties and the free will.

[Life] can be accidentally interfered with by poisonous food or by violent death, or by any number of ways: this is called by Aristotle "the inequality of matter", since it is through various accidents that there is a different disposition in what is moved by those orbs. So, men die more or less quickly of what should be according to nature, and the same also happens to all other animated things.[16]

The horoscope was drawn at birth theoretically, in order to calculate the length of an individual's life, "since the planets placed in the periodical circle, when they are stronger, they attribute more years of life"; and this is known,

since he who would attain the knowledge the virtues of signs and of those stars placed in them within the eclipse, when something is born, he will be able to make prognostications within the limits of celestial influences, concerning the entire life of the newborn; nevertheless, this act would not cause necessity, since, as we have said, the prognostication could accidentally be hindered.[17]

If lack of determination holds true for an event as natural as death, it would hold even more true as far as the intellective soul was concerned. This view was supported not only by theologians, but also by philosophers and astrologers.[18] Indeed,[19]

that the soul is especially subject to the motions of stars is against the doctrine of all the Peripateticians and of Ptolemy. The soul does, in fact, understand those things the highest in the sphere, and freely moves away from those things towards which the motion of stars would incline it, and it also perceives other things thanks to its wisdom and intelligence, as Ptolemy says.[20]

Albert's views on the issue of the relationship between astral influence and free will remain constant and coherent throughout all his writings. Even in the *Summa theologiae* he says that inclinations are derived from the stars, though those inclinations should never be considered necessary:

This quality of stars is capable of attracting bodies and of changing even the souls of brutes; but it cannot change nor attract with compelling necessity the soul and the will of man. These are created in freedom, after the image of God, and are the masters of their own actions and choices. Though, as a soul, it may be inclined towards its own body according to the powers proper to its organs (such as the powers of the sensitive and the vegetative souls), it can be attracted by these powers in terms of inclination, but not of compulsion.[21]

In the *Summa theologiae*, Albert maintains that this thesis could be reconciled with the anti-astrological polemic written by Augustine in the *De civitate Dei*, where the author was prepared to make allowances for natural causality: "if he sees the principle of meteorological variation in the movement of the sun ["in solaribus accessibus et decessibus"], as he sees it in the phases of the moon, he can equally see the principle of other phenomena, such as mollusks and shells and extraordinary oceanic tides". Yet, he "does not wish to submit the will of the soul to the positions of the stars".[22]

In the *De quatuor coaequaevis*, an early theological work where the question "On the effect of the movement of the heaven and of the stars in the lower world" is discussed at length, Albert starts from this distinction. He then proceeds to remark on the function of signs proper to heavenly phenomena.

The stars have the power to alter the elements, to change the complexions of men, to affect human mutations, and moreover, to provoke an inclination to action, and even to determine the issue of battles.[23]

Albert immediately raises an objection against the last two entries of this list arguing "that therefore actions and battles are among those things that are subject to free will, [the stars] seem to have power over free will".[24] He appeals to the authority of John of Damascus, who had vindicated the freedom of will with four theological-moral arguments and writes:

The stars are never causes of our actions, indeed we have been made free agents by the Creator and are the masters of our acts.[25]

This apparent contradiction is discarded by Albert, who in the *Super ethica* (a commentary on Aristotle he had dictated to Thomas Aquinas between 1250 and 1252) had already raised the question of "whether anyone becomes evil due to the stars rising at the moment of his birth", taking the lead from poetical utterances by Horace and medieval authors, and from two passages by Aristotle. If every kind of lower motion in nature must, according to Aristotle, be referred to a higher motion, which causes it to occur, man's affairs, that are "motions in the lower world [*motus in natura inferiori*], are caused by the motions of the superior world". Since the latter motions are necessary ("the rotations of the sun are necessary, and do not result from its deliberation"), it would thus seem that "even our actions are necessary, since we are produced by higher motions".[26]

Moreover, if we must believe what Aristotle said in the *De generatione animalium*, "man has the principle of his generation in the motions of the heavens", and if the principle of generation is also the principle of every mutation, then

all diversity which happens in the conception of each individual, can be traced to certain celestial principles. Hence, human actions too take their beginning from these principles".[27]

Amongst the arguments in favor of astral necessity, the most philosophical seems to be that which refers to the oneness of the principle of being and of knowledge. "In astrology the way in which human actions can be foreknown from the motions of the heavens is taught; therefore, they [human actions] have their effective principles in those motions".[28] Albert rejects some objections and in particular the opinion that "corporal motions follow those of the superior world, but the motions of the soul do not".[29] Albert is quoting here a passage from the *De causis*, where two levels and two means of causation are specified: "as the soul is impressed upon the body, so is the intelligence upon the soul".[30]

Albert's opposition to the thesis is drawn largely from the Fathers (Augustine and John of Damascus), but also from Aristotle's *De somno et vigilia*, where the fallibility of predictions[31] is upheld. He also cites Ptolemy's *Quadripartitum*:

In his scientific treatment [book 1] chapter 3 *On judgments* Ptolemy says: 'we must not believe that the superior bodies proceed in an inevitable manner, as far as their significations are concerned, as do things which inevitably happen through divine disposition [...]. Moreover, we must believe that the variation of terrestrial beings occurs in a natural, variable way, in an accidental way, and that these beings are affected by the first acts of higher beings, and that some of the accidents happening to men with general damage, do not occur through other properties of reality, as in the great perturbations of the air from which we can hardly find shelter, there are various casualties' [...]. From this work then we understand that the superior beings do not impose necessity on the inferior; for, many facts occur to men because of their own complexion or because of causes other than the effects of the stars.[32]

Even in his *Super Ethica* Albert refers to the inclination to solve the problem of moral freedom:

Fate, that is, the inflexions in the first causes do not impose necessity by affecting the will, but rather by giving it an inclination or a sort of disposition toward something [...] in the same way, dispositions cannot be completely drawn from the nature of man: for instance, someone can always be prone to anger; yet, habits contrary to such disposition can grow in the soul, if the same man will make every effort against it.[33]

Albert approaches this problem with the help of a distinction typical of the Aristotelian psychology. The soul can be regarded in two ways,

either it is considered as the actuality (*actus*) of the body, with regard to powers imprinted on the body and thus per accidens impressed upon the soul by the motions of the heavens, in so far as it follows the body's affections; or it is not considered as the actuality of any body, as far as the rational powers of the soul are concerned, and in this case no impression is made upon it by the motions of the heavens.[34]

This interesting example is clearly that of the rational soul, which is the actuality of the organic physical body of a living man. But also, in this case, there is no necessity.

The will, which is the principle of our actions, by which we are good or bad, is a power of the rational soul; and thus it is clear that we are not compelled by necessity to be good or bad by the disposition of our birth according to the effects of the stars, but rather only in so far as the stars implant certain dispositions in the nature of our body, which has some tendency toward anger or concupiscence; but it is not by necessity that the soul follow this tendency.[35] Albert does not say anything different in the *De mineralibus*, the most "astrological" of all his naturalistic production:

For, in man there is a two-fold principle of action, namely, nature and will: nature is ruled by the stars, will on the other hand is free. But unless will puts up a concrete defense, it is bound to be influenced by nature and hardened, so that it too, like nature, will be inclined to act according to the motions and the configurations of the stars. This Plato proves with reference to the actions of children, who are not yet able to resist nature and the inclination of the stars, by means of their free will ["libertate voluntatis"].[36]

Albert considers, and discards the contradiction between astral causality and freedom. It is worthwhile to stress Albert's consistency with respect to this problem, apparent both in his well-authenticated writings and the controversial *Speculum astronomiae*. It is equally important to stress that both the latter work, and the certainly authentic *De quatuor coaequaevis*, share the definition of the stars as signs rather than causes:

the stars have the force and the condition of a sign ["vim et rationem signi"] concerning those things that exist in transmutable matter, and even those that are connected to it.[37]

Matter, by definition, can be generated, corrupted and changed. "Things that are connected to it" indicate those sensitive and vegetative souls, "which by necessity follow the transmutation of matter".[38] But there is a third possibility:

There are some things that are in a certain, not absolute sense ["secundum quid et non sempliciter"], dependent on matter. One such thing is the soul of man. For this reason we say that blood rising around the heart inclines the soul of man to anger, though he does not by necessity become angry; and according to the degree to which the soul is inclined to matter and dominated by its own complexion, so that a constellation also has the power in a certain, not absolute sense. Otherwise, with free will and deliberation chance also will be taken away, if nothing, which is to be said about the future, would be contingent "ad utrumlibet" (as the Philosopher puts it very well).[39]

From these considerations Albert embarks upon a classification of the causes:

when it is said that the stars have power on inferior things, it is to be understood that they have such power as universal first causes moving immediate and "propter quid" causes; for that reason a constellation is not always followed by its effects in a necessary way.[40]

According to John of Damascus, the stars are not the causes of generation and corruption, but only the signs.[41] This is also true as far as simple meteorological phenomena are concerned (upon which the inclinations of temperament can also depend). Yet, these signs are merely indicative, and do not imply necessity.

These are great signs of rain and of meteorological changes. Perhaps, as someone would say, this happens because these are not causes, but signs of wars. But the quality of the air, produced in a way or another by the Sun, the Moon, and the stars, constitutes various complexions, habits, and dispositions. The aim of John of Damascus is to say that the sign is less than the cause: the cause indeeed, as Boethius says in his *Topics*, is what is necessarily followed by an effect; the sign, on the other hand, is a cause distantly removed, which does not act as a cause, without the conjunction of other causes.[42]

Albert repeatedly insists on the numerous mediations which occur between universal causes and their remote particular consequences:

Not the soul's [...], but only the body's motions can be reduced to the motion of the heavens; also these corporeal motions are not indeed compelled to follow by necessity celestial motions, for things are made subject to necessity not by first causes, but by proximate causes. Otherwise everything would occur by necessity, since the cause would be necessary either in the case of those things that occur frequently or in the case of rare events. Proximate causes are variable, as are the motions they produce. Hence, it is not by the power of the proximate causes that the superior world does not impose itself by necessity, but by a defect of the alteration that takes place in them.[43]

The aporia posed by the universal nature of astral causes was not a serious embarrassment to Albert. On the contrary, when much later Oresme and Pico in his *Disputationes adversus astrologiam iudiciarum* discuss astral causes, they deny that they are adequate to explain contingent and particular effects on account of this very aporia. Albert once again solves the contradiction by defining astrological prediction as probable, and by insisting on the Ptolemaic concept of "stellae secundae", with which he was very familiar.

Astrologers predict future events from first causes, which do not impose any necessity; and hence their judgment is not necessary, but –as Ptolemy says– conjectural according to variations in the secondary causes. It is for this reason that Ptolemy says that the prediction is more certain, if we can know the powers of higher causes as they are incorporated in the secondary causes: and he calls these secondary stars.[44]

It is difficult to analyze, and impossible to measure the way in which higher causes have an effect on secondary ones with any accuracy. This accounts for the fallibility of current astrological science and practice:

because all the principles of contrary dispositions are in the motion of the heavens; so, all the dispositions of inferior bodies have their principles in celestial motions, but no [disposition] happens by necessity.[45]

The mediation interposed by matter is also the principle of the efficacy of astrology and of its practical use. If astrological prediction were necessary in an absolute sense, it would be useless beyond its cognitive value. On the contrary, since astrology gives the probable prediction of mixed effects partly dependent on sublunar components, its cautious and sensible practice enables us to postpone, modify or even avoid some of the inauspicious effects of astral causality.

Those qualities, because of contrary natural properties and different dispositions found in matter, often exclude the effects of heavenly motions. For this reason Ptolemy says that "the wise men dominate the stars.[46]

This Ptolemaic saying was very dear to Albert, who used it both in the *Speculum*[47] and in well-authenticated works such as *De fato*, *De natura locorum* and the *Summa theologiae*.

The numerous points of theoretical agreement already examined make it superfluous to detail punctiliously all the themes and terms –in fact topics– which can be found in the undisputed Albertinian texts and in the *Speculum astronomiae*. It is sufficient to list the convergence of astrology and medicine, or, for instance, the treatment of quartan fever by preventing the "minoratio humorum";[48] the theory of the critical days;[49] the theological topics of the difference between twins;[50] the "elementary" qualities of hot, cold, dry, and humid stars, the mixture and constitution of primary qualities depending on planets,[51] and so on. It is even more interesting to note Albert's recurrent concern – in the *De proprietatibus rerum*, *De quatuor coaequaevis*, *De fato*, and in the *Summa theologiae* – with the explanation of the technical terms transliterated from Arabic. In the field of astrology, the author of the *Speculum* acknowledges the debt that a culturally deprived Latin world (*pauper latinitas*) owes to Arab science.[52] At this time, Arabic learning was available in Latin translations in which the terms were often kept in Arabic transliteration and were paraphrased and explained in Latin, as is the case with Alcabitius's work. This evidence – which Mandonnet could have stressed as similar to the linguistic concerns expressed by Roger Bacon[53] – is further proof of the close connection between the *Speculum* and the texts definitely written by Albert.

The same kind of consideration and conclusion can be reached with respect to the allusions to the necromantic Antichrist, which Semeria believed of Bacon. There is no doubt that the proximity of the Antichrist, and the necromantic character of his power was often announced by Bacon, as well as by other Franciscans worried by the inroads made by pagans and "necromantic" tartars. On the one hand, the theme was topical during the mid-twelfth century, witness the *De novissorum temporum periculis* by William of Saint-Amour (1255-56),[53] and Albert's *Summa theologiae*, where he appears to share this belief.

> At the time of the Antichrist, the power of demons will be bigger than now [...], as it is said in the Second Epistle to the Thessalonians, II, the coming of the Antichrist will be according to the activities of Satan, relatively to every power, prodiges, and mendacious signs.[54]

This belief explains why the author of the *Speculum astronomiae* – otherwise so cautious to avoid all judgment on black magic treatises – recommends the preservation of those texts in his conclusions, since in a near future they might have become very useful in opposing the Antichrist with its own weapons and deceptions.

Although we cannot exclude the possibility that the *Speculum* was the result of a collaboration between Albert and one of his pupils, it is important to draw attention to statements, based on the preparatory research Caroti, Pereira and Zamponi have undertaken for their edition of the *Speculum*. Readers may have noticed that the hypothesis suggested, that the *Speculum* might have been the result of a team effort, did not distinguish between the gradual compilation of the bibliographical information and the writing of the dense and rigorous theoretical parts of the work. Indeed, the main result of our investigation –based, in part, on the previous studies by Steinschneider, Cumont and Thorndike– consists of the identification of the literary models at the basis of these philosophical chapters. A few examples will suffice. Chapter 1, which defines the field and the problems of astronomy, reproduces word for word the headings and the initial passages of some of the sections ("differentiae") of al-Farghani's *De scientia astrorum* (*Numerus mensium Arabum*). A full understanding of this chapter is made considerably easier when one takes into account the full context of the Arabic treatise. The same observation holds true for chapter IV and various other passages, which are heavily indebted to the *Liber introductorius* by al-Qabī'sī. Chapter VIII, relies on Abû 'Alî's *De iudiciis nativitatum*. Chapter III, which introduces the reader to the foundations of astrology, "*secunda magna sapientia* (the second great wisdom)", is based on a source worth underlining, the *Introductorium maius* by Abû Ma'shar, quoted from the unpublished translation by Johannes Hispalensis. The latter work is mentioned again in chapter VI with great emphasis and ranked together with the *Quadripartitum*. Up to now we have mentioned more or less extensive borrowings employed to draft technical definitions. It is only necessary to mention only a few theoretical formulations, which are equally taken from the literature on astrology. Chapters XII and XIV, where, after the bibliographical sections, the author readdresses the definition and defense of judiciary science, are indebted to Abû Ma'shar's *Introductorium*. From the same author (as well as from the *Corpus Hermeticum* and from the Jewish astrologer, Māshā'allāh) come the recurrent expression "by order of God" (*"nutu Dei", "iussu Dei"*). This formula was designed to replace the Aristotelic notion of the "immobile motor" (with respect to astral motors) with the Plotinian idea of the stars as signs and instruments, and the Coranic (or generally religious) principle of a personalized Divine Will. "By the order of God" (*"iussu dei", "nutu dei", "dispositor iussu dei", "significare effectum et distructionem iussu dei"*) are expressions which frequently occur in the *Speculum* (though in the misleading reinterpretations of the vulgata, we find the expression "*motu dei*") and in other, definitely authentic Albertinian texts. The concept underlying those expressions is essential to the interpretation of heavenly motors as secondary causes, which are not independent, but instead embody and serve the Divine Will and Omnipotence. These terms are used in order to set astrology in a conceptual framework acceptable to Christian orthodoxy.

CHAPTER EIGHT

ARE "DEAF AND DUMB" STARS AND THEIR MOVERS AT THE ORIGINS OF MODERN SCIENCE? ANOTHER HISTORIOGRAPHICAL CASE-STUDY

I want to look at a question which seemed "curious" and "fatuous" to Albert,[1] and which may seem even more so to the modern reader. However, the Master General of the Dominicans, John of Vercelli thought that it called for high-level consultation and in 1271 submitted forty-three problems to three eminent Dominicans: Albert, Thomas Aquinas, and Robert Kilwardby.

Today this question reminds us of current discussions on the possibly creative character of artificial intelligence. Yet, above all it refers to the classical Aristotelian notion of the immobile motor, to the Platonic one of ideas-numbers, and to the discussions of unity of the first principle and the plurality of the worlds. H.C. Wolfson's fundamental comments on Thomas's texts from *De caelo* and *Metaphysics* confirm this. In addition to Aristotelian texts commented on by Jaeger, Wolfson discovers and comments upon texts of Avicenna, Averroes and Maimonides: "We shall examine here the explanations of Avicenna and Averroës and, in connection with them, we shall make reference also to Maimonides and St. Thomas".[2]

As Pierre Mandonnet observed in a 1930 review article on the two texts then available, "the list John submitted to St. Thomas and Kilwardby was merely the collection of the propositions he had collected while visiting the Lombardy province after the General Chapter in Milan in 1270".[3] However Mandonnet was reluctant to admit the importance given by Chenu to the doctrinal content in the most numerous group of articles, those concerning the Celestial Movers.[4] But paradoxically it was precisely this theme which divided and still divides the best Dominican historians. On one side, from Mandonnet to Van Steenberghen and Weisheipl, there is the traditional position governed by Neo-Thomist historiographical criteria. That is, wanting to combat the epistemological models of the proto-posivitists, the Neo-Thomists are conditioned by these models and acquire a type of inferiority complex which keeps them from admitting that Aquinas's science is not as good as Newton's (or even as good as that of the more advanced of his own contemporaries). On the other side and outside of these paradigms, there stand the more up-to-date and unbiased readings of three texts which have gradually come to light and which help to finish the puzzle: Thomas, whose text was always known, Kilwardby, known since 1930, and Albert, only found in 1960. It must be remembered that Chenu (and most of the historians later to be quoted) did not know Albert's answer...[5]

In this case, as in many others, Chenu had been extremely original: Thomas Litt followed his method, giving rise to widespread polemics. In 1930 Chenu made a statement about the forty-three problems which is worth re-reading today in the light of the current *histoire des mentalités* method:

> This controversy, in which such great personalities participated, lays bare intellectual preoccupations which will only surprise those who, in studying the Middle Ages, read nothing but speculative literature concerning lofty mataphysical questions. Actually, among the huge clerical population circulating through the universities, these humanly and religiously extremely interesting speculations are mixed with bizarre curiosities and infantile questionings; the religion of simple people and the scientific ideas of the period, even further debased by vernacular translations, offered many opportunities for these curiosities .... The cosmological problem at the center of most of the questions was a serious problem which the thought of the masters delighted in. Therefore the consultation is a document which helps us to know the world system which scientists and theologians were formulating at the time. It was precisely that mixture of theological views, metaphysical principles, cosmological theories, and celestial physics (i.e.astrology), which revealed the particular aspect of the medieval mentality to which the modern spirit, or at least the spirit of modern scientists, has become most extraneous.[6]

Thus "new life is conveyed to the documents which have come down to us from that small theological accident" of 1271:

> put back into their historical framework and their doctrinal context, they reveal much more than a superficial repetition of the main works of St. Thomas or Kilwardby; their expositions, in being more summary, are less *atemporal* than the long articles of the *Summas* and closer to the intellectual preoccupations of the thirteenth-century clerical community.[7]

When Litt asked ironically if "St. Thomas's philosophical system can be amputated from the pseudo-metaphysics of the celestial spheres without serious consequences",[8] Van Steenberghen had no hesitation in answering affirmatively. He was certain that "these pseudo-metaphysical and pseudo-scientific considerations are erroneous or imaginary applications of metaphysical principles and therefore do not condition them at all". In sum the metaphysical principles are untouched by the cultural framework in which they were formulated, and this framework does not disturb or interfere with the Neo-Thomist recovery of Aquinas's metaphysical and theological system which Leo XIII had proclaimed the only metaphysics capable of confronting contemporary sciences and equipped to furnish them with a metaphysical basis.[9]

Mandonnet had immediately and discreetly criticized his young Dominican brother for the attention he had dedicated to the articles concerning the nature of the Celestial Movers: "It is by no means clear why modern science should be interested in knowing what causes the movements of the stars, instead that it merely describes the heavenly bodies and determines the laws of their movement. What difference does it make to astronomy if the stars move by themselves or if they are acted upon by the separate intelligences or by the immediate action of God Himself?"[10] Mandonnet seems to be assuming a conventionalistic epistemology here or at least to assign purely descriptive aims to science. This is followed by a reference to Pierre Duhem and his well-known thesis that it was only with Buridan and the impetus theory that there was a qualitative leap from Ancient (anthropomorphic) science to Modern science. Several years later Chenu also came to consider this formula, on which according to Mandonnet, Duhem had changed his mind. [11]

There is a controversial historiographical point here, regarding Mandonnet's claim that Kilwardby had intuitively foreseen ideas not formulated until later periods. Were his distinctions between angels and Aristotelian intelligences (i.e. impersonal Celestial Movers) an anticipation of both Galileo's thesis (which introduced a celestial mechanics following laws of gravity which are homologous with earthly ones) and of the Cartesian idea of the "*chiquenaude*" (the small initial push which the creator gave to the world system especially the stars, after which, as a "*Dieu faitnéant*", he rested)? Is this the "anthropomorphism" which, according to Mandonnet, constitutes a milestone in modern science and philosophy?[12] On the contrary, a close look at Albert's and Kilwardby's answers to the 1271 consultation (and at the whole question of the Movers of the heavenly bodies) shows that they represented an attempt to eliminate anthropomorphism from the relationship between Heavenly Substances and the sublunar world. The scientific aspect of the question – I mean: the way in which it approaches or prepares for a celestial mechanics – can be summed up in Chenu's own words: "...two theories current at the time were trying to explain the circular turning of the celestial orbs. Some thought they were animated by an intellectual soul, having the function and properties of a form and, at the same time, the intellectual faculties of a spirit; others recurred to a 'separate Mover', a moving Intelligence which was neither act nor form of the mobile". This separate or angelic Intelligence did not have the same type of contact with the mobile, which was considered necessary in Aristotelian mechanics, but "a propulsive contact applied with a voluntary command sufficed" (implying that the moving Intelligence had will and other personal, anthropomorphic characteristics). However, Chenu observed, "in both these hypotheses heavenly bodies are denied a 'natural' virtue, which eliminates the need for a mover, and this contradicts the inviolable [Aristotelian] axiom 'everything moved supposes a mover distinct from the thing moved'", but contiguous to it for the duration of the movement. According to Duhem, Buridan and Ockhamist physics had, for the first time, " denied the difference between the two worlds, the heterogeneousness of the essence of the heavenly bodies, and consequently they had explained the movements of the stars with the same laws as those which governed terrestrial bodies, for which they had proposed the impetus theory", rejecting Aristotelian theory in this case as well. "This impetus continues until it is destroyed by the resistance of the means and by the action of gravity. In the case in which these resistances were not produced, the force of the impetus, proportionate to the volume and the density of the thing moved, would remain: the movement would continue indefinitely after the mover's action had ceased. This is precisely the case of the heavenly bodies, although they are not heavy. The law of inertia applies to them as it does on earth. It is useless to recur to separate substances and angels. God's general action alone on every living creature is sufficient".

In this conception Duhem had seen "a perfect application of the principle of economy which dominates nominalistic philosophy",[13] but Chenu, by highlighting Kilwardby's answers, which he had discovered, took this conception back to the thirteenth century. Kilwardby had in fact already maintained, "1. the identity of celestial and terrestrial mechanics in a regime dominated by gravity and by the law of inertia; 2. that every body, by virtue of its natural constitution in creation, possesses its own inclination, an energy proportionate to its own cosmic role, which impels it as though 'by the instinct of its own weight' (qu. 3); 3. that the theory of the animation of the stars is a philosophical opinion which has some basis; but that it does not go beyond a reasonable hypothesis ('*suppositio rationabilis*') which Christian doctrine can not admit (qu. 4)". On the other hand, Kilwardby considers "the case of miracles, which God can perform through the ministry of the angels, as obviously a case reserved for special authority; but here we go out of the natural order, and it is not different from God's miraculous works in the terrestrial sphere".

It is certainly very interesting that Chenu considers "the link which was established by Duhem between Ockham and Buridan's nominalistic method and the takeoff point for modern science" broken. However, no less interesting is the fact he noted that "Kilwardby does not present this solution as his personal property, but refers to it as an already existing opinion".[14] It seems worthwhile therefore to examine the prehistory of the question: who before Kilwardby had supported analogous theses, and hence ones which were different from or even contrary to those of Thomas Aquinas?

Against their common background of an Aristotelian-Ptolemaic cosmos, Chenu takes note of the "particular agreement between St. Thomas and Kilwardby concerning the problem of the animation of the stars [*anima ut forma*], and, in Christian terms, of the animation of the stars by means of the angels. Although the question was theoretically a free one for the individual believer, the common opinion of the Doctors had come up with a negative solution. St. Thomas, remaining faithful as usual to his method, makes his point by referring to the Christian tradition, called on St. John of Damascus, who explicitly denied animation, and St. Augustine, who was undecided", not to mention pseudo-Dionysius, while "Kilwardby, more personally, refused animation after citing the opinion of Aristotle and a few other philosophers". Both, "like most of their contemporaries, eliminated that *concordisme* which for a long time had brought the spirits of angels closer to Christian revelation".

They also agree in attributing influence on natural phenomena to movements of the heavenly bodies, which can be explained, according to Kilwardby, "influentia luminis atque virtutis".[15]

But soon their differences begin, both concerning the "presiding" of the angels over the heavenly bodies and the influence of the heavenly bodies on the terrestrial ones. Their agreement on the hierarchy of connected actions, which goes from angels to heavenly bodies, and from heavenly bodies to the sublunar ones, i.e. to the natural movements of the elements and of their compounds, is broken. Concerning the angels, Thomas says: "I don't remember ever having read that any of the saints or philosophers denied that celestial bodies are moved by a spiritual creature" (qu. 3),[16] whereas his "categorical answer", according to Kilwardby, "is not philosophical, nor do I remember that it has been approved as true and certain by any of the saints".[17]

The question which, from the thirteenth century onwards, had become commonplace, was referred to time and time again until the Renaissance.

While – much more than the variable theses left to us by Grosseteste, a "pivotal figure" in this question – the three answers to John of Vercelli are among the first documents to treat it extensively, it will crop up again in the works of other authors, especially some of Albert's disciples such as Giles of Lessines,[18] and, a few years later, Dietrich von Freiberg.[19] Quite similar was another question, which was asked in the same period, whether there was any type of matter in heavenly bodies and if so whether they had a potentiality similar to that of corruptible bodies, or a "potentiality referred to the sphere of the substance". i.e. a Thomistic "substantial form".[20]

More than half of the forty-three miscellaneous "problems" dealt with the identification of the angels (or, to be more precise, with the hierarchy of angelic *Virtutes*), with intelligences. They demonstrate the uneasiness which many Dominicans, more familiar with theological topoi used in preaching than with the commentaries and the *quaestiones* of the Peripatetics, had been experiencing ever since Aristotle's Celestial Movers were being made to combine with the Biblical figure of the angel. As rigthly Chenu observed with his usual frankness, the 1271 consultation was "a testimony to the agitation which an angelic physics inspired by Aristotle was provoking among theologians".[21] Along this line the most typical and extreme document is in another *Responsio de 6 articulis* which may also have been sent in 1271 by Aquinas to a certain Gerard, Dominican reader at Besançon:[22] according to Thomas the images used by Gerard for the creation of the stars and the form of the one seen by the Three Wise Men were too frivolous even to use for sermons.[23] The fact that such uneasiness was widespread is demonstrated by the frequency with which such questions were repeated. In the case of Thomas alone, many of the thirty questions which had been sent to him slightly earlier by Bassiano of Lodi, the Dominican reader in Venice, centered on the same theme. Aquinas's answer has two redactions: the *Responsio de 30 articulis* preceded (according to the chronology of the critical edition), while the *de 36 articulis* (or to the "articuli iterum remissi") came immediately after the 1271 answers to the Master of the Order, and in any case the three in the series overlap and complete each other. The *Responsio de 30 articulis*, together with the short works *De sortibus ad Jacobum de Tonengo* (from Orvieto and datable to the end of 1263),[24] the *De iudiciis astrorum ad quendam militem ultramontanum* (which could even be earlier)[25] and the *De occultis operationibus naturae* dating from his second stay in Paris (1269-1270),[26] could have been known to Albert and Kilwardby before they wrote their answers to the 1271 consultation. It is unlikely that priority could have been given to the youngest of the three prelates, as Mandonnet supposed.[27] However, when he proposed this hypothesis, he conveyed an impression of familiarity, which is quite justified. Thomas's positions remained quite constant, and Kilwardby could have already been familiar with the theses expressed in the works cited above, which were certainly available to an illustrious brother, even before he could have known of the 1271 *Responsio* and the still later (1270-1273) *De substantiis separatis*.

In truth, a number of authoritative scholars – from Pierre Duhem[28] to Mandonnet, from Chenu to Litt, and from Weisheipl[29] and Vansteenkiste[30] to Flasch[31] – have already recognized the presence and the pertinence of the issue under study at the metaphysical, cosmological, and scientific levels. What I would like to underline here is another aspect. What is the historical significance of the surprising circumstance which causes the most eminent Dominican theologians and philosophers to participate, at the request of their Master General, in a debate on Celestial Movers and on the nature of their causality? In questioning their identification with the angels, the three theologians were discussing their personal, voluntary, and hence anthropomorphic nature.

CHAPTER NINE

NOT THE HEAVENS, BUT GOD ALONE IS ENDOWED WITH LIFE AND THE STARS ARE SIMPLY HIS INSTRUMENTS

While important scientific ideas have already been noted, especially in Kilwardby's consultation, I would primarily like to show that the thesis of the non-anthropomorphic nature of the Celestial Movers is very important to the founding of a "natural" astrology within the Aristotelian system. This thesis forms the theoretical basis of the occult and yet still "natural" operations of magic. Whoever, like Albert, maintains this thesis also holds that magic is completely free of the evocations of necromancy and ceremonial rites or practices, which become impossible when there is no anthropomorphic and voluntary referent capable of responding to their requests. The refusal of the third type of talisman in the *Speculum astronomiae*[1] is not only a prudential measure to avoid being connected with some condemned practices, which can be summarily indicated by referring to the *Picatrix* and which were studied by Thorndike.[2] Its primary purpose is to maintain the coherence of this thesis. The *Speculum astronomiae*[3] assumes a decidedly negative position on this point, concerning the anthropomorphic animation of heavens, and it is extremely coherent with Albert's answer to the forty-three articles and with others of his later works which are surely authentic. It is interesting to see that this thesis in the *Speculum astronomiae* corresponds to the position taken by Albert in works definitely by him (for example, the commentary on the *Sentences* and the *De quattuor coaequevis* in the *Summa de creaturis*, and later the *De coelo*, the *Metaphysica*, the *Problemata determinata XLIII* and the *De causis et processu universitatis*), although one must admit that the statements in some of the later works are sharper than those to be found in his early theological writings. The *Speculum*'s formulation concerning "deaf and dumb stars",[4] inert instruments of a God who alone is endowed with life, is the most salient statement in the entire dialectic which was recorded in and also provoked by the 1271 consultation –certainly the most extensive and important among those promoted by John of Vercelli, who had sent at least another one to Thomas Aquinas in 1269.[4]

However, the definition of the nature of the Celestial Movers and their relationship to the world soul were problems which worried Albert from the very beginning of his career. While he was teaching theology in the 1240s, he returned twice to the question, and the succinct formulation to be found in the commentary to the *Liber Sententiarum* is a synthesis of the more articulated analysis already contained in the *De quatuor coaequaevis* from the *Summa de creaturis*.[5]

The two early theological writings show both that Albert was already very familiar with the Arab philosophers and astrologers and that he had introduced the distinction between three movers:

all of the philosophers of the Arabs said and showed in many ways that the heaven is moved by a soul which is conjoined to it. Aristotle, Avicenna, Averroes, Algazel, Alpharabi, Albumasar, and Maimonides say this, and they say that there is a threefold mover: that is, the first cause which is the object of desire of the first intelligence which contains forms which can not be explained by the movement of its own orbit. But since the intelligence is simple it can not direct its particular movement towards one or another destination; therefore, according to these authors, the third mover conjoined to the heaven is a soul, and the nature of the heaven consists in a disposition to motion: since its movement is circular and uniform by nature and it meets no resistance.[6]

When in his *De quatuor coaequaevis* Albert discusses the issue of "the soul of heaven" (*anima caeli*), he accepts the concept only in a metaphorical sense. The reading of a series of authorities ranging from Augustine and Gregorius to Avicenna and the hermetic *De sex principiis*, however, confirms the tenet that "stars are living beings", and establishes the concept of a "soul of the world" (*anima mundi*). At this point, Albert feels it appropriate to substantiate his Aristotelian conclusion with the usual methodological distinctions:

We said all this following the philosophers who do not contradict those Saints who deny that the heavens have a soul, apart from metaphor. They abhor the word "soul", and yet are willing to concede that some intelligences, that is angels, move the heavens following God's orders (*iussu Dei*).[7]

Even more interesting with regard to our discussion is a passage appearing in the commentary on the *Metaphysics*, where Albert distinguishes two roads which might be followed in order to discuss the problem of the superior or heavenly influence on inferior bodies. In Aristotelian terms, one has to consider the influence that is transmitted through the motion of animated bodies; yet, one has also to take into account another influence – one which Plato and the theologians acknowledged– which cannot be measured, since it consists of a kind of irradiation or direct inspiration:

This has not been said following the opinions of the Peripatetics, that there is no influence of the superior world on the inferior one, but through the motion of a body endowed with soul; since the first animated body has an influence on and desires the most undetermined and most universal forms, which become more and more determined to matter, according to their descent towards the one or the other matter.[8]

The influence of the superior bodies is also defined according to the matter which is subject to it (*materia recipiens*), following the important principle quoted above. Albert emphasizes it again in his commentary on the *De generatione et corruptione*. This principle would have momentous consequences as far as the practice of astrological prognostication was concerned: prognostication had to consider material factors, but was also granted the possibility of making indicative or possible predictions. This was not the case as far as the other kind of influence was concerned, and Albert limits his discussion to quoting the "theological" definition of it:

If there is some other radiation from the superior world on inferior things, as Plato said and as the theologians still say, this radiation cannot be investigated by reason, but, in order to study it, it is necessary to call upon other principles derived from the revelation of the spirit and religious faith; we must not talk of this irradiation in the terms of the peripatetic philosophy, since this knowledge cannot be put together with its principles.[9]

This influence works "per actum coniuncti", that is, it transmits the action of the "intelligentia" and of the first motor "as far as it is possible that the mobile can receive the form of the motor". This cautionary note is typical of Albert's interpretation of astral influence: the limitations inherent in contingent things explain the lack of determination, and therefore the relative predictability of the stars's effect, especially when the prognostication does not take into account particular physical facts. This lack of determination does not imply a limitation of God's omnipotence since, according to Albert, the stars are not gods, but only "instruments of superior motions", secondary causes, or (as we read in the *Speculum*) dumb and deaf executors of God's will. In comparison with man's intellect ("anima intellectiva") it is

in this manner [that] the agent intelligence which moves the orbit and the star or stars, transmits forms through the agency of the luminous stars, [as it is] by means of the light of the star that translates them into the matter which it moves; and by touching matter in this way, the intelligence leads it from potency to act. This is shown by the fact that astrologers, committed to these principles, which are the localization of stars, are able to prognosticate about the effects, induced in inferior bodies by the light of stars.[10]

In this way we can thus understand Plato's saying, that

the God of gods told the superior celestial gods that he himself was the one responsible for the semen of generation, and that he would have transmitted them that semen to accomplish it.[11]

Albert here attempts to conciliate Plato with the Ptolemaic concept of "secondary causes". It should also be emphasized that in the *Problemata determinata XLIII* Albert repeats an observation made in the *De quatuor coaequevis* and in the commentary on the *Sentences* and does not treat the identification of the heavenly intelligences with angels as demonstrated doctrine nor as being approved of by both Arab and Jewish philosophers.[12] It was merely a popular belief which they happened to mention:

No one has proved in an infallible way that angels are the movers of the heavenly bodies. But some Arabs and only a few Jews say that uncultivated people hold that angels are intelligences, and not even they proved that it was true nor did they approve of what was said by uninstructed people. On the contrary, according to what Holy Scripture says of the angels and Philosophy of the intelligences, the intelligences are not angels.[13]

If we compare the passage from the *Sentences* which Albert certainly had in mind when he was writing the *Problemata* a quarter of a century later, we see that the more cautious and mature author is careful to avoid giving the impression that the identification in the text between uninstructed people (*vulgus*) and "*loquentes in Leges*" is an opinion held by himself. The text of the *Problemata* continues with a few scriptural examples which may in a way remind us of the type of material Spinoza drew on in Maimonides. Even without insisting on cases like the angel of concupiscence who incited Judas and his daughter-in-law Thamar to incest[14] or the angel called Cerviel who helped David when he defeated Goliath,[15] Maimonides maintained that even Abraham's angel was a prophet or a good man.[16] If he had been an intelligence, which according to the neo-platonic conception is everywhere and pervades everything, he could not have come and gone with normal corporeal movements ("*motus processivi*"). Albert concedes that other cases do exist, nevertheless, which can be explained by a certain force originating in the heavens which moves men by natural instinct towards some goal. Other Aristotelian texts which Albert examined speak of the "motus processivus" of living beings which cannot be attributed to heavenly bodies. Certainly in orbs and stars we cannot see "wings or feet, or something of the sort!" After this ironical remark, he continues, man's "movements are in all directions, forwards, backwards, right, left, down, while a heaven's movement has one sole direction around the center".[17]

Angels are distinguished by their capacity to "carry out virtues of assistance and ministry", according to Pseudo-Dionysius and St.Gregory.[18] In Pseudo-Dionysius's definition, which constitutes an "auctoritas" in the case of the angels, Albert emphasizes:

1. that nothing can be "purified after its contact with the sphere of multiplicity ("purgatus a dissimilitudinis habitu") unless he is receptive and passive" ("recipit per modum passionis"), which does not coincide with the definition of an intelligence which is active by nature ("secundum seipsam totam activa est"), and
2. that it does not have the *modus* of passivity, but rather that of "*forma super formam fluens*", which gives it a new actuation thanks to a form which it adds to an inferior one ("perficit in virtute formali").[19]

Albert bases himself on the *Liber de causis*, on which he had shortly before written a commentary, for texts containing the thesis of philosophers and astrologers "that nothing which is superior ever endures influence,...or receives form from that which is inferior ("quod superior nunquam applicatur inferiori, quia superior non formatur ab inferiori"), but, on the contrary, the opposite is true. To be precise, "that which is superior is not determined by that which is inferior, but in that which is inferior" ("Superior determinatur non ab inferiori, sed in inferiori"), as, according to Ptolemy, Geber, Albategni, and Albumasar, takes place with Saturn, which can be "applied" to any other planets, while all others cannot "apply" to it.[20] Already in the second article of the *Problemata* – the first of the eighteen which, as Albert says, deal directly with the problem of angels and of their identification with intelligences – he gives an explicit denial "to the question whether all that which is moved naturally is moved through the ministry of the angelic movers of the heavenly bodies, [he answers that] there is no doubt that angels do not move heavenly bodies".[21]

However there are some [Christians] who in agreement with those Arabs and Jews, re-propose the identification saying that

> the *virtus* of the intelligences of an orb or a heaven which exercises an influence on inferior things is called an angel.[22]

Albert is very critical of this position and notes that one cannot reason in this way if he is expounding philosophy, nor can one have a discussion with a person who is so lacking in its principles. (Is Albert perhaps aiming here to criticize Aquinas?)

If it were said that by God's command the angels move the celestial spheres, that movement would be due to obedience, and not a natural movement. And it's impossible to give a *determinatio* for that in philosophy, because its principles, which are self-evident axioms, are inadequate to the task.[23]

Albert often criticizes the Pythagoreans' and Stoics' theses, as well as the Platonic one on the animation of the stars. In his commentary on the *Metaphysics*, Alberts places Plato together with Avicenna in the first of the groups he distinguishes among philosophers (or, as he said, among peripateticians). If there is no doubt that, according to a principle of the *Physics* ("all that moves by itself, has in itself the principle of motion, is animated, and the soul is its motor"), "all the peripateticians said that the heavenly orbits possess a soul", it is also true that they are divided on this issue according to three different standpoints. Albert has no sympathy for the most recent thesis, in which "Algazel and some other Arab philosopher", the Jews "Rabbi Moyses (Maimonides) and Isaac" claim that

the celestial circles possess souls, but apart from them there are distinct intelligences which act and preside over them; these intelligences are called angels, following the vulgar.[24]

In the *Metaphysics*, and more explicitly in the *Problemata determinata XLIII*, Albert leans towards the Platonic-Avicennian standpoint:

These souls are to be distinguished according to the intellect, imagination, and the desire or appetite [...]. It is necessary that they are intellectual in the terms of active intellect, since they produce forms through the motion of their orbit in the same way in which the artist realizes the form of the art through the instruments of his art.[25]

Those instruments cannot be considered to be sensible; in fact, it is only by using a metaphor that we can say that heavens have external instruments[26] superfluous "in view of the fact that the matter which go round after the heavens will obey them (*quia materia quae ambit oboedit ei*)", "thus they conceded the presence of desire in the soul, but they denied that it had senses, since the celestial circle cannot receive anything sensible".[27] The third point of view –"a middle way between us" defended by Alexander

Aphrodisias and his "summarizer" Themistius– denies the existence of separate intelligences or angels, and excludes that the heavenly souls possess imagination:

They said that those souls did not possess any of the virtues of the soul, but a universally active intellect, and desire or appetite.[28]

This third position presented with even greater firmness the thesis, already maintained by the first group of philosophers, that there are no superfluous faculties, and that the body of heavens is not as imperfect as that of man,

so that those virtues would be superfluous to the heavenly souls, since the virtues of the heavenly body are sufficient to the purpose [...]: there can be no disobedience between the motor and that which is moved.[29]

In his survey of opinions concerning the issue of heavenly souls, Albert disregards descriptive astronomers, whom he claims are irrelevant to the discussion,

since they do not consider the motion of the heavens according to the aims of their motion, but according to the number and measure of their quantity.[30]

Albert avoids all reference to theology when examining the Aristotelian text, and does not feel the need to justify his silence, as he had done with respect to astronomers. On the contrary when, as theologian, he wrote the *De quatuor coaequaevis*, he had to take the authority of the Fathers into account. Indeed, he had to follow them. When discussing the above mentioned solution, reached following the Saints' thesis and positing the existence of astral angels –denied by Albert as superfluous– he explains the contrast between theologians and philosophers by referring to the different terminology employed by the first group, which betrays theological preoccupations. Even though he seems unable to convince even himself completely, Albert maintains

that some angels could perform miracles, and concur with the laws of nature, it is not contrary to the faith: equally, it is not contrary to the faith that some angels help nature in moving and ruling the heavenly spheres, such as those moving angels or intelligences called souls by philosophers. But the Saints, fearing to be forced to say that heavens are animals, even when they concede that heavens have souls, deny that the heavenly motors are souls. And indeed it is clear that there is no contradiction between them [i.e. between philosophers and theologians]: in fact, the ancients used to call gods and angels the souls of the world.[31]

According to Albert, the theological preoccupations expressed by the Fathers, who wished to avoid all suspicion of pantheism and belief in souls' animation, did not therefore contradict the more coherent peripatetical theses, excluding that the celestial souls possessed affection and imagination. Later in the *De motibus animalium*, Albert criticizes this idea, a "deus ex machina" of Avicennian natural philosophy, which he considers "removed from truth". As far as men or stars are concerned, one has to take into account their real movements: it is the motion of the stars which causes influences, and therefore is the principle of astrology as a science:

If these events happened only because of the will of the soul or of the heavenly motors, without the motions of the orbs, there could be no science disclosed through practice, which might allow the prognostication of birth horoscopes concerning such [monsters]. We do have instead a science transmitted to us by many farmers, thanks to which we are to prognosticate from the positions of the stars and the motions of the orbs.[32]

In the commentary on the *Sentences* there is a brief mention of the movement of the stars, which "is said to take place quite effortlessly and painlessly and without causing fatigue as does the movement whereby our soul moves the body".[33] Here the concept has not been developed, but it had been-in order to show the difference between astral movement and the muscular and ambulatory movements of human bodies ("motus processivi") in the *Summa de creaturis*, and the declared source was Averroes's *De substantia orbis*.[34] A fundamental authority cited explicitly in both texts was John of Damascus, who, differently from Augustine,[35] had definitely asserted that the heavens were "inanimate and insensible". Albert's personal opinion was that "they only move by God's command" ["iussu divino"], a formula which he borrowed from the lexicon of astrologers and which turned up frequently in his other writings and in the *Speculum astronomiae*:

the above-mentioned reasons only go to show that the heavens are not moved by a nature which is the moving form of the body: others say this as well, both philosophers and astronomers, Ptolemy, Albategni, Albumasar, Geber and many many others.[36]

In the *Summa de creaturis* the discussion is developed much further, and Albert is already revealing his most characteristic methodological preoccupation, that of comparing the points of view of "saints" or theologians with philosophers and scientists. The "quaestio de motoribus orbium" begins with a series of texts from Aristotle: among the others quoted, Albert repeats the definition according to which "the prime mover is that which, moved by no cause and immobile both in itself and accidentally, moves heavens and earth", hence he is "God".[37] The "solutio" of this first article contains a more articulated distinction between the three movers[38] than the previuosly mentioned one in the commentary on the *Sentences*. Albert then asks if the second mover, "which is not divisible according to the quantity of the moving body, is or is not the soul of the heaven?" The eleven arguments "quod sic" are for the most part based on Aristotelian texts:

as can be seen in the second book of *De caelo et mundo*, where the Philosopher says that if the heaven has a soul and its soul has a moving principle, then it is certainly endowed with upward and downward movement and movement to the right and to the left. It follows that, as the cause of the movement to the right or to the left, the Philosopher puts forth the fact that a heaven has a soul.[39]

From Averroes's commentary on the *De caelo* and from his *De substantia orbis*, Albert quotes that "a heaven has an intellect, which is a form not divisible according to its quantity and that is its soul".[40] However, the Commentator differentiated this kind of soul from that of human beings:

the heavenly body is not determined in its being with the same necessary disposition which down here applies to the bodies of animals ... it is clear that the souls of these living beings are necessary to the being of their bodies and that they can only be saved by means of their sensitive and imaginative souls. The heavenly body, then, since it is simple and can not be changed by external factors, its being does not need sensitive and imaginative souls, but only a soul which a spirit moves from heaven, and a force, which is neither a body nor resides in the body according to its division, thus granting it an eternal duration and movement with no beginning or end.[41]

Citing both philosophers and religious texts, Maimonides also affirmed that "the heavens are rational animals, that is they know the Creator", they are "animals obedient to their creator who sing his praises".[42] But the most relevant authority for the "quod sic" argument is Avicenna: in his paraphrase of *De caelo et mundo* and in his *Liber sextus naturalium*, he

says expressly that a heaven has a soul and an imagination, and that all of the world's matter obeys it just as the body of an animal obeys the soul of an animal: thus just as an animal's body is modified by images of something pleasing or sad which its soul perceives, so does the matter of elements change according to the imagination of the Celestial Movers.[43]

In his "solutio" to this thesis Avicenna will object that "... he calls fantasy and imagination the application of the intellect to particular beings of nature".[44] However, other texts from Aristotle, significantly enough from the *De anima*, provided "contra" arguments: "a simple body can not be animated, ... heaven is a simple body, therefore it can not be animated; we do not see sense organs [in a heaven]; therefore it does not have a sensitive soul. If by chance it is said that, aside from the sensitive and vegetable soul, it has an intellectual soul and that that does not need organs, neither for itsef nor for its operations", in any case this intellect still does not proceed by a process of abstraction from "phantasmata" perceived by sense organs. If that were the case for the intellect of a heaven, it would either "never be actuated [in *actus*] or would be as though asleep", or "we would have to admit that it has imagination and a sensitive soul, and it is absurd to admit that of a heaven".[45]

The much-disputed identification of intelligences with angels was introduced by John of Damascus, and the same passage by John, which Albert used in the commentary on the *Sentences*, is also quoted here where he continues, remarking on a verse from the Scriptures which locates angels in heaven as men are on earth ("eos qui in coelis sunt angeli et qui in terra homines ad laetitiam vocant").[46] Here in the "solutio" Albert already assumes his characteristic position:

In agreement with the saints we profess that the heavens do not have souls and that they are not animals if we take 'soul' at its own proper meaning. But if we wanted that philosophers and saints agree, we might say that in the orbs there are some intelligences which cause it to move and that they are called the souls of the spheres, not in the same sense of men's intelligences insofar as heavenly intelligences do not operate by abstracting from sensory data (*phantasmatibus),* but they continuously return over their essence and, by means of their own essence over the other [things] with a complete return.[47]

Therefore,

those intelligences only have two faculties *(potentiae*), that is, intellect and the appetite which moves in places; and they have no relation to the orbs according to the definition of 'soul' which says that the soul is the entelechy of the organic physical body having life *in potentia* ... Its operation as regards the body is like that of the pilot as regards the ship, that is, as we read in the *De anima*, moving and directing it.[48]

Albert begins his concluding statement:

... let us say that the movers of the spheres bring about all differences with causal movement, which according to nature exists in lower things: and therefore knowing themselves, insofar as they are causes, they know the higher beings, because they are moved by them just as the thing desired moves the desire. It is clear that their knowledge is neither universal knowledge nor particular knowledge: by virtue of the very fact that they know themselves, they know the universal and the particular beings caused by their movers.[49]

He comes back to his basic methodological preoccupation concerning the "*duae viae*":

We have said all of this according to the philosophers, who are only nominally in contradiction with some saints who deny that a heaven has a soul, because they shrink from the name soul, but they nevertheless grant that certain intelligences or angels move the heaven according to God's command ("iussu dei"). Therefore, as we say according to the Catholic faith that certain angels do miraculous things and compete with the laws of nature, so it is not contrary to faith [to say] that some angels help nature in moving and governing the spheres of heavens: the philosophers call these moving angels souls or intelligences. Thus it is clear that there is no contradiction between them: in fact, the ancients called God and the angels world souls.[50]

Later commenting on the *De caelo* Albert writes a "Digressio declarans rationem naturalem de effectibus stellarum",[51] and takes a stand concerning the fundamental problem of whether the motors of stars are souls, endowed with will and passions, and capable of behaving with respect to their own body and the body of inferior things (always depending on them) in the same way in which the human soul behaves with respect to the human body; or whether they are pure "intellects". The problem did not question the legitimacy of astrology; it was nevertheless a complex issue, which required the analysis of various authorities. Among these authorities, for instance, was the criticism leveled by Averroes against Avicenna's theory of imagination as being tenable both for the particular and the universal. Albert's discussion of this question (which is impossible to examine here in detail) is perhaps less than rigorous. It is however important to stress that in the *De caelo* he offers a solution similar to the one put forward in the *Speculum*, and one which better defines his own views on the relationship between macrocosm and microcosm.

We must concede without doubt that stars, which are almost members of the heavens, are the first motors to which all the alterations, growths, and generations, of the universal matter of all bodies which are generated or corrupted are due. Stars have in themselves the strength of the intellectual motors, which are formal operational intellects, as is the intellect of the artificer with respect to the work he is producing; the actions of stars receive the form from them, in the same way in which the warmth of the complexion of the body receives its form from the virtue of the soul. And therefore the actions of stars through their motion transmit forms, as natural warmth induces in the food and in the body the form of the flesh and of the blood, when it derives from the virtue of the soul. It follows that we can make better conjectures concerning the production of things which are to be generated, and their length and their formation in general, when we know the virtues of stars from their positions and their motions.[51]

At a later time, this approach to the question will be a central aspect of Renaissance discussions on natural magic, and several authors cite both Albert's passages, as well as his quotations from the Aristotelian-Averroistic tradition (*Metaph.*, VII, comm.31), in addition to those from Plato (*Timaeus* 41CD; transl. Chalcidii p.35, 20ff.) and Avicenna ("Plato [...] said that all forms derive from the *dator formarum*"). Lastly, Albert closed his discussion of this issue by maintaining that it was possible to consider forms in two ways: that is, according to material and corporeal essences and according to intellectual and spiritual ones. So, for instance, a sculptor can give his statue only the intellectual form, and not the material:

The sculptor *per se et essentialiter* makes statues, but these are not disposed in a material way according to the form of statue. In the same way, stars do give form essentially, but only the spiritual and intellectual ones, since they are instruments of the moving intelligence.[52]

The theoretical developments quoted above are extremely important as methodological justifications of natural magic. As theoretical justifications of the foundations of astral influence, the position Albert defends in the *De Caelo* is virtually identical to what is poignantly said in the *Speculum astronomiae*:

the highest God has ordered this world according to His supreme wisdom, so that He, being the living God of heavens, which does not live, will act through deaf and dumb stars as His own instruments upon things that are created and are to be found in the four inferior elements.[53]

CHAPTER TEN

DIVINE PROVIDENCE AND THE MEANING OF "INTERROGATIONES"

In order to refute the accusations that some "friends" leveled against astrology, in particular against two authors, Al Qabisi and Abu Ma'shar the author of the *Speculum* borrowed word for word the polemical passages against various interpretations of astrology and of the notion of contingency from the *Introductorium maius*. Contrary to what Thorndike believed, this was not "some juggling with the terms necessity and possibility".[1] The author followed Abu Ma'shar in the analysis and refutation of the "third sect of opponents", which had reminded him of the tenets of his contemporaries:

> They contradicted the science of stars, and they maintained that planets do not have signification for those things which happen in this world. They have used this argument and they have said that stars do not signify that what is possible, but only what is necessary and what is impossible.[2]

The Arab critics of astrology followed "the reasoning of some ancients" (the megaric theses criticized by Aristotle, or the Stoic ones?) and offered an interpretation of the *Metaphysics* (Alpha 5; Theta 4) and of the *Periermeneias*, 9, which excluded contingency and therefore astrology. They maintained that "the Philosopher said that the being of things which are in the world occurs in three ways:

1. a necessary one, such that fire is hot;
2. an impossible one, such that man can fly;
3. a possible one, such that man can write.

And the stars signify two principles: the necessary and the impossible; they do not signify the possible. Hence, the science of stars is false". Even though Aristotle never said so, Abu Ma'shar is convinced that he was bound to question the foundations of free will and of the "electiones", maintaining that "when indeed a person was walking, he was deprived of the possibility, and he was brought to the definition of necessity". Therefore, "man cannot choose but that which has been signified by planets, since his choice is directed towards something and to its contrary thanks to the rational soul, which in the individual is connected with the vegetable soul due to the signification of planets". The author of the *Speculum* – much more than the historian Lemay[3] – believes that the foundation for contingency provided by Abu Ma'shar is adequate, and that this *Introductorium* offered a different definition of Aristotelian potency, one that provided free will with better safeguards against determinism. Alhough he rephrases the inelegant expressions of John of Seville with a terminology derived from an Aristotelian vocabulary, the line of argument remains untouched and recognizable:

For it is not the same thing to be necessarily as to be simply by necessity. It can not be, therefore, before it is; and yet, it will be because it is not necessary that potentiality [must] be reduced to actuality.[4]

The above philosophical expressions show clearly and correctly the revision of the Aristotelian doctrine of potency and act, and the influence of Abu Ma'shar. The author of the *Speculum* not only repeats word for word the arguments offered in the *Introductorium*, and concisely evaluates the arguments given by Aristotle and Abu Ma'shar, but also expresses his choice between the two:

Similarly, regarding that about which it is signified that it will not be at a determined time, and about which it is true to say that it will not be then [at that time], nevertheless, it can always be before that [time], and [up until that time when] it finally reverts back to the nature of the impossible. And this is the opinion of Albumasar, from which the famous Aristotle seems to depart to some extent, since he [i. e.: Aristotle] does not concede that it may be true to say [something will or will not be] beforehand. I do not regret having said this.[5]

In chapter XII, the author of the *Speculum* elevates the opinions of Ptolemy and Abu Ma'shar over those of Aristotle, whom he praises but does not take as an absolute authority, exactly as Albert does. This when discussing the rather harmless issue of "the revolutions and changes of time" ("revolutiones et temporum mutationes"), one that did not imply the issue of free will, but only the "will of his Creator" ("voluntas sui conditoris"): because by God

the motion [of the heavens] will be stopped, as it began by His command. In this matter alone, we find the [normally] useful Aristotle has erred (nevertheless, he is to be thanked for a million other [ideas]).[6]

The preference for Abu Ma'shar and his "elegant [...] testament of faith and of eternal life, not acquired save by faith"[7] is expressed even more strongly when the author discusses the definition of the "domus fidei", the horoscope of Christ, and the divine providence:

In those things which God operates by means of the heavens, the indication of heaven is nothing other than divine providence. In those things, indeed, which we initiate, nothing prevents [the fact] that there is not a cause in heaven, but a signification. For of the two sides of a dilemma from which man can choose one or the other, God knew from eternity which of these he [i. e.: the man] would choose. For which reason, in the book of the universe, which is the vellum of heaven (as was said before). He was able to configure, if He wished, what He knew; [but] if He did this, then the compatibility of free will with divine providence or with the indication of an interrogation is the same.[8]

We should not be shocked by this conflation of the sublime issue of divine providence and the question of the foundation of the "interrogationes". On the other hand, the author of the *Speculum* also added that we should not mistake the celestial signifier for the Supreme Will. We must not believe that "which ever of those things that are not hidden by known divine providence might also be recognized in heavens".[9] This was the reason for contradictory and unclear prognostications. In those cases, "the counsel of the profession of the stars is to abandon [the interrogation], since God wished to keep it hidden from us".[10]

The author of the *Speculum*, who was well acquainted with Arab astrological doctrines through their Latin translations did nevertheless consciously and exactly transfer them into the vocabulary and the philosophical universe of thirteenth-century Christian theology and philosophy.

As d'Alverny has pointed out, following the translation of two Arabic texts both condemned in 1277 (the *De radiis stellicis*, or *Theorica artium magicarum*, by Al Kindi – explicitly criticized by Giles of Rome – and the pseudo-Aristotelian *De causis proprietatum elementorum*) a kind of astrological determinism became popular, based on, and justified by, a line in the Koran (Sura 60), "stars will kneel in front of God".[11] Thanks to this doctrine, Albert himself (whom d'Alverny was inclined towards considering as the author of the *Speculum astronomiae*), "took away the sting from astrological determinism, by claiming that the conjunctions of celestial bodies are only an instrumental cause of the Divine Will".[12] Indeed, Albert offered a thorough paraphrase of the *De causis proprietatum elementorum*, an Arabic text which had been included in the pseudo-Aristotelian corpus (which he integrated with his own writings on the topics he felt relevant, even when he could find no Aristotelian or pseudo-Aristotelian text in his support).[13] In Chapter IX of the tr.II of the *De causis proprietatum elementorum*, Albert innocently attempts "to identify the causes of deluges", which he classifies under the headings "universal deluges" (such as the ones narrated by Noah and Deucalion which Albert compares) and "particular deluges", which happen "more often" ("saepius"). Up to this point, Albert does not approach the questions which will be condemned. He limits his discussion to the distinction between "universal causes", "a cause less universal (in which there is a convergence of celestial and terrestrial)", and, lastly, "the cause truly particular [...] (in which there is convergence either of celestial elements only, or only terrestrial)". The discussion became more delicate and dangerous when Albert, in a text of undoubted authenticity, summarizes "the opinion of some Arabs: [...] prodigious things occurring on earth which are caused by the intelligence which moves the sphere of the moon".[14]

Even more dangerously, he later on discusses the doctrine of conjunctions in depth. Albert does not limit himself to listing the number and kind of the more or less "great" conjunctions, but claims that their study constitutes one of the primary tasks of philosophical investigation:

> That conjunction signifies great accidents and great prodigies, and changes in the general state of the elements and of the world: the cause of this must be said by the natural astrologers, since he knows astrology, and for this reason the Philosopher says that astrology is the second part of philosophy, and Ptolemy says that the judicial astrologer, the one who practices 'elections' and the observer of stars will err if he be not a philosopher.[15]

From the *Speculum*, full of references to astrology and to related scientific and philosophical doctrines, we can move to consider two better known works, to which the author himself refers as already completed (2921; 297b; 328a): the *De natura locorum*, a geographical work based on the astrological doctrine of climate; and the *De caelo et mundo* (293a). The above mentioned treatises allow us to reconstruct Albert's beliefs at a given time, as if they were expressed in a single work. In the *De causis proprietatum elementorum* Albert does, in fact, announce that he is about to embark upon commentaries on the *De generatione* (294a; 296a; 197), on the *Meteorologica* (382a), on the books on animals (295b) and to write a "history" –which Aristotle had forgotten – *De mineralibus* (328a). With regards to the *De caelo*, Paul Hossfeld, the editor of the critical edition, has suggested a date for the work between 1248 and 1258 (this year is the *ante quem* for the *De animalibus*, the last part of the naturalistic series), during the period in which Albert was teaching at Cologne. He has hypothetically proposed 1251 as the actual year. In the autographed Köln manuscript of the *De caelo*, it immediately precedes the *De natura locorum* and the interesting treatise on *De causis proprietatum elementorum*. As a result of the critical edition of the *De caelo*, however, it is now possible to list a few elements which seem to echo the problems and terminology present in the *Speculum astronomiae*. In the *De caelo* we find that "the stars generate and move the matter of generated beings".[16] We also find an interpretation of the chain of being which clarifies how in the inferior bodies, subject to generation and corruption, and in all bodies in general "both simple and composed, the decor of the celestial body cannot fully express itself".[17] There is, in the *De caelo* "an extremely difficult investigation" and one that "must not lead anyone to believe that we are presumptuous". It is the author's "intention and willingness to research the causes of occult things, which are either impossible or difficult for man to know". The reader of the *Speculum astronomiae vel de libris licitis et illicitis* will be particularly struck by the distinction Albert introduces in his judgement of this kind of research: "this is an investigation of a very difficult subject, which it is sometimes to be vituperated, as sometimes to be praised".[18] After what we have read above, it is not surprising to find the revealing expression as "nutu dei",[19] or a pointed definition of the scientific role of judicial astrology, particularly of the branch concerned with the "elections", in the *De caelo*:

> In philosophy there are two issues concerning the various effects of the stars, that is, what is the effect of each star, and when and where does it occur? To study this problem is the task of those who practice 'elections', and make divinations through stars, who can choose the hours and know them, hours thanks to which the events in the inferior world are related to the configurations of the stars. And this belongs to the science of those who practice elections, or, in other words, to the *genethliaci*, since what they investigate through the figure of stars is mainly the nativity of what is generated and the events of those which are born.[20]

The *De caelo* also provides concise bibliographical information on two masterpieces by Ptolemy, "one of which deals with the great universal accidents in the world" ("quorum unum est de accidentibus magnis universalibus in mundo"). Albert does not hesitate to maintain the legitimacy of investigating the great conjunctions, probably the most scandalous intellectual enterprise from the point of view of theological orthodoxy:

The great accidents are changes of kingdoms from one nation to another, and the introduction of sects and doctrines of new religions, and so on.[21]

Not even the study of birth horoscopes, which posed so many difficulties to the doctrine of free will, bothers Albert, who talks about Ptolemy's investigation "in the small particular accidents, such as the events of a single man born in this or that constellation".[22] The field was one of purely natural investigation, and the distinction Albert introduces between philosophy and theology makes it impossible for natural research to hinder theological concerns.

According to what has been investigated concerning the effects of stars, there is a natural cause for which a star is said to have this or that effect, and this is to be determined by those astrologers who practice birth horoscopes [*geneatici*] and 'elections'.[23]

Even the use of astrological premises for magical purposes is not completely excluded. Albert says only that this task is particularly difficult and of dubious efficacy with respect to artificial objects:

What is then more difficult to know, is according to what nature stars can produce luck or bad luck and can infuse strength not only to what derives from nature, but sometimes also to what is produced by art, such as images or clothes which have been recently cut, or to buildings recently built, and so on. All these things are produced by mutable causes, and therefore they can be or not be. It seems then that their destiny does not depend upon some nature or virtue of stars.[24]

Equally, the reservations expressed in the *De caelo* on the variable, non-systematic and yet acknowledged efficacy of talismans find their counterpart in the *Speculum* in Chapter XI, "De imaginibus astronomicis et earum auctoribus licitis atque de imaginibus superstitiosis et earum auctoribus". Here, talismans are listed under "three ways": the first is considered as "abominable", the second "somewhat though less abominable, [is] nevertheless detestable", and "the third type is [that] of astronomical images, which eliminates this filth, does not have suffumigations or invocations and does not allow exorcisms or the inscription of characters, but obtains [its] virtue solely from the celestial figure [...] and it will have a [good] effect from the celestial virtue by the command of God [*nutu Dei*]".[25] The *Speculum* does, in fact, offer –though written with worry and openly acknowledged regret– a unique survey of this kind of literature in this chapter.

After what we have seen so far, it is not suprising to note that Albert announces his project of specifically writing on astronomy and astrology in several passages of the *De caelo*:[26] "It is clear that astrology, of which Alpetrauz Abuysac claimed paternity, is false according to Aristotle. It will be the subject of investigation in my *Astrologia*", where "we will make a comparison [of doctrines] in our astrological science", in view of the fact that "all those things which are to be said in my *Astrologia* will be sufficiently determined through mathematical principles".[27]

Albert does not intend, however, to write a mere study in descriptive astronomy or a comparative commentary *In Almagestum Ptolemaei* (a work mentioned in the ancient catalogues of his writings, and attributed to him by a still unstudied manuscript of less than certain authenticity).[28] But the plan of work Albert expressedly outlines in another passage (154/87-89) of the *De caelo* is not limited to the investigation of the first of the "duae magnae sapientiae", and announces the discussion of astrological "elections".[29]

If, when he presented the expositio on the *De caelo*, Albert included in his working program the plan of freely discussing a very controversial part of astrology, why should we exclude the possibility that he had devoted to that topic, and to the consequences on the view of nature and of history it implied, the precious introduction and bibliographical guide represented by the *Speculum astronomiae*? In the *Speculum* (which does not belong to the genre of the "expositio", nor of the "quaestio", and, in my view, is not part of his Aristotelian cursus)[30], the author, adopting a methodology positing the autonomy of philosophical investigations, nevertheless appears aware of the possible conflicts with theological doctrines and authorities. Clearly, this attitude reflects the "occasion" he repeatedly refers to as the impetus for writing the work, that is, the censorship to which those disciplines were submitted not only in 1277, but also in previous years. After this date, no case of analogous criticism is registered, which could at the end of the thirteenth or at the beginning of the fourteenth century give occasion to a pseudo-Albertinian *Speculum*.

PART THREE

Traditions, Collections and Heritage

CHAPTER ELEVEN

ALBERT'S *BIBLIONOMIA*

In the thirteenth century, readers must have experienced great difficulties in finding their way through the many Arabic astrological works which had recently been translated into Latin, the compilations from Arabic sources prepared by Latin scholars, or through the questions this tradition had left open. If even today, after the meritorious work of generations of specialists, we find it difficult to identify one or another treatise, and to explain to our students the distinctions between the various parts of astronomy and astrology, we can well appreciate the historical importance of the work known as *Speculum astronomiae*, or, more rarely, as the *De libris licitis et illicitis*. The treatise helps the reader to identify various works and distinguish between the various kinds and parts of astrology. It soon became a classic in schools, because it offered a comprehensive view of the structure and problems of the "duae magnae sapientiae" astronomy and astrology (which, from then on, were almost universally referred to with the expression quoted above, following the ordering by Ptolemy and Abu Ma'shar) as well as an introduction to their vast literature.[1] Clearly, the aim of the author was to offer a description and outline the philosophical foundation of the two disciplines by eliminating their questionable features. He also wanted to list

> almost all of the praiseworthy books which latin culture, impoverished in this [subject], has begged from the riches of other languages by means of translators.[2]

The desire to exhaust all possible information on Arabic scientific writings is also expressed elsewhere in the *Speculum*. And even though the author does not guarantee that he has achieved a complete knowledge of the literature, he makes a valiant effort as far as those works which had been translated ("and possibly [...] there are others, as in the case of those [mentioned] above, but they are not translated").[3] As we have seen, however he does add that he has at his disposal a translator who will clarify some of the Arabic technical terms still present in the translations, for instance, of al-Qabī'sī.

If there are names in an unknown language in his text, their meanings are immediately added to the text itself; but if perhaps the meanings of some [of those words] should be missing, [there is a] man prepared to supply them.[4]

The author of the *Speculum* does not claim to be familiar with the Arabic language, but only of having accurately reviewed all the scientific patrimony accessible to the Latin world. This preoccupation with technical terms finds its counterpart in the *De quatuor coaequaevis*, the *De fato* (in parallel passages to the *Summa theologiae*), the *De caelo*, as well as in the *De causis proprietatum elementorum*, where Albert clarifies and paraphrases those technical terms which had entered the Latin language in a transliterated Arabic form.[5] Equally consistent in all Albert's work is the care he takes to provide accurate bibliographical orientation. His aim is to offer complete and exact entries. All this confirms the affinity between the *Speculum* and Albert's well authenticated works. It is important to remember that commentators have pointed out those features of the *Speculum* discussed above in order to attribute the work to Bacon, famous for his interest in Arabic culture. In the *Opus tertium* however, he does express his concern over the difficulty of Arabic terms which, he says, is one of the reasons for the ignorance of his contemporaries:

In the texts of philosophy and theology, numberless words from foreign languages are introduced, which cannot be written, said or understood save by those who know those languages, and it was necessary that this happened as it did, since these sciences were written in their original languages, and the translators did not find sufficient words for them in Latin.[6]

But it is precisely this familiarity Bacon exhibits with Arabic science – not only by means of translators, whom he quotes in this as well as in other texts[7] – which gives rise to the supposition that he was able to read those works in the original, as he had done with Greek and Hebrew texts.[8] This is totally different from the attitude we have described as characteristic of the author of the *Speculum*. On the other hand, those who at that time took an interest in science could not have avoided coming to terms with Arabic textbooks and their vocabulary. But it is equally true that the bibliographical concern we have highlighted in the *Speculum* was neither new, nor unique.

Medieval bibliographies or library catalogs offer other, though less complete, examples of analogous citations. I will briefly examine two cases: the well known *Biblionomia* by Richard de Fournival, and the *De originibus* by Guglielmo of Pastrengo. Our choice is not made by chance, in view of the fact that the description of these two libraries, offered by the two treatises bear dates which are not very far from the decades when the *Speculum astronomiae* could have been written. Both libraries were probably not just ideal ones, as A. Birkenmajer and P. Glorieux have shown for the first case,[9] and R. Sabadini has maintained, perhaps rashly, for the second. It is commonly agreed by scholars that Richard de Fournival's guide was composed for the readers he admitted to his private library at Amiens, afterwards inherited by Gérard d'Abbeville and transferred to Paris, where in 1271 it constituted one of the first public libraries, as well as one of the earliest parts of the library at the Sorbonne. The *Biblionomia* was certainly written before 1243. The testimony by Arnold Gheylhoven (1424), according to whom Richard was the author of the pseudo-Ovidian *De vetula*, has been re-examined by P. Klopsch the editor of that astrological work, who doubted the attribution, but could not exclude it.[10] Even without supposing that Richard was the author of the *De vetula* (quoted by Bacon on the subject of the horoscope of religions for the verses inspired by Abu Ma'shar),[11] there is no doubt that he was very interested in astrology, witness the fact that in 1239 he calculated his own horoscope.[12] His general familiarity with astrology also emerges from the section of the *Biblionomia* devoted to the subject,[13] and from the manuscripts Birkenmajer and others have traced in the holdings of the Sorbonne now at the Bibliothèque Nationale in Paris.

In Richard's work, the "*Tabula quinta areole phylosophyce*", which follows the section devoted to geometry and arithmetic, was divided between music and astrology. Yet, astrology took the lion's share: on a total of sixtheenth entries which composed the chapter, dealing with the artes reales of the quadrivium, astrology was represented in the description of eight, often highly articulated miscellaneous manuscripts. The examination of entries fifty-three through sixty leaves the reader with a curious impression of *déjà vu*. Since Birkenmajer could not directly examine all the codices, he had to rely on other scholars' descriptions concerning the first item listed at number fifty-four, the so-called *Almagestum minus*. Birkenmajer described this work as a compilation from Ptolemy and al-Battani made by a Christian author.[14] He also referred to the *Speculum astronomiae* where, in the *Biblionomia*, there are certain allusions to the author of the *Almagestum parvum* or *minus* ("A book by Galterus de Insula [i. e.: de Chatillon] extracting the elements of astrology from Ptolemy's *Almagest*, going up to the end of the sixth book") which, however, are not pursued. The author of the *Speculum*, as has already been emphasized, instead embarks upon a very detailed description, certainly based on first hand acquaintance with the work (possibly, on collaboration with its author):

Also from these two books someone has compiled a book in the style of Euclid, whose commentary contains the opinions of both Ptolemy and Albategni, and it begins like this: "Omnium recte philosophantium etc." (Of all of those who philosophize correctly etc.).[15]

This coincidence between the titles described in the *Biblionomia* and in the *Speculum* could be regarded as due to chance. The systematic reference by the author of the latter work to items listed in the former, however, completely excludes this hypothesis. The beginning of chapter II, "On the astronomical books of the ancients (*De libris astronomicis antiquorum*)", reflects the general structure and the aims of the *Biblionomia*: "therefore, amongst the books found by us written on these [matters], after the geometrical and arithmetical books [...]" (II/1-2), the first item on the list, a rare text seldom quoted in the thirteenth century, confirms the similarity of the bibliographical lists, as well as the peculiarities of the *Speculum*. The author of the *Biblionomia* tries to find a secure attribution for item 53, "The book by Mercurius Trismegistus concerning the motion of the inclined celestial sphere, which bears the title of Nemroth to Joanton". The author of the *Speculum* does not take up the hermetic identification, but instead justifies the fact that the work, which he appears to have examined personally, opens the list, by calling upon chronological arguments:

the first in time of composition is the book written by Nemroth, the giant, for his disciple Iohanton, which begins thus: "Sphaera caeli etc." ("The sphere of heaven etc."), in which there is not much that is useful and quite a few falsehoods, but nothing that is against the faith, as far as I know.[16]

After the description of the above archetype of which great antiquity was presumed (it is now well known thanks to studies by Haskins, Lemay, Nardi, Livesey and Rouse),[17] the relationship between the *Speculum* and the *Biblionomia* becomes even more apparent when the *Almagestum* is described. Richard did not possess a codex of that work, but he knew of its existence through the above quoted *Almagestum minus* and through the *Elementa astronomiae* by Jabir ibn Aflah, quoted as item 56,

The book of Geber of Seville on the science of the form of movements of superior bodies and on the knowledge of their orbits and to avoid some mistakes found in the books of Claudius

Ptolemy Phudensis [sic] which bears the title *Elmegisti* or *Megasinthasis*, which is corrupted in *Almagesti.*[18]

The author of the *Speculum astronomiae*, as he had already done on several occasions, checked and completed the indications in the *Biblionomia* concerning this fundamental text with a thorough reference to the relevant codices. His bibliographical skills ("making a list of both types of books, showing their number, titles, incipits and the contents of each in general, and who their authors were, so that the permitted ones might be separated from the illicit ones")[19] and historical accuracy led him to revert the order of the listing. He correctly places the comment after the original text, and makes them both precede the Tabulae and all the other texts described in the *Biblionomia* (nn. 54-55). None is deleted, but some were simply deferred to later in the chapter. Thus, the reference by Richard to the corrections introduced by Geber and other Arab commentators in Ptolemy's work was only modified by the author of the *Speculum*. Up to this point, it has been possible to argue that the two bibliographical surveys were made independently from each other, and that all coincidence has been due to the very nature of the subject-matter. In the *Speculum* the description of the *Almagestum*, the masterpiece of Hellenistic astronomy, however, is sufficient to abolish all idea that the two lists were independent and that the coinciding parts between them were there by chance:

But what is found [to be] more useful concerning this science is the book by Ptolemaeus Pheludensis called *Megasti* in Greek, *Almagesti* in Arabic and *Maior perfectus* (The greater perfect) in Latin, which begins in this manner: "Bonum fuit etc." (It was good to know etc.).[20]

The *Speculum* always employs the title *Almagestum*. It is clear that his Greek is weaker than Fournival's. But his preoccupation with listing all the forms of this title, as well as, later on, of the *Quadripartitum*, betrays his dependence on the *Biblionomia*. To limit our discussion to the major examples of coincidence between the two texts, in the *Speculum* the indications concerning the commentaries by Jabir and Al-Battâni, Mâshâ'allâh, Al-Fârghanî and Al-Bitrûjî, are all taken from the *Biblionomia*, but later checked on codices. A recent and important paper by David Pingree has throughly confirmed the relationship I suggested between the *Speculum* and the *Biblionomia*. This proves beyond any doubt that the author of the *Speculum* was inspired by Richard's work and took it as his model, especially as far as the astronomical section was concerned. Nevertheless, the borrowing was not mechanical nor did it exclude other sources. The author of the *Speculum* revised the data he took from Richard following his own, more refined and more sophisticated, historical and bibliographical criteria. Moreover, he also added several comments. It is only natural to ask oneself whether the author of the *Speculum* knew the *Biblionomia* and the manuscripts described there because he had been granted access to Richard de Fournival's library, either during Richard's lifetime, or when it became the property of Gerard d'Abbeville and later, the Sorbonne. These hypotheses are not crucial, in view of the fact that the reading of the rare catalog compiled by Richard alone was sufficient to the task.[21] Even without discussing the issue of the date and the place of composition of the *Speculum*, the similarities between this work and the library described by Fournival in his catalog confirm that the anonymous treatise was written after 1243.[22] More importantly, the *Speculum* was clearly indebted to the holdings of a very rich specialized library, a true rarity at the time. In the middle of the thirteenth century, the eight miscellaneous codices Richard described are unanimously considered to represent a considerable collection, though they were limited to descriptive astronomy, and, contrary to what Klopsch has pointed out, they did include works by Abu Ma'shar, the astrological authority so influential in the *De vetula* and the *Speculum*. The author of the *Speculum* added six chapters of entries to Richard's bibliography, thus enormously expanding the scope of the *Biblionomia*. This amount is not surprising, especially if one remembers that Pierre de Limoges, an author active in the same Parisian intellectual milieu, and a commentator on Richard's horoscope, died in 1306 leaving, according to the obituary of the Sorbonne, a specialized collection of one hundred and twenty astrological volumes.

Pingree, after confirming the relationship *Biblionomia* – *Speculum*, has mentioned other very interesting contributions from three manuscripts known to former scholars, but never before seen under his clarifying light. In fact, the first one (Paris. lat. 16204), "copied by one of Fournival's scribes furnishes the basis of a great part of the catalogue of the *Speculum*, i. e. "70%" of it. The second (Oxford, Corpus Christi 248) contains a short catalog extracted from the former manuscript and prepared by a person collaborating with the writing of the *Speculum*, since a note in it confesses the sole lack of an astrological section missing in the bibliography of the *Speculum*, i. e. – as Pingree has emphasized – the lack of "the entire subject of anniversary horoscopes";[23] the third manuscript (the well known Laurentinaus Pl. 30,29 copied in the 1280s) contains in a collection of occult writings "one after the other, neither with author or title indicated, the complete copy of *De arte alchemica* of Richard of Fournival" and the oldest copy of the *Speculum*; this manuscript was copied about 1280 and according to Pingree "presumably in Paris, from one that Fournival has used", so that "it clearly comes from the same circle of Parisian scholars to which Fournival and Albert belonged".[24]

It is difficult to say if the group collaborating to prepare the *Speculum astronomiae* (and mainly – in my view – its bibliographical sections) was based in Paris or at the papal curia or elsewhere; for a "Doctor universalis" like Albert, it is difficult to suppose that his drafts and notes should have been autographs, when he had many secretaries. Unfortunately they have not yet been studied as deeply as have Thomas Aquinas's scribes.

All these data – Albert's knowledge of the *Biblionomia*, his use of some of Richard's codices, seen in France or more likely at the papal court, given the fact that Richard and Albert visited there, and Campanus and Moerbeke etc. also stayed there between 1256 and 1264, i. e. at the very period of the discussions on the legitimacy of astrology between Dominicans, and on the comet of this last year – suggest the following conclusion: a group of scholars, some of them probably Dominicans, collaborated to prepare the *Speculum astronomiae* and among them was Campano, responsible for the Orientalist and technical problems, while the main writer and sponsor of the enterprise was Albertus Magnus. If my hypothesis is correct, it could explain why this booklet is found under his name in the older manuscripts, being conceived mainly as a memorial to be diffused among prelates in order to prevent a condemnation of astrology. To bear the signature of such an *auctoritas* was certainly useful. The polemic itself on behalf of unrestricted readings in the *libri naturales*[25] and the *travail en équipe* are two favorite features of Albertus Magnus.

CHAPTER TWELVE

THE LITERARY TRADITION OF THE *SPECULUM* AND ITS ROLE AS A REFERENCE BOOK

The second bibliography we are going to examine briefly seems to reflect, at first sight, an analogous collection. Remigio Sabbadini was the first to study Guglielmo da Pastrengo (1290-*ca.*1362), a politician and judge from Verona who acted as ambassador to Vicenza, Venice and Avignon (where he met Petrarch, and began a life-long friendship with him). Sabbadini maintained that Guglielmo's *De originibus rerum libellus, in quo agitur de scripturis virorum insignium* (*A booklet on the origins of things, concerning writings by famous men, ca.* 1346-1350) represented the discoveries of a group of manuscript-hunters who were responsible for the birth of humanism in Verona and in Europe in general.[1] According to Sabbadini, Guglielmo's culture was "firm and wide-ranging", covering the whole spectrum of classical and medieval literature, and particularly deep as far as astrology was concerned. Guglielmo "had a special passion for astrology, and was able to collect an astrological library of twenty-five authors". It is very difficult to determine which of the works he lists "he had under his eyes" and which "he found quoted in his sources". Sabbadini, who, being a humanist philologist of classical training, was not acquainted with the history of science, and was convinced that the astrological library was exceptional, in view of the fact that "these are the only works of which he gives the incipits".[2] In fact, a cursory comparison between the *De originibus* and the *Speculum astronomiae* immediately destroys Sabbadini's belief. The astrological entries are more numerous, more precise, and provided with incipits, only because Guglielmo was able to take advantage of the exceptional bibliographical survey offered by the *Speculum*, as is shown even if one considers only those authors listed under "A" ("Archesel known as Albategni, the astrologer Alboali, Almansor, Abrazath, Alchindus, Alpetragius, Aldilazith, Alfraganus", etc.). The abandonment of the chronological order in favor of the alphabetical one and the numerous misspellings of names and titles show that Guglielmo consulted his codex of the *Speculum astronomiae* without either care or competence. He does not deny that he knows the work. Indeed, after mentioning "Albertus the German, of the Order of the Preachers, a man of very sharp intellect (*acerrimi vir ingenii*)"[3] and one whom he places amongst the "doctors of law" following Albertino Mussato, Guglielmo claims that Albert is the author of the *Speculum*. This passage was later censored by Michelangelo Biondo, the sixteenth century editor of Guglielmo's work. Carlo Cipolla published the astrological entries quoted by Pastrengo, and re-established some of the pseudo-Aristotelian titles: "*De iudiciis in astrologia* (On astrological judgments) which starts as follows: *Signorum alia* ('some of the signs') and later *De imaginibus*, which is the worst of all books concerning images, dedicated to Alexander, which starts as follows *Dixit Aristoteles Alexandri* ('Aristotle said to Alexander'). This book also bears the name of *Mors animae* (Death of the soul), as Brother Albertus writes".[4] Both titles are in fact taken from the *Speculum* (VI/12-14; XI/95-98). It would be redundant to add to the list of evidence proving that Guglielmo di Pastrengo used the *Speculum* and attributed it to Albertus.

It should be added that this attribution is not the first. Of the earliest manuscripts, only three gave a contrary indication, all naming Philippus Cancellarius as the author. One of these, however, was copied by Thomas Allen at the end of the fourteenth century from another manuscript. The third bears the name of two authors: "*Albertus or Philippus Cancellarius*". Numerous older manuscripts, the earliest catalogs of Albertus's works, and a note found by Lynn Thorndike in a manuscript of the beginning of the fourteenth century, confirm that the majority of the earliest readers of the *Speculum* attributed it to Albertus.

In order to substantiate his thesis, Thorndike offered precious philological contributions, such as his commentaries on chapter XI, which provided the foundation for his exhaustive bibliography of *Traditional mediaeval Tracts concerning engraved astrological Images*,[5] integrating the previous commentary sketched by Steinschneider and updated by Cumont. The *Speculum astronomiae* was always on his mind. On many occasions he verified its enormous utility as a bibliographical guide (as Carmody has done),[6] and often collected data to establish its origin. In the manuscript Paris BN Latin 16089, Thorndike identified a bibliography on the occult sciences compiled in the year 1300 which is "largely indebted to the *Speculum*".[7] This manuscript proves that the work enjoyed a brisk circulation before the survey by Guglielmo da Pastrengo, and was used as a basis for further bibliographies. Of greater interest however is the manuscript Vat. Ottob. lat. 1826, which Thorndike alluded to in 1958. In its marginal notes datable to 1333 the work is without hesitation attributed to Albertus.[8] Predating this manuscript, the attribution is present in the Parisian catalog of the *grande librairie*.[9]

The recent publication and analysis by R.A. Pack of a pseudo-Aristotelian *Chyromantia*, allows us to add a thirteenth-century testimony concerning the fortune of the *Speculum*. This apology for chiromancy, written soon after the death of Albertus in 1280, was undoubtedly based on the *Speculum*, a work as firmly attributed to the deceased scholar as the well authenticated *De animalibus* was. The influence of Albertus on this work is profound. It is particularly evident in the definition of physiognomy as non-deterministic:

The natural affective inclinations of men can be clarified following the signs of their natural members [...]. This science [*physiognomy*] does not impose necessity on the customs of men, but shows their inclinations which derive from the blood and from physical spirits, which can be controlled by reason.[10]

Several literal, but unavowed quotations from the *Speculum* have recently been detected in an introduction to *Philosophia* by its forthcoming editor: in every case these quotations have been made extremely early and have to be seriously considered when dating the *Speculum*. If the author of the *Philosophia* attributed to Oliverius Brito, could still be identified, as it was traditionally, with the Dominican Olivier Lebreton (Tricorensis, Armoricensis) his strong use of the formulas found in the *Speculum* would strongly confirm the early circulation and authority of this text amongst Albert's contemporary dominicans.[11]

These testimonies provided by the manuscript tradition, by ancient quotations and by the earliest biographies and bibliographies of Albertus could have been completed and fully confirmed, if the fire which destroyed the "most famous" library of the Dominican friars in Cologne in 1659, had not also burned what might have been a decisive document.[12] We have thus lost the ancient holdings of the convent and of Albertus' "Studium generale", but we can still quote the testimony of the *Legenda coloniensis*:

In the Monastery of the Dominicans in Cologne is preserved one of his [Albertus's] works, the autograph commentary *On Matthew*; there is also another autograph work, *De naturis animalium*, and equally by his hand there is a *Speculum mathematicae*.[13]

The testimony of this biography written on the basis of the lost archetype of the *Legenda prima* (that is, of a contemporary biography), is confirmed by the list of works by Albertus contained in the biography by Peter of Prussia. Peter, a Dominican living in the fifteenth century, commented on the listing of the *Speculum* amongst Albertus's works by writing "I saw and we have it".[14] These testimonies acquire even higher credibility if we reflect that the other autographs worshipped in Cologne during the Renaissance were works such as the *De animalibus* and the *Super Matthaeum*, which have reached us in manuscript autographs from Cologne.

The *Speculum*'s critical fortune is not only attested to in bibliographies or, as mentioned above, in old lists of works by Albertus.[15] We must also mention the reception it found with various authors immediately following Albertus. I can hardly presume to give a complete list, which would undoubtedly be very long indeed. An old and important example has already been suggested by Bruno Nardi;[16] that is, the *Lucidator dubitabilium astronomiae* written by Peter of Abano in 1303.[17] This work[18] was written a mere quarter of a century after Albertus's death. Nardi had already pointed out that Petrus of Abano based his astrology on the *Speculum astronomiae*, and this has been confirmed by several recent studies which analyze the manuscript tradition of the *Lucidator* and others of Peter's minor works in relation to Albumasar, his disciple 'Sadan' and other Arabs. "Like the author of the *Speculum astronomiae*, which seems to have influenced many passages of the *Lucidator dubitabilium astronomiae*, Peter of Abano believed that God acts on the sublunary realm through the heavenly bodies". As in the *Speculum*, "Peter of Abano felt impelled to found a science which would be capable of investigating that which lies between being and not-being, between the necessary and the possible". Peter postulates an "astral causality which happens '*nutu dei*' ('at God's command')"; "this concept receives a certain amount of support throughout Peter's writings and it had already been influentially espoused by Albert the Great (see especially the *Speculum astronomiae*)".[19] Writing in the generation following the *Speculum*, Peter "wrote his *Lucidator* in order to defend astrology from the attacks of those whose ignorance of philosophical ideas prevented them from understanding Ptolemy".[20] Although he attenuated the distinction between astrology and astronomy which, following Greek and Arab sources, Albert had sought to maintain,[21] Peter discussed both disciplines and drew on the *Speculum* as his preferred source: "The first is fully expounded and demonstrated in Ptolemy's *Almagest*, the second is to be found", Peter declares, "in the works of Alfargani, Azarquiel, Thabit ben Qurra, Alkindi".[22] "Like the author of the *Speculum*, Peter provides an excursus on necromancy and the magical arts", as well as on "astronomical, astrological and magical *imagines*, 'as [...] described by Ptolemy, Thabit ben Qurra and Zahel in their books of seals'".[23]

In his *Expositio Theoricae Planetarum Gerardi Cremonensis*, fruit of his university teaching at Bologna in 1318, Taddeo Alderotti of Parma not only distinguishes between the two types of *mathesis*, but also between permissable and necromantic *imagines*, "presumably in much the same way as Albertus Magnus had in the *Speculum Astronomiae"*.[24] Thorndike observes that in the *Expositio* "many of the works on *imagines* and nigromancy listed there are identical with titles mentioned in the *Speculum Astronomiae*".[25] Taddeo transcribes "a few chapters literally" from the *Speculum*, thereby revealing his poor preparation.[26]

Although more knowledgeable and rather original in his criticisms of astrology, Nicolas Oresme had no hesitation in drawing on the *Speculum astronomiae* in an analagous way. This can be seen in the introduction to his *Livre de divinacions*, where he "expounds a division of astrology into its various parts", a section which represents an addition to his Latin writings on the same subject. It doubtless appears in this French treatise because it was a work for the general public, and "it is not improbable that the philosopher was influenced by Albertus Magnus's *Speculum astronomiae* which contains a systematic subdivision [...] and which must have been regarded as a bibliographical guide for students of this discipline".[27] Albertus is also cited in Nicolas's *Livre des Ethiques*, where he appears precisely as a critic of illegitimate occult disciplines. He is quoted as maintaining that "some sciences are held to be bad either because they have bad principles, such as some auguries and spells, or because they deal in bad matters, such as necromancy".[28] The distinction can surely be traced back to the *Speculum astronomiae*.

In this connection, the contemporary testimonies of Pierre d'Ailly and Jean Gerson are interesting because they are very explicit and are the end result of a polemic directly concerning the legitimacy of astrology. The two prelates were in an appropriate position to evaluate with precision the aims and the authority behind a work like the *Speculum astronomiae*. It has to be stressed that neither doubted its Albertian authorship. Pierre d'Ailly explicitly mentions[29] this text in his *Elucidarius*, even though he does so in order to say that "Albertus is mistaken in his *Speculum*.[30] Pierre makes use of it again in the *Apologia defensiva astronomiae* in response to the *Tricelogium* written by Gerson in 1419. In the conclusion to this apology Pierre draws on arguments from Aquinas and the *Speculum*, a text whose usefulness he recognizes. He did not consider Albertus to be too favorable towards astrology, and he himself favored that discipline, while adapting it to a theological context.[31] He wrote in the *Elucidarius* that he aimed "to bring into concord true astrological science with sacred theology, since it [astrology] also has to serve it [theology] as a handmaid her lady, even more so than the other sciences ... it [astrology] is called appropriately natural theology".[32] In fact, just as real theology ("superior") leads to the knowledge of God by means of faith, so does astrology ("theologia naturalis"), the handmaid, lead to the knowledge of God by means of natural reason. According to Pierre d'Ailly, those theologians err who "do not adequately understand the aims of astrology". Its function consists in approximating with its own *iudicia* "the sayings of the prophets and in applying the meanings of the great conjunctions to the tenor of the prophetic messages" [3]. Pierre's distinction between true and false astrology is based on three points, which are also underlined in the *Speculum*:

1. that the influence of the stars is not binding,
2. that mixed together within treatises of astrology one also finds "detestable superstitions of magical art", and
3. that the power of the stars' influence is limited in the presence of free will. Such distinctions have been put forward not only "by holy theologians, but also by truthful astrologers".[34]

Not only Augustine, but also Ptolemy wrote against a deterministic interpretation of the influence of the stars. On the second point, set forth by Pierre with unusual emphasis, "many expert astronomers have excluded magic from their books, as a detestable and abominable art".[35] This was the promise and the program of both Leopold of Austria and of the *Speculum*. Pierre has no doubts as to the identity of its author:

> Albertus Magnus published also a very useful treatise in which he distinguished, according to their beginnings and ends, books concerning true astrology and those concerning the magical art, in order to separate astrological truth from magical vanity.[36]

Pierre, moreover, used literal quotations from the *Speculum* in one of his sermons. He had found himself in opposition to Jean Gerson because their conceptions regarding the legitimacy of astrology were profoundly different [7]; nevertheless, they were in agreement on the authenticity of the *Speculum*. Both Mandonnet[38] and Pangerl[39] have indicated a passage from Gerson's *Tricelogium astrologiae theologizatae* that took issue directly with Pierre, author of the *Vigintiloquium*.

Albertus Magnus wrote a short work on this subject entitled *Speculum Alberti*, relating that in his time some persons wanted to destroy books by Albumasar and several others. Preserving honour to so great a Doctor, it nevertheless seems that just as [Albertus], in expounding books of natural science, especially those written by Peripatetics, took too great care, more than was appropriate for a Christian Doctor and without adding anything concerning Christian piety, so also in his approving some astrological books, especially those on images, on birth-horoscopes, on engraved stones, on characters, and on interrogations, he came down on the side of irrational superstitions.[40]

This passage demonstrates that Gerson's position towards astrology was less favorable than has been stated by Liebermann[41] and it should be emphasized that he thought the *Speculum* contained "determinations", although its author was careful to deny this on many pages. Actually, Gerson applied to astrology the old criterion of its subordination to theology, which, as the sovereign science, must not only "cleanse", but "amputate" what is superfluous, harmful and indecent in astrology.[42] Many formulas used in the *Speculum* for the foundation and defense of astrology reappear, however, in Gerson's pages, such as calling celestial phenomena the "instruments of God",[43] even though he differs from the *Speculum* by refusing to see them as "signs"[44] Gerson concludes by more or less recognizing incommensurability in heaven ("eas ab omnibus comprehendi non posse").[45] The influence of the heavens is universal and remote and does not therefore provide explanations of particular effects, while God's actions operate in a "very singular and close way" ("singularissime and propinquissime").[46] Not only is divine operation to be valued infinitely more than that of the stars, but the effects of the latter are mediated in various and sometimes opposing ways according to the diversity of the receiving matter ("pro diversitate materiae").[47] Gerson was an attentive reader of Oresme, whom he mentions explicitly to confirm the incommensurability noted by him of the movements of the stars and of their influence.[48]

The *Tricelogium* contains the same problems as the *Speculum*, but provides more rigorous "solutions": "the heavens receive different powers from God according to their various parts, stars, planets and movements". This is a very early appearance of the idea of God as celestial clockmaker. The "powers" should be thought of as the internal mechanisms of a "very beautiful clock put together by the supreme arteficer", or as a "book full of many wise sayings copied into an exemplar, which is the infinite and eternal book of life, called world-archetype".[49] However, "the heavens obey the glorious bidding of God and are inferior and subject to works of human re-creation or repair": thus Gerson brings in a polemical note against those who see creation as a system regulated by necessity. "Many astrologers and philosophers erred in this respect by stating that God's actions spring from natural necessity".[50]

In a work of 1428 against a doctor from Montpellier who turned to talismans to effect his cures, Gerson is more severe than Albertus and holds that

> the making and use of *images* called "astrological" smacks highly of superstition and idolatry or magical ceremonies. [...] If such characters possess, or are thought to possess, a certain efficacy, the cause must be a spiritual one rather than one which is purely natural and corporeal like heaven and its influence on the body [...] they contain characters, letters, figures and phrases which do not produce any natural or purely corporeal effect for curing a disease of the kidneys and the like [...] as was noted especially by St. Thomas who, following the example of his teacher Albertus Magnus, conceded to astrology everything that was rationally possible, but in keeping with the Catholic faith".[51]

In the *Tricelogium*, Gerson had treated Albertus less favorably, and perhaps this was what attracted Mandonnet's attention. In the passage quoted above it can be seen that Gerson, a century and a half later, is reproving Albertus for excessive faithfulness to the astrological and Aristotelian texts which he had expounded too carefully and whose incompatibility with "Christian piety" he had not sufficiently emphasized. However, Gerson's criticisms place Albertus's Aristotelian commentaries on the same level with the *Speculum Astronomiae*, a work composed by Albertus Magnus in order to recount how, in his time, some persons wanted to destroy Albumasar's books and those of several others. There were reasons for the *rapprochement* which Chancellor Gerson made without hesitation, and they were not only cultural but institutional. This is important because it shows that he was well-informed concerning the interests, documents, and conflicts existing in earlier generations of Parisian Scholastics. This *rapprochement* is a strong argument in favor of the previously uncontested authenticity of the *Speculum*, and it is hard to see how Mandonnet found an argument for the opposite thesis in the passage just quoted.[53]

CONCLUSION

The *Speculum astronomiae* is relevant to the scientific and philosophical history of the Latin Schools in the thirteenth century. It illustrates the totality of material in Latin translation available during Albertus's lifetime and it could only have been written by a person who had access either to an exceptionally rich library or to a very wide-ranging network of bibliographical data. In the first case, Richard de Fournival's library, or at least his *Biblionomia*, come to mind; and for the second, we must imagine the collaboration of many people who sent *incipits* and descriptions of astronomical and astrological manuscripts from various Dominican convents.

Such a systematic and complete effort of bringing oneself up to date should not be played down, as it tends to be by the internalist historian of science mentioned at the beginning of this book.[1] After the long phase of rediscovery of Greek and Arabic scientific texts which extended from the twelfth century to William of Moerbeke, an effort at bibliographical and theoretical classification like that of the *Speculum astronomiae* represents a conspicuous achievement. Although it never became a standard schoolbook like Sacrobosco's *Sphaera* and the *Theorica planetarum*,[2] its manuscript and printed circulation was vast. It has been said that although Campanus of Novara showed no scientific originality, his place is assured in the history of medieval science on the strength of his dutiful interpretation of Euclid and of Ptolemaic astronomy.[3] This granted, there is surely no reason to attribute less importance to the author of the *Speculum astronomiae*.

The *Speculum* shows, on the one hand, an astrologer's ability to find his way through a vast specialist literature and, on the other, the thinker's awareness of the main problems which that discipline poses for theology and for philosophy itself. The solutions proposed for these problems do not differ from solutions appearing in other authentic works by Albertus, especially the *De fato* recently given back to him. After the convincing attribution of this *quaestio* to Albertus in the *editio coloniensis*, any controversy concerning the *Speculum*'s authenticity becomes a paradox. The *De fato* and other works declared authentic all attest to such a clear and consistent conviction on the part of the scientists' Patron Saint concerning the influence of the stars on nature and history, individuals and groups, that restoring the *Speculum astronomiae* as the canon of his works hardly constitutes a problem, nor changes an existing situation. The *De fato* and other writings reveal that Albertus was as much an astrologer as Roger Bacon, and recognizing his paternity of the *Speculum* does not change this point.

Within the jungle of uncertain and frequently changing attributions, of anonymous, pseudonymous, doubtful and spurious works, which constitutes the scientific and philosophical writings of the Middle Ages, the enigma of the *Speculum*'s authenticity is threatening or exciting only to a certain type of Dominican or, rather, Neo-Thomist historian. It is an interesting enigma, however, in that it represents a remarkable cultural crossroads – if we accept the attribution to Albertus (which it seems to me should be maintained) – in an author who stands between Alexander of Hales and Thomas Aquinas, perhaps working and collaborating with the scientific group assembled at the papal curia in the 1260s, and possibly with Campanus of Novara, in opposition to the attacks on astrology advanced by such theologians as Bonaventure of Bagnorea and the Dominican Gerard of Feltre.

The *Speculum astronomiae* is written in an unusual literary form. It is distinct from the type of encyclopedia often called, similarly, *Speculum*, or from *Summae*, *Expositiones*, *Commentarii* and *Quaestiones*, and could, instead, be considered one of the first examples of something resembling a modern encyclopedia entry which intelligently combines bibliographical information with a succinct theoretical discussion. Similar contemporary texts are hard to find, indeed, the pseudo-Albertian *Libellus de alchimia*, which has been likened to the *Speculum* because of the way in which its contents were organized, is in no way comparable.[4] We come closer, instead, in a few pages from the *De caelo* or others of Albertus's works on natural philosophy, which try to orient the reader concerning Ptolemaic or man-centered cosmic systems. However, very little can be said pertaining to a specific style in Albertus Magnus. He is an author who has more than one style, and his choice does not depend entirely upon whether he is writing theology or philosophy. He tends to use the texts he is paraphrasing as models, and consequently the style of his pseudo-Dionysian expositions is quite distinct from that of his expositions of the Aristotelian corpus. In the latter case it has been noted not only that many of the authors cited and references given by Albertus come literally from Averroes's commentary, but also that in many passages "Albertus seems to have identified himself with Aristotle and Averroes to such an extent that he speaks with the same voice", and sometimes attributes their literary projects to himself.[5] Such identification probably sprang from the habit of teaching by the "lectio" of texts to students. Albertus was, therefore, most himself stylistically in some *quaestiones* and in those works of natural history, the *De natura locorum* and, even more so, the *De mineralibus*, which he could not base on an 'Aristotelian' text.[6] It is not insignificant that these are precisely the works which bear the greatest stylistic resemblance to the *Speculum*: the two just mentioned, the commentary on the pseudo-Aristotelian Arabic *De causis proprietatum elementorum*, and the *De fato* (a *quaestio* the Scholastic tradition of which goes back at least as far as the *Summa halensis*). But it is difficult to find true resemblances since the *Speculum* contains literal transcriptions and occasional summaries from Ptolemy, Haly, Alfarghani, Thebit, Alkabitius and Albumasar. In our 1977 commentary on the *Speculum* (republished here), which was well received,[7] full passages from these authors not only throw light on Albertus's vast store of information and the careful search for significant definitions which constitute the backbone of the *Speculum*, but they also facilitate an understanding of the work itself. Albertus's tendency to use summarizing quotations often gives his text too dense a texture, and only the reading of the contexts behind the brief quotations clarifies the meaning. The *Speculum*'s style is influenced by these calques from various Arab authors and Arab-Latin translations, and is therefore not very similar to expositions of the *Sentences*, the *Scriptures*, or of pseudo-Dionysius (but it is more similar to his Aristotelian commentaries). Even the *Speculum*'s theoretical chapters have the same characteristic, since they too are based on calques and quotations, from, for example, texts by Albumasar on contingency. All this notwithstanding, it has been possible to note above the many correspondences of terminology and style between the *Speculum* and Albertus's works of natural history.

As for the *Speculum*'s tradition, it is certainly surprising to read in a study published for the recent centenary that "during years of debate about its authorship the work has been attributed to, among other figures of the High Middle Ages, Albertus Magnus".[8] But this attribution was generally accepted long before the debate, a debate which began when an attribution to Roger Bacon was produced which finds no confirmation in the manuscript or literary tradition. There is no time here for the codicological analysis which will have to be performed by whoever wants to prepare the definitive edition of the *Speculum astronomiae*.[9] However, from the data published in 1977 it can be deduced that the work appears in fifty-three codices, a number which exceeds that of even the most popular of Albertus's works on the natural sciences, and that the attribution to Albertus predominates. Alongside the attributions to Albertus there is only one attribution to

Thomas (in a codex containing others of the latter's works on similar subjects) and one in two old manuscripts and a late copy to Philippus Cancellarius. This fact is convincingly explained by the institutional functions of the chancellor as recipient of the *Speculum*'s peroration and guarantor of the law forbidding astrological readings. Together with the numerous mentions of Albertus as author (in which can be discerned no response to an objection as has been strangely asserted), we must also add that the *Speculum* appears several times in miscellaneous codices not only of translations of Arab treatises on astrology, but above all of works by Albertus on the natural sciences.[10] Finally, the presence of the *Speculum* in old catalogs of Albertus's works and the tradition which held that an autograph manuscript of this work existed until recent times in the Dominican library in Cologne, all further confirm his authorship, as do those announcements made in the *De caelo* of his intention to write on astrology and "electiones". The literary tradition is unanimous, from the generation immediately following him (Peter of Abano) up to the period of Campanella and Naudé.

All of these reasons, therefore, lead me to maintain the attribution of the *Speculum* to Albertus. With the sole exception of Giovanni Pico – an exception which, as we have seen, was due to polemical concerns and which was not expressed unequivocally – the attribution to him of this work has been virtually unanimous. Unanimous, that is, up to the time of Mandonnet's ingenious but poorly-documented hypothesis, which was proposed in order to safeguard, at the beginning of our century, Albertus's prestige as a scientist and thinker, when the relevance of astrology and its indubitable presence in the medieval world view had not been fully understood. While analyzing the themes and questions present in the *Speculum*, I have often had to state that these themes and questions were commonplace both in the thirteenth century and in all discussions concerning astrology. This shows how widespread and commonly accepted they were, even by Bonaventure of Bagnorea when he treated the general concept of the influence of the stars. The world system in the two variant versions of Ptolemy and al-Bitruji, the theory of the planetary properties of the elements and of the correspondences between the heavens and the sublunary world, which lay behind the action of natural and talismanic magic, astrological medicine with its emphasis on critical days for illnesses, phlebotomies or purges – none of these can be understood without turning to the notion of celestial influence and its consequences. Many aspects of medieval science and some practices, which were not only medical, were based on these principles. To provide a sketch of the fundamental outlines of this discipline in a theoretical-bibliographical introduction was no small achievement, and the *Speculum astronomiae* was consequently greeted with great interest and used as a standard reference book for many centuries. Many later astrological treatises, beginning with that of Peter of Abano, follow its scheme and draw on it abundantly; and such a fact is also a confirmation of the importance of this guide to astrology. It is interesting that the *Speculum* was born in a climate of accusation and censure formulated by friends of the writer. The background of discussions taking place in the second half of the thirteenth century, and especially in the 1260s, within the Dominican order and in the Arts Faculties also dispels all doubt that the *Speculum* was written by Albertus, perhaps with some collaborators, among whom could have been Campanus of Novara. This is not the most important point: what concerns us here is an appreciation of the clear and by now classical arrangement provided by this work for a series of problems which definitely cannot be removed from the panorama of medieval thought.

NOTES*

CHAPTER ONE

* See Abreviations at p. 204.

1. 'Further consideration of the *Experimenta*, *Speculum astronomiae* and *De secretis mulierum* ascribed to Albertus Magnus', *Speculum*, XXX, 1955, p. 427.
2. J. D. North, *Horoscopes and History*, London, 1986, p. 172. On the "complacency towards astrology, or positive acceptance it" from the times of Hildegard of Bingen to those of Bartholomeus Anglicus, Vincent of Beauvais, Roger Bacon, and Etienne Tempier, *cf.* Richard of Wallingford, *An edition of his Writings with Introductions, English Translations and commentary* by John D. North, Oxford, at the Clarendon Press, II, 1976, p. 84 ff.
3. M. H. Malevicz, ed., '*Libellus de efficacia artis astrologiae*', *Mediaevalia philosophica polonorum*, XX, 1974, pp. 3-95; *cf.* pp. 47, 53, 92. Among the arabs Abu Ma'shar is the more certain of the sources for the *Libellus* (possibly together with al-Farghani and al-Qabisi).
4. P. Glorieux, *Répertoire*, Paris 1933, I, p. 75, on Albert; I, p. 392 on Philippe de Thory; II, p. 73, on Roger Bacon. See below, for full citation of the works by Mandonnet and Geyer here quoted. *Cf.* Albertus Magnus, *Speculum astronomaie*, ed. S. Caroti, M.Pereira, S. Zamponi, P. Zambelli, Pisa, Domus Galileana, 1977, Appendix II, which lists its mss., the codexes Bodleian, Digby 81, (13th Cent.) and Digby 228 (14th Cent.). This ed. will be quoted hereafter as *Speculum* and reproduced below p. 203 ff.
5. P. Mandonnet, 'Roger Bacon et le *Speculum astronomiae*' (1277), *Revue néoscolastique de philosophie*, XVII, 1910, pp. 313-335; *cf.* further comments by Mandonnet in the second edition of his classic study *Siger de Brabant et l'Averroisme latin au XIIIe siècle*, Louvain 1911, I, pp. 244-248 and in the *Dictionnaire de théologie catholique*, I, Paris 1930, col. 673, s. v. Albert-le-Grand. See also P. Mandonnet, 'Roger Bacon et la composition des trois *Opus*', *Revue néoscolastique de philosophie*, XX, 1913, pp. 52-68, 164-180. Before Mandonnet, few had approached the problem: amongst others, see L. Choulant, 'Albertus Magnus in seiner Bedeutung für die Naturwissenschaften', *Janus. Zeitschrift für Geschichte und Literatur der Medizin*, I, 1846, p. 138 ("nicht in der gewöhnlichen Schreibart Alberts verfasst, vielleicht unecht"); M. Steinschneider, 'Zur Geschichte der Übersetzungen aus dem Indischen ins Arabische', *Zeitschrift der Deutschen Morgenländischen Gesellschaft*, XXV, 1871, p. 386 registered this doubt concerning the attribution to Albert, but confirmed the dating of the *Speculum* to Albert's time; Steinschneider did however abandon all reservation in his study 'Zum Speculum astronomiae des Albertus Magnus, über die darin angeführten Schriftsteller und Schriften', *Zeitschrift für Mathematik und Physik*, XVI, 1871, pp. 357-396, where the identification of the eastern sources and of their Latin translators offered the first philological contribution for the study of this treatise; the research had been undertaken at the request of Jessen, the editor of the Albertinian text *De vegetalibus*. Jessen was then preparing a critical edition of the *Speculum*, and had sent a ms. copy of the work to Steinschneider, who in the article quoted above listed some of the manuscripts. J. Sighart, *Albertus Magnus*, Regensburg 1857, p. 343 did not doubt the

authenticity ("So zeigt sich Albertus auch in dieser Schrift als Forscher, der selbst geprüft und sich über den Wogen des Aberglaubens seiner Vorgänger hierin glücklich zu erhalten gewusst hat"), contrary to what is reported by F. von Bezold, *Aus Mittelalter und Renaissance*, Berlin 1918, p. 403 n. 351. Ch. Jourdain too did not doubt the authenticity of the work in his 'N. Oresme et les astrologues', *Revue des questions historiques*, XVIII, 1875, p. 139.

6. A. Birkenmajer, *Études d'histoire des sciences et de la philosophie du Moyen Age*, Wrocław-Warszawa-Kraków 1970, pp. 143-145; *cf.* Ch. H. Haskins, *Studies in the History of Mediaeval Science*, Cambridge, Mass. 1924, p. 69n., 164, 288, 338n., in which is reprinted a study written in 1911 (immediately after the article by Mandonnet) quoting Albert as the author of the *Speculum*.
7. See *Speculum, Prooemium* / 1-5: "apud quos non est radix scientiae, [...] verae sapientiae inimici, h. e. D. N. Iesu Christi [...] catholicae fidei amatoribus merito sunt suspecti".
8. L. Thorndike, *A History of Magic and experimental Science*, II, New York 1923 [hereafter TH, II], pp. 11, 55-56 and *passim*; see also some distinctions I deployed in my essay 'Il problema della magia naturale nel Rinascimento', *Rivista critica di storia della filosofia*, XXV, 1973, pp. 271 ff. An interesting discussion about whether it was right to "class astrology with the occult sciences" has recently been published by J.V. Field, 'Astrology in Kepler's Cosmology', in *Astrology, Science and Society. Historical Essays*, ed. by P. Curry, Woodbridge, Suffolk, 1987, p. 143 ff.
9. *Speculum, Prooemium* / 6: "placuit aliquibus magnis viris, ut libros quosdam alios et fortasse innoxios accusarent".
10. *Speculum, Prooemium* / 7-8: "quia plures ante dictorum librorum necromantiam palliant professionem astronomiae mentientes". *Cf.* Thomas Aquinas, *De sortibus* [1268-1272], in *Opuscola theologica*, ed. R. Averardo, I, Torino-Roma 1954, pp. 161-162 and 642-650; *Opera omnia*, Roma 1976, vol.XLIII (*Opuscula*,IV), pp. 229-238, 239-241: according to Thomas "patet quod sors proprie in rebus humanis locum habeat" and he put forward a classification of all forms of divination in order to distinguish the natural, legitimate ones from those considered to be supernatural or evil.
11. In a critical discussion which appeared separately from Mandonnet's preoccupation with the connection between Tempier and the *Speculum astronomiae*, R. Hissette, 'Etienne Tempier et ses condamnations', *Recherches de théologie ancienne et médievale*, 48, 1980, p. 265, thought it was "certain que la seconde condamnation de Tempier fut bien davantage prise au sérieux que la première [du 1270]. D'aucuns l'ont critiqué. Ainsi Gilles de Rome et un maître qui pourrait être Jacques de Douai; ceux-ci ne cesserent pourtant de s'y soumettre".
12. E. Grant, 'The condemnation of 1277, God's absolute power, and physical thought in the late Middle Ages', *Viator*, 10, 1979, p. 239. Taking up again the important critical remarks Koyrè deployed against Duhem (pp. 212-213), Grant focused on articles 34 e 49 (*cf.* pp. 139-141), the ones which provoked the physical discussions on the concept of emptiness, of center, of weight and of natural places, of celestial bodies' rectilinear or circular motion, and on the consequences of such discussions – that continued up to Suarez, Campanella, Hobbes, Gassendi, and Locke – for the re-definition of the power of God ("The God of the Middle Ages, who could do anything he pleased short of a logical contradiction, was replaced by a God of constraint, who, having created a perfect clock-like universe, rested content merely to contemplate his handiwork ever thereafter",

p. 244). In Grant there are interesting remarks on Tempier's articles 204 and 219 referring to the problem of intelligences –where even Thomas Aquinas was censored– and concerning article 6, on the subject of the great year (pp. 235 ff., 238). To the bibliography he listed (p. 211, n. 1), one should add the summary of his own article Grant offered in *Cambridge History of Later Mediaeval Philosophy*, Cambridge 1982, pp. 537-539; R. Hissette, *Enquête sur les 219 articles condannés à Paris le 7 mars 1277*, Louvain-Paris 1977; J.P. Wippel, 'The Condemnations of 1270 and 1277 at Paris', *The Journal of Mediaeval and Renaissance Studies*, VII, 1977, pp.169-202; R. Wielockx, 'Le ms. Paris. lat. 16096 et la condemnation du 7 mars 1277', *Recherches de théologie ancienne et médievale*, XLVIII, 1981, pp. 227-232; K. Flasch, *Aufklärung im Mittelalter? Die Verurteilung von 1277, Das Dokument des Bischofs von Paris übersetzt und erklärt* v. K. Flasch, Mainz, DVB, 1989; L. Bianchi, *Il vescovo e i filosofi. La condanna parigina del 1277 e l'evoluzione dell'aristotelismo scolastico*, Bergamo, Lubrina, 1990.

13. M.-Th. d'Alverny, 'Un témoin muet des luttes doctrinales du XIIIe siècle', *Archives d'histoire doctrinale et littéraire du Moyen Age*, XXIV, 1949, p. 226; *cf.* Godefroid de Fontaines, *Quatuordecim Quodlibeta*, ed. M.De Wulff- A.Pelzer- J.Hoffmans, Louvain-Paris 1904-1935, *Quodl.* XII, q. 5, where articles 96, 124, 36, 215, 204, 219, 129, 130, 160, and 163 of the condemnation of 1277 are re-examined.
14. Raymundus Lullus, *Declaratio per modum dialogi edita*, hrsg. v. O. Keicher, Münster 1909 (= Beiträge z. Geschichte d. Philosophie und u. Theologie d. Mittelalters, VII, 4-5), p. 95 ff., *cf.* J. N. Hillgarth, *Ramon Lull and Lullism in fourteenth century France*, Oxford 1971, pp. 230-31, 251.
15. Aegidius Romanus, *Errores philosophorum*. Critical Text with notes and introduction by J. Koch, English Translation by J. O. Riedl, Milwaukee 1944, pp. XXIX-XL, LV-LIX, 3-67; *cf.* also the first ed. (as an anonimal text) in Mandonnet, *Siger cit.*, II, 2nd. ed. 1911, pp. 3-25.
16. *Speculum*, III / 28-30 "universi ordinationem nulla scientia humana perfecte attingit, sicut scientia iudiciorum astrorum"; *cf.* XIII/54-60: "Quod si propter hoc condemnetur ista scientia, eo quod liberum arbitrium destruere videatur hoc modo, certe eadem ratione non stabit magisterium medicinae: numquid enim ex eius magisterio iudicatur quis secundum causas inferiores aptus ad huiusmodi vel ineptus? Quod si magisterium medicinae destruatur multum erit utilitati reipublicae derogatum, eo vero stante non videntur habere quid contra partem nativitatum allegent".
17. *Speculum*, II / 76-84: "non est lumen geometriae cum evacuata fuerit astronomia". The tone is not unusual for Albert: see for instance "quidam qui nesciunt omnibus modis volunt impugnare usum philosophiae... blasphemantes in iis quae ignorant", a passage taken from his commentary *In Epistolas Dionysii*, a text which dates back to the first period in Cologne (1248-1249). This passage, already quoted by Mandonnet, in *Siger cit.*, I, pp. 35-36n, now is also to be found in Van Steenberghen, *La Philosophie au XIIIe siècle*, Louvain-Paris 1963, p. 275 n. Further, highly polemical passages by Albert are to be found in *De somno et vigilia*, I, i, 1 e III, ii, 5, ed. Jammy, V, pp. 65a, 106 and are quoted in TH, II, 585.
18. *Speculum*, XII/1-8. "Quoniam autem occasione eorum, ut dictum est, multi libri praenominati et fortassis innoxii accusantur, licet accusatores eorum amici nostri sint, veritatem tamen oportet, sicut inquit Philosophus, honorare, protestor tamen quod si aliquid dicam quo velim uti in defensione eorum, quoniam *determinando* non dico, sed potius *opponendo*

vel *excipiendo* et ad determinationis animadversionem determinatoris ingenium provocando". The expression "veritas salvari" is usual for Albert, and for him carried epistemological implications, see *De caelo*, ed. P. Hossfeld, *Opera omnia*, t. V, I, Münster 1971, p. 129/10; bk. II, tr. 2, ch. 2, and especially p. 168/31, where an importat discussion of the various astronomical systems ended with a remark and an hypothesis "salvantes Aristotelem et veritatem, quam invenimus diligenti astrorum inspectione". The use of the "quoniam" instead of the "quod" was not unusual in the *Speculum*, XIV/61, 71-71, and reflected an imitation of the patristic style.

19. R. Lemay, *Abu Ma'shar and Latin Aristotelianism in the Twelfth Century. The Recovery of Aristotle's Natural Philosophy through Arabic Astrology*, Beirut 1962; T. Gregory, 'L'idea di natura nella filosofia medievale prima dell'ingresso della fisica di Aristotele', in *La filosofia della natura nel Medioevo. Atti del III Congresso Internazionale di Filosofia Medievale* [La Mendola 1964], Milano 1966, pp. 27-65; M.-Th. d'Alverny, introduction to 'Al-Kindi *De Radiis', Archives d'histoire doctrinale et littéraire du Moyen Age*, XLI, 1974, p. 139 ff. Mandonnet, 'Roger Bacon', p. 329 believed that only one of the propositions condemned in 1277 and analysed in the next paragraph (namely nr. 167, numbered by Mandonnet 178) concerned the divinatory sciences, an assumption he took as confirmation of the hypothesis that the teaching of Aristotle in Paris was developed independently of Arabic influences in this field. TH, II, 709-713 listed, but did not analyze, several propositions. In his analysis of the condemned theses, E. Gilson, *La philosophie au Moyen Age*, Paris 1952, 2nd ed., pp. 560-561, did not discuss astrology, but saw in the tendency to necessitarianism the decisive element that provoked Bishop Tempier's decree.
20. H. Denifle and E. Chatelain, *Chartularium Universitatis Parisiensis*, I, Paris 1899, p. 552, art. 156: "Quod si celum staret, ignis in stupam non ageret, quia Deus non esset"; R. Hissette, *Enquête sur les 219 articles cit.*, p. 142 (followed by L. Bianchi, *Il vescovo e i filosofi cit.*, but not by K. Flasch, *Aufklärung im Mittelalter? cit.*) has proposed an interesting emendation to the last words of this article, namely: "quia natura deesset". Two mss. give this reading, but the traditional form cannot be so easily dismissed being not only printed by Du Plessis d'Argentré or Mandonnet, but also being the subject of discussion by contemporaries such as Ramon Llull and John of Naples. Hissette, pp. 70 ff., 117 ff. has fully commented on the nature and role of intelligences and their influence on inferior substances. He traces the sources of the theses condemned mainly to Siger. Bianchi does not consider intelligences, but makes interesting comments on Mandonnet's historical method. See also in Denifle, art. 21: "...quod nichil fit contingenter considerando omnes causas"; art. 38: "Quod Deus non potuit fecisse primam materiam nisi mediante corpore celesti"; art. 59: "Quod Deus est causa necessaria motus corporum superiorum et coniunctionis et divisionis continentis in stellis"; art. 92: "Quod corpora celestia moventur a principio intrinseco, quod est anima; et quod moventur per animam et per virtutem appetitivam, sicut animal; sicut enim animal appetens movetur, ita et celum".

CHAPTER TWO

1. J. A. Weisheipl, 'The *Problemata determinata XLIII* ascribed to Albertus Magnus', *Mediaeval Studies*, XXII, 1960, pp. 323-327; *cf.* now the edition, based also on a new ms., by Weisheipl in *Opera omnia*, t. XVII/1, Münster 1975, p. 48 ff.: "Sciendum autem quod

non inveniuntur antiqui perypatetici aliquid de angelis tradidisse, sed novi quidam et tantum quidam arabes et quidam judaei... Concorditer autem isti dicunt quod intelligencie sunt substancie quas vulgus angelos vocat... Non est dubium quod corpora caelestia non movent angeli... Patet igitur quod intelligencia nec angelus est, et si esset, non adhuc esset motor proximus alicuius spere celestis. Et si sic est, quod certissime probatum est, tunc angeli per ministerium non movent corpora celestia et sic ulterius sequitur quod nec alia inferiora corpora moventur ab ipsis". Before the discovery of these *Problemata*, Fritz Pangerl, 'Studien über Albert den Grossen', *Zeitschrift für katholische Theologie*, XXXVI, 1912, p. 800n., remarked: "fast in allen seinen Werken wendet er sich gegen die auch nach ihm nicht vermiedene Verwechselung der Intelligentiae (*Substantiae separatae*) und der Engel. Vgl. [*Opera*, ed. Borgnet, Paris 1890-1899] I, 189-190; X, 431, 45; XXXII, 368". The same views were expressed in TH., II, 502n. *Cf.* also K. Flasch, 'Von Dietrich zu Albert', *Freiburger Zeitschrift für Philosophie und Theologie*, 32, 1985, p. 23. The historiographical debate and many texts by Albert will be discussed below, ch. 8.

2. *Chartularium cit.*, I, p. 551, art. 133: "Quod voluntas et intellectus non moventur in actu per se, sed per causam sempiternam, scilicet corpora celestia".
3. *Chartularium cit.*, I, p. 549, art. 112: "Quod intelligentie superiores imprimunt in inferiores, sicut anima una imprimit in aliam, et etiam in animam sensitivam; et per talem impressionem incantator aliquis prohicit camelum in foveam solo visu"; see also art. 142: "Quod ex diversitate locorum acquiruntur necessitates eventuum", art. 143: "Quod ex diversis signis caeli significantur [ed.:signantur] diversae conditiones in hominibus tam donorum spiritualium quam rerum temporalium"; see also art. 206 and 132.
4. *Chartularium cit.*, I, p. 552, art. 162: "Quod voluntas nostra subiacet potestati corporum celestium".
5. *Chartularium cit.*, I, p. 547, art. 74: "Quod intelligentia motrix celi influit in animam rationalem sicut corpus celi influit in corpus humanum".
6. *Chartularium cit.*, I, p. 553, art. 167: "Quod quibusdam signis sciuntur hominum intentiones et mutationes intentionum, et an illae intentiones perficiendae sint et quod per tales figuras sciuntur eventus peregrinorum, captivatio hominum, solutio captivorum, et an futuri sint scientes an latrones".
7. *Chartularium cit.*, I, p. 555, art. 207: "Quod in hora generationis hominis in corpore suo et per consequens in *anima, quae sequitur corpus*, ex ordine causarum superiorum et inferiorum inest homini dispositio inclinans ad tales actiones vel eventus. Error, nisi intelligatur de eventibus naturalibus et per viam dispositionis."(Italics mine).
8. *Cf.* A. Bouché-Leclercq, *L'astrologie grecque*, Paris 1899; F. Boll, C.Bezold and W. Gundel, *Sternglaube und Sterndeutung. Die Geschichte und das Wesen der Astrologie*, Darmstadt 1965.
9. Mandonnet, *Siger cit.*, I, p. VII; II, pp.27-52; now see the critical ed. by B. Geyer, in *Opera omnia*, XVII/1, Münster 1975, pp. xix-xxiii, 31-44. Through a careful comparison of the above quoted answers with the thirteen articles condemned in 1270, as well as with their "more elaborated formulation" in 1277, Geyer showed that Albert's pamphlet was composed shortly before the condemnation of 1270, and not during the interval between the first and the second condemnation (1273-1276). The latter thesis had been put forward by F. Van Steenberghen, "Le *De XV problematibus* d'Albert le Grand", in *Mélanges Pelzer*, Louvain 1947, pp. 438-39, and was supported by the critical remarks on the chronology of the commentaries on Aristotle elaborated by Pelster, now deeply revised if not

abandoned. The date 1273-1276 was still accepted by T. Schneider, *Die Einheit des Menschen*, Münster 1973 (= Beiträge zur Geschichte der Philosophie und Theologie des Mittelalters, N. F., 8), p. 71.

10. Albertus Magnus, *De XV Problematibus*, ed B.Geyer in *Opera omnia*, t. XVII/ 1, Münster 1975, p. 36: "quod fatum, quod ex constellatione est, necessitatem non imponit propter tres causas. Quarum una est, quia non immediate, sed per medium advenit, cuius inaequalitate impediri poterit; secunda autem, quia per accidens et non per se operatur in natis; operatur enim per primas qualitates, quae non per se virtutes stellarum accipiunt; tertium est, quod operatur in hoc in quod operatur in diversitate et potestate materiae natorum, quae materia uniformiter et prout sunt in caelis recipere non potest coelorum virtutes".
11. *Op. cit.*, p. 36: "quamvis allatio Solis et planetarum in circulo declivi sit causa generationis inferiorum et recessus eorundem in eodem circulo sit causa corruptionis, et sint aequales periodi generationis et corruptionis, tamen inferiora periodi aequalitatem et ordinem non assequuntur propter materiae inaequalitatem et inordinationem. Quis autem dubitet propositum hominis magis inaequale et inordinatum esse quam naturae? Multo minus propositum necessitati subiacet quam natura".
12. *Op. cit.*, p. 37: "Necessitatem ergo in inferioribus superiora non imponunt. *Nec unquam hoc aliquis dixit mathematicorum.* Si enim hoc esset, periret liberum arbitrium, periret consilium et periret contingens secundum omnem ambitum suae communitatis, quod est valde absurdum".
13. *Op. cit.*, p. 36: "anima humana secundum philosophos est imago mundi; propter quod in ea parte quae *imago* intelligentiae et causae primae est, impossibile est eam motibus coelestium subiacere. In ea autem parte quae in organis est, quamvis sidereis moveatur scintillationibus, tamen necessitatem et ordinem superiorum non assequitur, et sic nec illa parte necessitati subiacet vel subditur superiorum".
14. Cfr. Siger de Brabant, *De aeternitate mundi*, first ed. in Mandonnet, *Siger cit.*, II, pp. 139-140 and *cf.* I, pp. 171-172; now see the critical ed. by B. Bazán, Siger, *Quaestiones in tertium de anima*, Louvain-Paris 1972, p. 132: "Ex hoc autem quod semper est movens et agens, sequitur quod nulla species entis ad actum procedit quin prius praecesserit, ita quod eadem specie quae fuerunt circulariter revertuntur, et opiniones, et leges, et religiones, et alia, ut circulent inferiora ex circulatione superiorum, quamvis circulationis quorundam propter antiquitatem non maneat memoria. Haec autem dicimus opinionem Philosophi recitando, non ea asserendo tanquam vera".
15. *Chartularium cit.*, I, p. 544, art.6: "Quod redeuntibus corporibus coelestibus omnibus in idem punctum, quod fit in XXX sex milibus annorum, redibunt idem effectus, qui sunt modo".
16. On the circulation and the attribution of this text (PG, 40, 729 ff.) in the Middle Ages, *cf.* R. Klibansky, *The Continuity of the Platonic Tradition during the Middle Ages*, London 1950, and E. Garin, *L'età nuova*, Napoli 1969, pp. 41-42. See Nemesius of Emesa, *De natura hominis*, trad. Burgundio Pisanus, éd. crit. G. Verbeke-J. R. Moncho, Paris-Louvain 1975 (= Corpus Lat. Comm. in Aristotelem Graecorum), p. 142, ch.xxxvii: the text is literally copied by Albert in the passage reproduced in footnote 17.
17. Albertus Magnus, *Summa theologiae*, Pars I, tr. 17, q. 68, ed. Jammy *cit.*, XVII, p. 388: "Stoici aiunt restitutos planetas in idem signum secundum longitudinem et latitudinem, ubi in principio unusquisque erat cum primum mundus costitutus est, in dictis temporum

circumitionibus incendium et corruptionem eorum quae sunt perpetrari et rursus a principio in idem mundum restitui, et rursus unumquodque astrorum in priore circuitione figens [recte: fiens]secundum longitudinem et latitudinem, inde similiter alium mundum perfici. Futurum rursus esse Socratem et Platonem et unumquemque hominem cum eisdem amicis et civibus, et eadem suadere et cum eisdem colloqui et omnem civitatem et municipium et agrum similiter instaurari ut prius." The discussion which immediately follows the one on fate or the temperamental disposition in individuals was introduced by a harsh theological judgment: "Addunt etiam, quod maxime horribile est, quod quia talis ordo causarum certus est, fundatus motu regulari et uniformi caelestium, ideo certus et necessarius est cursus fati et etiam cursus fortunae. Unde Homerus: 'Inveniunt sibi fata viam'. Et hoc est contra fidem, sicut patuit in praecedentibus de fato". [Those works that are not yet included in the critical *Opera omnia* ed. by the Albertus-Magnus-Institut, are quoted from the ed. Jammy, Lyon 1651, in preference to the Borgnet ed., Paris 1890-1899, which is in fact a reproduction of the earlier one.]

18. *Ibidem*: "Addunt etiam deos sive corporeos, sive incorporeos, sive caelestes, sive terrestres, sive infernales, qui non subiiciuntur corruptioni huic quae mortalium est, cum assecuti fuerint unam circumitionem, hoc est perfecte cognoverunt, ex hac cognoscere omnia quae sunt futura in his quae deinceps sunt circumitionibus. Nullum enim extraneum futurum esse dicunt, praeterquam ea quae facta sunt prius, sed omnia similiter et immutabiliter esse in una circumitione sicut in alia etiam usque ad minima. Tempus autem unius circumitionis dicunt esse XXXVI millia annorum, quod vocant magnum annum, in quo, sicut dicit Aristoteles in primo *Primae Philosophiae*, dii caelestes iureiurando informaverunt ad idem principium circumitionis se redituros et similem circumitionem ut prius se perfecturos. Et quia sic fatum et fortunam in diis caelestibus radicaliter posuerunt, ideo fatum et fortunam pro diis colebant supplicationibus et sacrificiis". The *quaestio 68* is briefly analyzed in TH, II, 589-592. Albert took it for granted that the period of revolution of the heaven carrying the fixed stars was 36.000 years exactly: see *De quattuor coaequaevis*, tr. III, q. xii, a. 2; ed. Jammy, XIX, p. 64; *Metaphysica*, bk. XI, tr. ii, c. 22; ed. B. Geyer, *Opera omnia*, t. XVI, 2, p. 510/55 ff.

19. This Arabic geomancy translated by Hugo of Santalla in the twelfth century is different from the one he himslef wrote, which is preserved in several manuscripts (CLM 588, ff. 58va-77vb; Bodley 625, f. 54; Paris. lat. 7354, ff. 2r-55v and – together with *Speculum* – Vindob. 5508, ff. 182r-200r); partial ed. in P. Tannery, *Mémoires scientifiques*, IV, Paris 1920, pp. 373-401 (but *cf.* pp. 324-328, 339-340, 402-411): *Super artem geomantiae*, inc. prologus: "Rerum opifex Deus qui sine exemplo nova condidit universa"; inc. op.: "Arenam limpidissimam a nemine conculcatam et de profundo ante solis ortum assumptam". Hugo of Santalla worked under Bishop Michael of Taragona (1119-1150), preparing versions of the *Centiloquium*, of the *De pluviis* by Albumasar, of pseudo-Aristotelian occult treatises, etc. Concerning him see C. F. S. Burnett, 'A Group of Arabic-Latin Translations', *Journal of the Royal Asiatic Society*, 1977, pp. 62-118. *Cf.* the recent fundamental study by T. Charmasson, *Recherches sur une technique divinatoire: la géomancie dans l'occident médiéval*, Genève-Paris 1980, and also: Haskins, *Studies cit.*, p. 78ff.; TH, II, pp. 86-88, 118-119; P. Meyer, 'Traités en vers provençaux sur l'astrologie et la géomancie', *Romania*, XXVI, 1897, who at pp. 248-49 published the prologue to the *Estimaverunt Indi* from the ms. Laurenziano pl. XXX, 29, f. 1ra; an explanation for the condemnation is easily found in the extreme necessitarianism at the basis of this divinatory practice.

"Incipit liber geomancie nove magistri Ugonis Sanctiliensis editus ab Alatrabuluci translatione. Estimaverunt Indi quod quando lineantur linee absque numero et proiciuntur pares et eriguntur ex eo quod remanet figure quatuor, deinde generantur et concluduntur ad inveniendam intentionem, significat illud quod erit anime et facit ea necessitas orbis ad illud quod rectum est, et interpretatur de eo quod in anima est... circulus erit secundum querentis et cognitionem cordis eius ad illud quod est rectum et interpretatur de eo quod est in anima". Albert spoke of geomancy in objective terms, and did not feel compelled to censor it, in *De mineralibus*, bk. II, tr. iii, ch. 3, ed. Jammy, V, p. 240(b).

20. Mandonnet, 'Roger Bacon' *cit.*, pp. 317-8.
21. See ch. 5, pp. 46–47 and footnotes 5–9.
22. Th., II, 708-709.
23. Mandonnet, 'Roger Bacon' *cit.*, p. 319; but *cf. Siger cit.*, 1909(2), II, p. IX, where he describes this very manuscript CLM 8001, dating to the first half of the fourteenth century, without identifying Chap. XVII of the *Speculum astronomiae*, there reproduced with a rubrica (15th) which gives it the title: "Epistola Thomae de aliquibus nominibus librorum astronomiae" (f. 145r) and followed by "Thomas, An licet judiciis astrorum uti" as well as by four further writings by Thomas, many by Albert and by Giles of Rome. In Dondaine's edition of Thomas Aquinas, *De iudiciis astrorum*, in *Opera omnia*, Roma 1976, vol. XLV [= Opuscula, IV], p.192 the codex is listed, but the attribution to Thomas of this fragment of the *Speculum astronomiae* is not discussed or even mentioned. So not persuaded by some critical reviewers I still believe, that because of this and other new observations some use can be found in the description of CLM 8001 we gave – as for all the other mss. – on the basis of microfilms in our edition *Speculum cit.*, pp. 154-157; *cf. ibid.*, pp. 179-181 for a summary of the attributions to Albert found in 29 mss., and pp. 93-175 for a description of all the known mss. To the list, we should now add a 53d ms.: London, British Meteorological Association/ Institution of Electrical Engineers, Thompson Collection 5 (XV Cent., membr.), ff. 1-43(2), indicated by L. Sturlese, in his review of *Speculum cit.*, *Annali della Scuola Normale Superiore di Pisa, Cl. di Lett.*, S.III, vii, 1977, p. 1616.
24. Mandonnet, 'Roger Bacon' *cit.*, pp. 313 and 320; *cf.* F. von Hertling, *Albertus Magnus. Beiträge zu seiner Würdigung*, Münster 1914, 2nd ed., p. 26.
25. Though I share the severe reservation put forward by F. Pangerl against the reliability of such ancient lists in his 'Studien über Albert den Grossen', *Zeitschrift für katholische Theologie*, XXXVI, 1912, pp. 514 ff., I have collected all the data they offer, and they are invariably favorable to the authenticity of the *Speculum*, which is indicated as *Contra libros nigromanticorum* in the *Tabula Stams*, at n. 85 (see both eds. of this older list of dominican writers in H. Denifle, 'Quellen zur Gelehrtengeschichte des Predigerordens im 13. und 14. Jahrhundert', *Archiv für Literatur und Kirchengeschichte des Mittelalters*, II, 1886, p. 236; *Laurenti Pignon catalogi et chronica. Accedunt catalogi Stamensis et Uppsalensis Scriptorum O.P.*, ed. G. G. Meersseman, Roma 1936, (MOPH, XVIII), pp. 22–33; in the catalog by Pignon (*ca.* 1412) at nr. 8, *cit.* ed. *ibid.*, p. 22; in the one by Henrichus Herfordiensis (1370) at n. 49; in the *Liber de rebus memorabilibus*, ed. Potthast, Göttingen 1859, p. 202, in the *Legenda I* according to B. Geyer, 'Der alte Katalog der Werke des hl. Albertus Magnus', in *Miscellanea G. Mercati*, Città del Vaticano 1946, II, pp. 398-413. Geyer made an effort to eliminate "so zahlreiche Pseudoepigraphen" (besides *Speculum*, a title which appears together with *Speculum astrolabicum* and *Contra libros nigro-*

manticorum, one can find titles such as *Alchimia*, *Secretum secretorum Alberti*, and *Almagestum et quosdam alios mathematicos*). See also the list in ms XL. C. 1, Prague, University Library, ed. by P. Auer, *Ein neuaufgefundener Katalog der Dominikanerschriftsteller*, Paris 1933 (= S. Sabinae Dissertationes Historicae, II), p. 89, at n. 47. *Cf.* P. Simon ed., 'Ein Katalog der Werke des hl. Albertus Magnus in einer Handschrift der Lütticher Universitätsbibliothek', in *Zur Geschichte und Kunst im Erzbistum Köln. Festschrift für W. Neuss*, Düsseldorf 1960, pp.79-88; J. A. Weisheipl, 'The Problemata', quoted above, pp. 309-311; lastly, see the fifteenth-century lists by Luíz de Valladolid 1414, Rodulphus Noviomagensis 1488, and those offered in the *Legenda coloniensis* 1483. Very useful also is the synopsis established by H. C. Scheeben, 'Les écrits d'Albert le Grand d'après les catalogues', *Revue Thomiste*, XXXVI, 1931, pp. 260-292 (pp. 290-291, nn. 63, 71, 72 and 73): almost all the catalogs mention the *Speculum astronomiae* starting with the most ancient one – which in some parts dates back to the end of the thirteenth century, and in any case is not later than 1310 – perhaps written by Gottfried von Duisburg, Albert's last secretary. The *Tabula*, found by Denifle in the Abbey at Stams, was edited in 1886 by its discoverer, and was reproduced by Scheeben also with the help of a second copy found in Basel and compared with all the other lists and biographies. Numbers 19 (*Contra librum nigromanticorum*), and 25 (*Speculum astrolabicum*) of the Stams' catalog reappeared in Henry of Herford (nn. 50 and 54), in Albertus de Castello (Jacobus de Soest) (nn. 44 and 48). In the second group of texts (Bernardus Guidonis, Tolomeo da Lucca and Johannes Colonna) there are only a few titles, insufficient to consider them as catalogs. The third group of sources was characterized by Scheeben as independent from the *Legenda I* from Soest, the lost archetype of the entire first group. Luíz de Valladolid, the first author of the group, who in 1414 compiled a *Tabula* following an official request by the University of Paris, on the basis of its library holdings in that year, recorded several titles which can be identified with the *Speculum astronomiae*: n. 70. *Speculum astrolabicum*; n. 73. *De imaginibus astrologorum*; n. 75. *Librum ubi improbavit scientias magicas nigromanticorum*. Peter of Prussia, Albert's biographer, wrote in 1486, when in Cologne, and, showing remarkable critical judgment, made use of the rich collection of ancient sources there preserved. At nn. 88, 91, and 94, Peter not only reproduced the indications offered by Luíz de Valladolid, but volunteered a significant comment for the last one: "Item fecit Albertus *Speculum astronomiae* in quo reprobat scientias magicas". Peter devoted eight chapters of his *Vita Alberti* to the analysis of this work; this is particularly important in view of the fact that, thanks to his fine critical discrimination, Peter had already excluded *Semita recta* and the *De secretis mulierum*, two works which were certanly pseudoepigraphic. The repetition of titles can be explained (here as well as in previous lists) as resulting from the comparison of several lists when there was no original work available, with the few exceptions of those works for which the incipit was given. Peter in fact declared: "Notandum tamen quod isti libri hic enumerati feruntur ab Alberto editi... non sum ita certus de omnibus hic enumeratis, quemadmodum de illis quos vidi... ideo volui notare signanter quos vidi vel quos habemus in nostro conventu coloniensi"; the sentence we quoted above, referring to the *Speculum*, was followed by the mark which indicates "vidi, habemus". Rodulphus of Nimegen, a compiler from Peter who pursued popularizing goals, reproduced all the data discussed above; he was however bereft of critical judgment, and added less reliable information. See Rodulphus de Noviomago O.P., *Legenda b. Alberti Magni* (ca. 1490), 2nd ed. by H.C. Scheeben, Köln 1928, pp. XV-95.

26. G. Pico della Mirandola, *Disputationes adversus astrologiam divinatricem*, ed. E. Garin, Florence 1946, I, pp. 64-67: "Tum si mihi forte obicias librum *de licitis et illicitis*, in quo reicit quidem magos, astronomicos autem probat autores, respondebo existimari quidem a multis esse illud opus Alberti, sed nec ipsum Albertum, nec libri inscriptionem usquequamque hoc significare, cum auctor ipse, quicumque demum fuerit, nomen suum consulto et ex professo dissimulet. Quid? Quod in eo multa leguntur indigna homine docto et bono christiano... Quae utique aut non scripsit Albertus aut si scripsit, dicendum est cum Apostolo: 'in aliis laudo, in hoc non laudo'". *Cf. ibid.* p. 24, where, after having discussed Bacon's authorship, he examined other dubious attributions, and referred to Albert in a different way: "qua temeritate vel ignorantia Eboracensis cuiusdam opusculum multi referunt ad Albertum".
27. Mandonnet, 'Roger Bacon' *cit.*, pp. 321-322.
28. *Ibid.*, pp. 323-324.
29. Th. Litt, *Les corps célestes dans l'univers de saint Thomas d'Aquin*, Louvain-Paris 1963, pp. 21-22.
30. Mandonnet, 'Roger Bacon' *cit*, p. 324.
31. *De animalibus* (XVII, tr. 2, c.4, n. 72), ed. Stadler, Münster 1916-21 (= Beiträge zur Geschichte der Philosophie u. Theologie des Mittelalters, 15-16), p. 1183/27-29: "De natura tamen et dispositione Lunae considerandum est in alia scientia quae est *altera pars astronomiae*, in qua quaeruntur effectus caelestium in terrenis corporibus"; *In Dionysium de divinis nominibus*, ed. P. Simon, in *Opera omnia*, XXXVII, Münster 1971, p. 155/55: "dicit Avicenna, quod prima pars astrologiae, quae est de dispositione superiorum corporum, est *demonstrativa*, quia illa semper eodem modo sunt, sed pars altera quae est de dispositione inferiorum per superiora, est coniecturalis. Et per hoc patet solutio ad obiecta, quia quamvis superiora determinent inferiora, haec tamen non consequuntur determinationem illam necessario, ut dictum est, nec illa sunt principia istorum propinqua et essentialia, nec iterum est mensura quam necesse sit sequi suum mensuratum, quia quamvis vita alicuius sit determinata ad determinatum tempus secundum propriam periodum, potest tamen plus vel minus vivere, secundum quod disponit se ad hoc vel illud". In the same theological text (pp. 155/11-13) Albert insisted again on the influence of coelestial bodies on temporal entities ("caelum est principium determinans et continens temporalia, ut dicitur in Littera; ergo ista inferiora possunt cognosci in superioribus"), and further emphasized on the conjectural nature of this kind of knowledge (p. 155/20-44): "Dicendum quod inferiora non possunt cognosci in superioribus corporibus certitudinaliter, sed tantum coniecturaliter; et huius duas causas assignat Ptolomaeus et tertiam Aristotelem. Prima est, quia certitudo effectus caeli non haberi posset nisi per experimentum pluries acceptum unius effectus ab eadem dispositione stellarum secundum eandem imaginem. Hoc autem non convenit accipere, quia quamvis una stella redeat ad punctum idem, non tamen similiter redeunt omnes stellae ita, ut efficiatur eadem imago omnino quae fuit, sed redit aliquid simile illi, eo quod plures illarum stellarum redeunt ad situm priorem, licet in aliquibus deficiat, et ideo non erit idem effectus, sed diversus, qui certitudinaliter determinari non potest, sed per coniecturam secundum similitudinem prioris effectus. Secunda causa est, quia caelum non influit tantum per stellas, sed etiam per spatium. Et quamvis omnes stellae redirent ad eandem imaginem, tamen non posset cum hoc computari, ut redirent in eodem spatio per tempus trium mundorum, et propter hoc etiam non habent eundem effectum. Tertia causa est inaequalitas materiae propter dispo-

sitionem contrariam inventam vel inductam in suscipientibus actionem caeli; unde non necessario sequitur effectus".

32. Albertus Magnus, *De fato*, ed. P. Simon, in *Opera omnia*, XVII/1, Münster 1975, pp. 73/36-44: "Dicendum quod duae partes sunt astronomiae, sicut dicit Ptolomaeus: una est de sitibus superiorum et quantitatibus eorum et passionibus propriis; et ad hanc *per demonstrationem* pervenitur. Alia est de effectibus astrorum in inferioribus, qui in rebus mutabilibus mutabiliter recipiuntur; et ideo ad hanc non pervenitur nisi per coniecturam, et oportet astronomum in ista parte secundum aliquid physicum esse et ex signis physicis coniecturari". Paul Simon, the editor of this work, pointed out some terminological similarities between this page of the *De fato* and the *Speculum astronomiae*, both quoting, moreover, the verbum 22 of the *Centiloquium*: "Nova vestimenta facere vel exercere Luna in Leone, timendum". Further, equally topical similarities are to be found at p. 70/23-24 and 76/42, where the author is discussing the legitimate third kind of imagines: "figurae imaginum magicarum ad aspectum stellarum fieri praecipuuntur". At pp. XXXIII-XXXV of his *Prolegomena* Simon lists seven codices bearing Albert's name, against seven others, plus a fragment favoring Thomas' authorship. Yet, despite the testimony of the catalogs of Albert's works, which do not include this work, and of the contrary data of the *Tabula Stams* and the lists by Bernard Guy and Tolomeo da Lucca, which attributed the work to Aquinas, father Simon relied on the decisive testimony, of the ms. Vaticanus Chigianus E. IV. 109, entitled "quaestio disputata a fratre Alberto apud Anagniam de fato". Among scholars, Thorndike and Pelster had already shown a preference for Albert and pointed out its similarity to several of his other texts; to that list, Simon added the *Speculum astronomiae*, though he thought that this treatise was pseudoepigraphic: "Hoc autem opusculo... agitur de quaestionibus, quid stellae in vita hominum efficiant, num libero arbitrio necessitatem imponant, quomodo effectus stellarum cognoscatur vel etiam prohibeatur. Similiter Albertus in *Summa de homine* [ed. Paris, t. 34, p. 448 ss.] in *II° Sententiarum*, d. 15.a.5 [t. 27, pp. 276-77]... in commento *Super Dionysii librum de divinis nominibus* [*Opera omnia*, Münster, t. 37, p. 153/22-155/72]... *Super Ethica* [*Opera omnia*, Münster, t. 14, p. 174/42 ss.]... *Physicae* 1.2, tr.2, c. 1920 [ed. Paris, t. 3, pp. 153-155]... *De animalibus*, l. 20, tr. 2, c. 2 [ed. Stadler, p. 1308/38 ss.]... in libro I *De causis et processu universitatis* [tr. 4, c.6; ed. Paris, t. 10, pp. 421-423]... *De XV problematibus* et *Problemata determinata*. *Summae theologiae* denique pars prima continet prolixam quaestionem de fato [q. 68; ed. Paris, 31, pp. 694-714]". It is very interesting that Simon saw the issue of fate as connected to the more important and general question of interpreting Aristotle in a way that avoided the offenses to Christian dogmas coming from the Arab tradition: "Sic erat cur Albertus, tum iam magister famosus, de hac re coram Curia Romana disputaret... Nam, sicut Stephanus Gilson animadvertit, iam saeculo duodecimo Iohannes Saresberiensis demonstravit Aristotelem determinismum astrologicum, quo liberum arbitrium escluderetur, docuisse. Quare quaestio a doctore operum Aristotelis peritissimo disputata, quid de fato, idest de effectu et potestate siderum, veris philosophis esse sentiendum, a multis procul dubio attentis auribus excipi potuit, praesertim cum de hac re illis temporibus saepius dissereretur".

33. The topical nature of the Ptolemaic partition of astronomy finds confirmation in the marginal gloss that Lemay published from a thirtheenth century ms. (*Abu Ma'shar cit.*, p. XXXVI n.): "Astronomia duas habet partes. Una que considerat situs planetarum... traditur in Almagesti... Alia est que considerat planetas secundum suam naturam, cuius

complexionis sint et cuius operationis in inferioribus... traditur in libro Albumasar".

34. *De fato cit.*, in Opera omnia, t. XVII/1, p. 73, 45-56: "Coniecturatio autem, cum sit ex signis mutabilibus, generat habitum minoris certitudinis, quam sit scientia, vel opinio. Cum enim huiusmodi signa sint communia et mutabilia, non potest haberi ex ipsis via syllogistica, eo quod nec in omnibus nec in pluribus includunt significatum, sed quantum est de se, sunt iudicia quaedam multis de causis mutabilia... Et ideo saepe astronomus dicit verum, et tamen non evenit quod dicit, quia dictum suum fuit quoad dispositionem caelestem verissimum, sed haec dispositio a mutabilitate inferiorum exclusa est". For the repeated attempt to define the epistemological status of astrological forecasts: *De fato*, p. 73/61-64: "dicit Ptolomeus, quod elector non nisi probabiliter et communiter iudicare debet, hoc est per causas superiores communes, quas propriae rerum causae frequentissime excludunt"; p. 74/8-15: "dicendum quod via syllogistica sciri non potest conclusio coniecturalis; sed tamen imperfectio scientiae non impedit, ut dicit Ptolemeus, quin hoc inde sciatur, quod inde sciri potest, sicut etiam est in pronosticatione somniorum. Non enim habitudo syllogistica est inter imaginem somnialem et interpretationem somnii; et sic est in omnibus existimationibus coniecturalibus" (*cf. Quadripartitum*, tr. 1, c. 1); p. 70/7-11: "Cum tamen dispositio fatalis exclusibilis sit et impedibilis ab oppositis dispositionibus inventis in materia... [et] in anima sensibili".
35. *Cf.* G. Paré, *Les idées cit.*, p. 277; R. Lemay, *Abu Ma'shar cit.*, pp. 38-39; T. Gregory, 'La nouvelle idée de nature et de savoir scientifique au XIIe siècle', in *The Cultural Context of Medieval Learning*, eds. J. E. Murdoch and E. D. Sylla, Dordrecht 1975, p. 204.
36. *Cf.* J. Agrimi-C. Crisciani, 'Albumazar nell'astrologia di Ruggero Bacone', *ACME. Annali della Facoltà di Lettere e Filosofia... Milano*, XXV, 1972, p. 321; for Albumasar's translation by Johannes Hispalensis Albert and the *Speculum* made use of, see Thorndike, 'Further consideration' *cit.*, pp. 424-25. Some texts quoted by T. A. Orlando, "Roger Bacon and the *Testimonia gentilium de secta christiana*", *Recherches de théologie*, 43, 1976, p. 210, suggest an additional observation: Bacon, well informed of both translations of Abu ma'shar's *Introductorium*, when using the famous passage XII,76–83, on the Virgo decane, preferred to cite from the version not used in the *Speculum astronomiae*.
37. Mandonnet, 'Roger Bacon' *cit.*, p. 327; *cf.* E. Charles, *Roger Bacon, sa vie, ses oeuvres, ses doctrines*, Paris, 1861, pp. 45 ss.
38. C. Schmitt and D. Knox, *Pseudo-Aristoteles Latinus: A Guide to Latin Works falsely attributed to Aristoteles before 1500*, London 1985, pp. 1, 3, 5.

CHAPTER THREE

1. See the undoubtedly more penetrating remarks by Etienne Gilson and Paul Simon concerning the astrological determinism that John of Salisbury already attributed to Aristotle, quoted above, ch. 2, n. 32.
2. 'Docteurs franciscains et doctrines franciscaines', *Études franciscaines*, XXXI, 1914, fasc. 1, pp. 94-95.
3. B. Vandewalle, 'R. Bacon dans l'histoire de la philologie: IV. Roger Bacon et le *Speculum astronomiae*', *La France franciscaine*, XII, 1929 , pp. 196–214.
4. *Cf. above*, pp. 18-19. See however F. Pangerl, *Studien über Albert cit.*, pp. 325-326. Without mentioning the then recent standpoint taken by Mandonnet, Pangerl re-examined

Albert's defenders (Petrus de Prussia, Trithemius, Aventinus, Martin Delrio, Athanasius Kircher and Naudé) as well as two of his critics, namely Pico –who tried to demonstrate that the *Speculum* was a spurious work– and Gerson –who reproached Albert precisely for the opposite reason: he considered him the author of the treatise. Pangerl was surprised by Gerson's criticism, in view of the fact that "Albert had repeatedly and at length said that false astrology was a 'diffamatio stellarum', and precisely in the *Speculum* had taken a stand against the discipline". T. Witzel expressed his agreement with the thesis put forward by Mandonnet, and used the *Speculum* as a Baconian work in the entry 'Roger Bacon', *The Catholic Encyclopedia*, XIII, New York 1913, pp. 111-116. Equally in favor of Mandonnet was P. Robinson, 'The Seventh Centenary of R. Bacon', in *The Catholic University Bulletin*, 1914, fasc. I; *Roger Bacon Essays*, ed. by A. G. Litle, Oxford 1914, p. 25; R. Carton, *L'expérience physique chez Roger Bacon*, Paris 1921, p. 14 ("jusqu'au P. Mandonnet... il était attribué [à Albert] et continue d'ailleurs de l'être encore par d'autre médievistes") and p. 172 ff., where the work is unhesitatingly attributed to Bacon. Also P. Duhem, *Le système du monde*, VIII, Paris 1958, p. 390: "il semble que cette attribution [de Mandonnet] soit légitime, car, de Roger Bacon, on trouve dans le *Speculum Astronomiae*, certaines locutions coutumières, certaines métaphores habituelles, certaines pensées favorites". A critical standpoint was taken by Ch. V. Langlois, who reviewed Mandonnet's *Siger* in the *Revue de Paris*, VII, 1900, p. 71.

5. *Opera R. Baconis hactenus inedita*, ed. R. Steele, V. Oxford 1920, p. 26.
6. G. G. Meersseman, *Introductio in Opera omnia Alberti Magni*, Bruges 1931, p. 132.
7. F. Tinivella, 'Il metodo scientifico in S. Alberto Magno e Ruggero Bacone', *Angelicum*, XXI, 1944 [= Serta albertina], p. 76: this Franciscan scholar maintained, "against Mandonnet and the authors who took from him, the Albertinian authenticity of the *Speculum*" and concluded: "The very fact that the *Speculum astronomiae* has been at times attributed to the Universal Doctor, and at times to the Admirable Doctor, proves the two authors' almost identical views on the matter". In his *L'expérience physique cit.*, pp. 23-25, R. Carton too establihed a relationship between the observational methods used by Albert and by Bacon. A. G. Little, another editor of Bacon, was less positive in the attribution, and in his introduction to Bacon's *Opus tertium*, Aberdeen 1912, p. XX, he mentioned the *Speculum astronomiae* as a work "generally ascribed to Albertus Magnus, but attributed by Father Mandonnet to Roger Bacon".
8. Mario Brusadelli [Giovanni Semeria, pseud.], 'Lo *Speculum astronomiae* di Ruggero Bacone', *Rivista di filosofia neoscolastica*, VI, 1914, pp. 572-79; *cf.* also Fleming, 'R. Bacone e la Scolastica', *ibid.*, p. 541. Semeria regarded Roger Bacon as a victim of Church authorities (who imprisoned him because of his modern opinions) and turned Mandonnet's arguments over. The *Speculum astronomiae* is consistent also in its more delicate passages (as when dealing with the horoscopes of religions or with Antichrist and necromancy): just because it is consistent, the *Speculum* is not to be excluded from the work of Semeria's hero, Roger Bacon.
9. *Cf. De retardatione accidentium senectutis* in *Opera R. Baconis hactenus inedita*, IX, ed. A. G. Little and E. Withington, Oxford 1928, pp. 34, XXIV-XXV. This edition contains a passage from the shorter version given by two ms., "hanc [epistolam] incepi ad suasionem duorum sapientum Parisiensium, sc. Joh. Castellionati et Philippi cancellarii Parisiensis". The second figure mentioned in this passage could be Philippe de Grève, chancellor from 1218 to 1236, the year of his death, after having been condemned for corrupted habits

and heretical doctrines (*cf.* Thomas Cantimpré, *Bonum universale de apibus*, I, cap. 19; ed. Douai 1597, p. 59). The dates of Philippe de Grève do create some difficulty also concerning the Baconian pamphlet.

10. Brusadelli, 'Lo *Speculum astronomiae*' *cit.*, p. 575.
11. *Ibid.*, pp. 577-78.
12. *Ibid.*, p. 575. Compare similar topical statements in the *Speculum* with relevant Baconian passages: "quamvis loquantur [astrologi] de sectis, et sectae dependant ex libertate rationis, tamen non imponunt aliquam necessitatem libero arbitrio dicentes planetas esse *signa* innuentia nobis ea quae Deus disposuit ab aeterno fieri sive per naturam, sive per rationem humanam, sive per rationem propriam secundum beneplacitum suae voluntatis" (*Opus maius cit.*, p. 646).
13. Brusadelli, 'Lo *Speculum astronomiae*' *cit.*, p. 575 f.
14. B. Geyer, 'Das *Speculum astronomiae* kein Werk des Albertus Magnus', *Münchener Theologische Zeitschrift*, IV, 1953; [*cf.* the same paper printed in *Studien zur historischen Theologie. Festgabe für F. X. Seppelt*, hrsg. v. W. Düring u. B. Panzram, München 1953], p. 97.
15. *Cf. Speculum*, pp. 93 ff., Appendix I and II: only 9 out of 51 mss. are anonymous, and one of these (ms. Vat. Borghesiano 134), as well as the CLM 8001 reproduced the *Speculum* – here attributed to Aquinas – as part of a collection of Albertinian works on natural history. When the *Speculum* is included in astrological miscellanies, it is usually attributed to Albert.
16. Ueberweg-Geyer, *Die Geschichte der patristischen und scholastischen Philosophie*, Berlin 1928, p. 406.
17. Geyer, 'Das *Speculum astronomiae*' cit., p. 98 n. 58.
18. Geyer, 'Das *Speculum astronomiae*' *cit.*, p. 99: "Es ist ferner schwer vorstellbar, dass Albert sich mit dem Schleier der Anonymität umhüllt und sich selbst im Prolog als 'vir quidam zelator fidei et philosophiae' bezeichnet habe. Er hat stets mit offenem Visier gekämpft und seine Person hinter der Sache zurücktreten lassen. Für jeden, der mit dem Schriftum Alberts vertraut ist, steht ohne weiteres fest, dass er diesen Prolog nicht geschrieben haben kann".
19. *Ibid.*, pp. 99-100.
20. *Cf.* S. Dezani, 'S. Alberto Magno: l'osservazione e l'esperimento', *Angelicum*, XXI, 1944 [= Serta albertina], pp. 43-47. Dezani pointed out that Bacon, "wholeheartedly devoted to the exact sciences", theorized but did not practice experiments; on the contrary, Albert cultivated experimental concerns, even though "his concept of the experiment amounted to mere observation". *Cf.* below ch. 4, n. 11 and *passim*, the remarks and the texts adduced by Grabmann, here taken up by Dezani. Tinivella too emphasized the role of experiments in Albert, see 'Il metodo' *cit.*, p. 73.
21. Geyer, 'Das *Speculum astronomiae*' *cit.*, pp. 99-100.
22. *Super Matthaeum*, ch. II, ed. B. Schmidt,in *Opera omnia*, Münster 1987, p.46/36 ff. (already cit. on the basis of the autograph by Geyer 'Das *Speculum astronomiae*' cit., p. 100 n.): "Si quis enim pronosticatur per stellas de his quae non subiacent nisi ordini causarum naturalium, et sua pronosticatio est de his secundum quod ordini illi subiacent, et non extendit se ad illa eadem, nisi eatenus quo inclinat ad ea primus ordo naturae, qui est in situ stellarum et circulo, *non malefacit*, sed potius utiliter a multis cavet nocumentis et promovet utilitates. Qui autem non consideratis omnibus praenuntiat de his *quae futura*

sunt aliter quam dictum est, trufator est et trutannus et abiciendus" (Italics mine). *Cf. De sommo et vigilia* (III, tr. ii, c. 5), ed. Jammy, V, p. 106, where Albert concluded his interesting discussion of the distinction between necessary and purely probable forecasts – the only possible ones for the human mind – in this way: "universaliter dicendo non omne contingens fieri in futurum eveniet... Et haec est causa quare non deceptus videtur decipi astronomus et augur et magus et interpres sommiorum et visionum et omnis similiter divinus: omne enim fere tale genus hominum *deceptionibus gaudet*, et parum literati existentes putant necessarium esse quod contingens est, et pronuntiant tanquam absque impedimento aliquid futurum, et cum non evenit, facit scientias vilescere in conspectum hominum imperitorum, cum defectus non sit in scientiis, sed potius in eis qui abutuntur eis; propter quod etiam Ptolomeus sapiens dicit nihil esse iudicandum nisi valde generaliter et cum protestatione cauta, quod stellae ea quae faciunt faciunt *per aliud et per accidens*, ex quibus multa in significatis suis occurrunt impedimenta: frustra enim poneretur studium ad scientias vaticinantes si ea quae futura previdentur impediri non possent; ad hoc enim praevidemus ut mala impediantur et bona expediantur ad actum, sicut faciunt periti medicorum in suis prognosticationibus" (Italics mine). On this text, see TH, II, 585.

23. *Super Matthaeum*, cap. II, ed. cit., p. 46/21 ff.: "Magus enim et mathematicus et incantator et maleficus sive necromanticus et ariolus et aruspex et divinator differunt. Quia magus proprie nisi magnus est, qui scientiam habens de omnibus ex naturis et effectibus naturarum coniecturans, aliquando mirabilia naturae praeostendit et educit". *Cf. In Danielem*, cap.I, 20; ed. Jammy, VIII, p. 8b (*cit.* by TH, II, 554 n.). Among previous authorities, going back to Varro and Isidore of Seville, we also find the Liber introductorius to astronomy (a source often referred to in the *Speculum*), where Michael Scot distinguished between true and legitimate mathesis (astronomy and astrology) and *mathesis* (forbidden magic); *cf.* C. H. Haskins, *Studies in the History of Medieval Science cit.*, pp. 285-286; Thorndike, *Michael Scot*, London 1965, pp. 118-119; on Bacon, see Opus tertium *cit.*, pp. 26-27; *Secretum secretorum*, in *Opera hactenus inedita*, V, Oxford 1920, ed. R. Steele, pp. 3-7; on other texts, see TH, II, 668-69.
24. *In II Sententiarum*, d. vii, art. 9: "An daemon in suis operationibus constellationibus iuvetur an non? et utrum scientia imaginum sit operatione daemonum an non?"; ed. Jammy, XV, pp. 87-88: the answer is "videtur quod [fiat] operatione daemonis, quia talis scientia prohibetur; non autem prohiberetur, si fierit operatione naturae...".
25. It is not easy to identify a *Liber de mansionibus Lunae* among the numerous still unpublished homonym or similar works (*cf.* TK, cols. 834-84, 139, 819), many of which contain the attribution to Aristotle. Albert's text, however, gives a lection which is open to some doubt, in the absence of a critical edition of the *Commentary to the Sentences*. The more probable identification is however with the *Liber Lunae*, a work usually recorded as Hermetic, also in the *Speculum*, XI / 47; in some ms. of this work, such as the one preserved in Copenhagen (Gl. kg. S. 3499, ff. 92v-95v) we find the following paragraph: "Et nota quod Aristoteles plenior artibus dicit Selim, idest luna...". *Cf.* TK, 819; M. Steinschneider, 'Die europäischen Übersetzungen aus dem Arabischen', in *Sitzungsberichte der K. Akademie der Wissenschaften in Wien*, Ph.-Hist. Kl., 151, 1906 p. 6; F. Saxl, *Verzeichnis astrol. u. mythol. illustrierter Handschriften der National-Bibliothek in Wien*, Hamburg 1927, p. 102; Zinner, 8225; Thorndike, 'Traditional Medieval Tracts concerning engraved astrological Images', in *Mélanges Auguste Pelzer*, Louvain 1947, pp. 238, 255.

26. *In II Sententiarum cit.*, p. 88a: "videmus nativitates variari secundum substantiam et secundum operationes consequentes a constellationibus, sicut dicit Ptolomeus in *Quadripartito*, quod Sole existente in quadam parte et minuto Arietis, non fit generatio humana, et si cadat tunc semen in matricem, monstrum nascetur. Et ut credatur, ego probavi *experimento* hoc in duabus matronis probis et bonis, a quibus ego percepi quod monstra pepererunt, et quaerens tempus ab eis et *aequans stellas*, inveni quod Sole existente circa eundem gradum et minutum secundum suas aestimationes conceperunt" (my italics). It is noteworthy that in this passage Albert declares that he performed an experiment in astrological measurement; shortly afterwards, we find his interesting discussion of a thesis by Avicenna, where Albert denies the origin of such monsters from a combination of semen of various animal species, but calls upon an astrological cause, defined as equally natural: "ergo videtur quod hoc sit naturale: non differt autem scientia imaginum ab illa impressione, nisi *sicut ars et natura*: quia si natura tunc efficeret imaginem talem qualis fit per artem, facilius et melius haberet ista mirabilia, quae imprimit aspectus stellae, quam imago facta per artem: ergo videtur, quod ibi nihil sit de opere daemonum, sed tantum opus artis et naturae". The discussion of the relationship between art and nature as far as demons' actions were concerned, had already taken place *ibid.*, dist. VII, articles 6-8, pp. 84-87.
27. *Ibid.*, p. 88b: "absque dubio, sicut etiam supra per auctoritatem Augustini probatur, ortus et aspectus stellarum magnum habent effectum in operibus naturae et artis, sed tamen super nostrum liberum arbitrium non habent, ut dicit Damascenus. Sed imaginum ars ideo mala est, quia inclinans est ad idolatriam per numen quod creditur esse in stellis et quia non sunt inventae imagines nisi ad vana vel mala".
28. *Summa theologiae* (Pars II, tr. XI, q. 61), ed. Jammy, XVIII, p. 322: "Hoc autem quod dicit Albumasar error pessimus est et vituperandi sunt qui hoc adducunt *quasi pro testimonio* quod philosophi nobis testificentur de partu virginis" (Italics mine). *Cf. Speculum*, XII/60-61. For an even more blasphemous formulation, expressed within a discussion of spontaneous generation –which is considered as caused by celestial influence– *cf. ibid.*, p. 321: "Albumasar dicit in *Introductorio*, quod Virgo, in cuius facie prima oritur constellatio quaedam ad similitudinem virginis in gremio habentis puerum, quae tantae virtutis est ut fecunditatem quibusdam virginibus afferat sine commixtione virili. Et dat examplum, quod *penitus haereticum est*, quod beata Virgo sic conceperat Iesum quem gens Christianorum adorat" (Italics mine). *Cf. Speculum*, XII/78ss and 83.
29. Geyer, 'Das *Speculum astronomiae*' cit., p. 100. In a letter sent to me on November 30th, 1973, Lemay pointed out: "Que dans ses ouvrages propres Albert ait parfois pensé différemment que dans le *Speculum*, il n'y a pas raison d'en être surpris. En particulier, la difference de pensée remarquée par Mgr. Geyer entre le *Speculum* et les ouvrages authentiques d'Albert, concernant la prédiction de la naissance virginale du Christ s'explique (en plus de l'argumentation importante de Thorndike répondant à Geyer) par le fait que dans le *Speculum* d'Albert ne rapport pas sa pensée definitive, mais trace les limites permises à la spéculation des chrétiens sur ce sujet délicat. Dans ses ouvrages plus personnels, par contre, Albert prend partie pour ou contre la validité scientifique de certaines de ces doctrines des *Libri Naturales* dans le cadre de son propre système de pensée. Il peut très bien alors avoir rejeté de sa synthèse philosophique ce morceau astrologique, que dans le *Speculum* il ne jugeait pas opposé à l'orthodoxie. Faire, comme Mgr. Geyer, de ces différences un argument historique pour rejeter la paternité d'Albert dans le

Speculum c'est méconnaître aussi bien le milieu culturel d'alors que les véritables circonstances de la production intellectuelle d'Albert le Grand". Thorndike, 'Further consideration' *cit.*, p. 426 conceded that "Undoubtedly the two passages in *Summa* and *Speculum* are contradictory, but that in the *Summa* may be an interpolation, or Albertus may have changed his mind upon this point... Albertus like others, wrote from a different standpoint in his theological and natural writings. In the one he was apt to reflect the views of the Church Fathers, in the other not merely those of Aristotle and Avicenna, but of Aaron and Evax, Hermes and Albumasar". As readers will easily deduce from my own insistence on the Albertinian method allowing the so-called "duae viae", I find the second of the two remarks by Thorndike quoted above as more convincing.

30. *Speculum*, XII / 60-100.
31. *Cf.* Meersseman, *Introductio cit.*, p. 112: "Omnes conveniunt Albertum hoc opus scripsisse in ultimo suae vitae decennio et a continuatione eius impeditum fuisse memoriae lapsu, morbo et morte. Doctrina ibi magis evoluta et generaliter magis aristotelica est quam in priori *Summa* vel in *Sententiis*. Imo, influxus *Summae theologicae* S. Thomae non omnino excludendus videtur. Major tamen et indubitabilis videtur influxus *Summae theologicae* quae Alexandri Halensis dicitur". To the bibliography on this issue listed by Meersseman we should now add H. Neufeld, 'Zum Problem des Verhältnisses der *Theologischen Summe* Alberts des Grossen zur *Theologischen Summe* Alexander von Hales', *Franziskanische Studien*, 27, 1940, pp. 22–56, 65–87, where some texts are confronted on parallel columns.
32. O. Lottin did not agree with Meersseman's Aristotelic-Thomistic characterization of the *Summa*, and saw in that work the presence of Franciscan tendencies; see his review of R. Kaiser, 'Die Bedeutung proklischer Schriften durch Albert den Grossen', in *Bulletin de théologie ancienne et médievale*, IX, 1963, pp. 387-88; the sentences we have quoted have been taken from that review.
33. This information still awaits confirmation; on the contrary, it is certain that Albert attended the Council of Lyon in May 1274, which means that his vitality was unimpaired at least until that date: *cf.* P. von Loe, 'Albert der Grosse auf dem Konzil von Lyon', *Literarische Beilage der kölnischen Volkszeitung*, LV, 1914, Nr. 29, pp. 225-226. The earliest biographers who mention the trip to Paris in 1277 agree that Albert's sight and memory remained unimpaired until three years before his death, that is, until 1277; the trip to Paris is endorsed by H. C. Scheeben, *Albertus Magnus*, Köln 1955 2nd ed., pp. 172-173; by W. A. Wallace, *s. v.* 'Albert' in *New Catholic Encyclopedia*, I, 1967, and *s. v.* 'Albert' in Dictionary of Scientific Biography, I, 1970; J. A. Weisheipl, 'The Life and Works of St-Albert', in *Albertus Magnus and the Sciences*, Toronto 1980, pp. 43-46, has denied that the trip took place.
34. Some of the Baconian texts concerning the horoscopes of religions have been quoted and discussed by D. Bigalli, *I tartari e l'apocalisse*, Firenze 1971, pp. 179-188, and p. 110 n. 46 where an interesting passage is quoted from the *Chronica magistri Rogeri de Hoveden*: "Habebit autem Antichristus magos, maleficos, divinos et cantores: qui eum, diabolo inspirante, nutrient et docebant eum in omni iniquitate et falsitate et nefaria arte".
35. *Summa theologiae*, Pars I, tr. XVII, q. 68, m. 4, ed. Jammy, XVII, p. 387 a-b: "Christus assumpsit nostros defectus indetractabiles, ut dicit Damascenus. Unus autem et praecipuus nostrorum defectuum est subiacere fato et fortunae. Ergo illum assumpsit Christus. Adhuc, omnibus mobilibus adhaeret dispositio quae est fatum. Christus mobilis fuit

secundum corpus, hoc constat. Mobilis etiam fuit secundum electionem... Ergo videtur quod secundum corpus et animam fato subiacuit et fortunae".

36. *Summa theologiae* cit., ed. Jammy, XVII, p. 389b: "Ad id quod ulterius quaeritur, utrum Christus secundum corpus vel secundum animam subiacuerit fortunae, dicendum quod non. Cum enim ipse sit conditor dispositionis quae in rebus est, vel ex ordine causarum vel ex positione siderum, non potest subiacere dispositioni tali: nec qui Deus est gubernans unumquodque ad debitum ordinem et finem providentia sua, ab alio quodam potest gubernari et suo ordine necti. Et hoc est quod dicit Augustinus in sermone de Epiphania, quod de Christo verum non esset, quod sub decreto stellae nasceretur, si etiam alii et alii homines sub decreto stellae nascerentur. Nam Christus Dei filius propria voluntate homo factus est: alii homines nascuntur conditione naturae. Ad id quod obicitur in contrarium, dicendum quod Christus defectus nostros assumpsit voluntate et non contraxit naturae vitiosae necessitate, et ideo non subiicitur ei, sed supponitur quod passus fuit quae voluit et quando voluit et a quibus voluit. Ad aliud dicendum, quod Christus mobilis fuit secundum corpus, sed mobilitas voluntati suae subiacuit et ipse non ei; secundum animam autem non fuit mobilis. Et quod dicitur, quod proficiebat sapientia et gratia tropice dicitur, tropo illo quo res dicuntur fieri quando innotescit, ut dicit Ambrosius".

CHAPTER FOUR

1. See above ch. 1, n.1. For very recent endorsements of the authenticity of the work, see T. Gregory, 'La Filosofia medievale: i secoli XIII-XIV', in *Storia della filosofia*, ed. by M. Dal Pra, Milano 1976, VI, pp. 227-230, and E. Garin, *Lo zodiaco della vita*, Bari 1976, p. 42. C. Crisciani-C. Gagnon, *Alchimie et philosophie au Moyen Age. Perspectives et problèmes*, Montréal, L'aurore-Univers, 1980, p. 27; R. Kieckhefer, *Magic in the Middle Ages*, Cambridge U. P. 1989 (= Cambridge Medieval Textbooks), p. 117: "the *Mirror of astronomy* ascribed (probably rightly) to Albert the Great distinguished the disciplines and dealt with both" astronomy and astrology. It is worth mentioning a series of articles by A. Cortabarria Beitia,O.P.: especially 'Fuentes arabes de San Alberto. Albumasar', *Estudios filosoficos*, XXX/84, 1981, p. 284: "Es sabido que la paternidad albertina de esta obra [i.e. *Speculum astronomiae*] ha sido puesto en duda. El P. Mandonnet la attribuyó a R. Bacon, pero sin que su opinión lograra la unanimitad entro los historiadores interesados en la cuestión. Por mi parte, no dejaré de recoger las referencias que el *Speculum astronomicum* nos da de Albumasar". *Cf.* p. 288 n.15 where Cortabarria mentions Mandonnet's criticism by Litt, and Meersseman, who "justifica a continuacion la partenidad albertina del *Speculum astronomicum*, pero segnala tambien algunas difficultades de vocabulario" and also the problem – to be discussed in the following pages – of the sentence of ch. 12 mentioning books 12 and 13 of Aristotle's *Metaphysics* "qui nondum sunt translati". Quotations to be found in the *Speculum* as well as in works certainly by Albert are used by Cortabarria in his papers 'Al-Kindi vu par Albert le Grand' *cit.*, p. 125, and 'Deux sources arabes de S. Albert le Grand', *Mélanges de l'Institut d'Etudes Orientales* [MIDEO], 1982, pp. 40, 43, 46.
2. TH, II, 522, 578.
3. TH, II, 578.

4. TH, II, 621 compared the famous statement in the *Opus tertium* concerning Bacon's ten years "exile" from the schools, with a similar one by Albert (*Mineral.* III, 1, 1: "Exul... longe vadens ad loca metallica"), and asked why one should take the first statement at its word to support the conjecture that Bacon had been subjected to censorship and imprisoned. "Perhaps, however, Father Mandonnet would infer from the passage and from the favorable attitude of the treatise on minerals towards astrological images that Bacon was really the author". *Cf.* 'Further Consideration' *cit.*, p. 427.
5. TH, II, 531.
6. M. Grabmann, 'Zur philosophischen und naturwissenschaftlichen Methode in den Aristoteleskommentaren Alberts des Grossen', *Angelicum*, XXI, 1941 (= Serta albertina), pp. 51-52.
7. *Siger cit.*, I, p. 36 n. 1 reproduces, besides the passage here quoted (*In Sententias*, II, d. XIII, a. 2: "Sciendum quod Augustino in his quae sunt de fide et de moribus plus quam philosophis credendum est, si dissentiunt. Sed si de medicina loqueretur, plus ego crederem Galeno vel Hippocrati, et si de naturis rerum loquatur credo Aristoteli plus vel aliis experto in rerum naturis"), a more explicit one taken from the *Physica* (IV, tr. III, c. 4: "Nec Galenus, nec Augustinus sciverunt bene naturas rerum").
8. Cfr. Grabmann, *op cit.*, pp. 54-55.
9. *De causis proprietatum elementorum* (I, 2, 9), ed. P. Hossfeld, in Opera Omnia, V, p. 77/44-47: "non enim sufficit scire in universali, sed quaerimus scire unumquodque secundum quod in propria natura se habet: hoc enim optimum et perfectum est genus sciendi".
10. Ibid., pp. 51-52. On this methodological issue, see TH, II, 536.
11. Albertus Magnus, *De mineralibus*, II, 2, 1; ed. Borgnet, V, p.30a: "Scientia enim naturalis non est simpliciter narrata accipere, sed in rebus naturaliter inquirere causas", quoted in Grabmann, art. cit., p. 52. On the same Albertinian text, see TH,II, 545 and D. A. Callus, 'S. Tommaso d'Aquino e S. Alberto Magno', *Angelicum*, XXXVI, 1960, p. 144.
12. Albertus Magnus, *Physica*, VIII, 5, 2 (quoted by Grabmann): "Accipiamus igitur ab antiquis *quaecumque bene* dicta sunt", "Conclusio quae sensui contradicit est incredibilis".
13. Albertus Magnus, *Ethica*, VI, 2, 25; ed. Borgnet, VII, pp.442b-443a: "Multitudo enim temporis requiritur ad hoc, ut experimentum probetur, ita quod in nullo modo fallat... Oportet enim experimentum non in uno modo, sed secundum omnes circumstantias probare, ut certe et recte principium sit operis".
14. Albertus Magnus, *De vegetabilibus*, VI, 1, 1; ed. Jammy, V, p. 430: "Earum autem, quas ponemus, quasdam quidem ipsi nos *experimento probamus*, quasdam autem *referimus ex dictis eorum*, quos comperimus non de facili aliqua dicere nisi probata per experimentum. Experimentum enim solum certificat in talibus eo quod de tam particularibus naturis syllogismus (ed.:simile) haberi non potest" (Italics mine).
15. Albertus Magnus, *De caelo*, ed. by P. Hossfeld, in *Opera omnia*, V, 1, Münster 1971, p. 156/62 ff.: "debent sufficere solutiones topicae et parvae in his quae sunt de caelo quaesita, eo quod ad ipsa cognoscenda perfecte non sufficimus"; see also *Physica*, II, tr. 2, c. 11, and *De fato*, quoted *above* ch. 2 nn. 32 and 34.
16. TH, II, 540 n. 4 and 548. *Cf.* Pangerl, 'Studien' *cit.*, p. 305; Dezani, 'S. Alberto Magno: l'osservazione e l'esperimento' *cit.*, p. 47; A. Walz, 'L'opera scientifica di Alberto Magno secondo le indagini recenti', *Sapienza*, V, 1952, p. 443. See also the papers by Y. Congar, 'In dulcedine' *cit.* below ch. 5 n. 5 and by P. Hossfeld, 'Die eigenen Beobachtungen', *cit.* below, ch. 5 n. 14. Cfr. below ch. 10, n. 5.

17. TH, II, 618-619: "one is impelled to the conclusion that Bacon's writings, instead of being unpalatable to, neglected by, and far in advance of, his times, give a most valuable picture of medieval thought, summarizing, it is true, its most advanced stages, but also including much that is most characteristic and even revealing some of its back currents".
18. TH, II, 534.
19. TH, II, 530.
20. TH, II, 577.
21. *Cf.* B. Nardi, *Studi di filosofia medievale*, Roma 1960, p. 119 ff.; T. Gregory, 'Forme di conoscenza e ideali di sapere nella cultura medievale', *Giornale critico della filosofia italiana*, LXVII (LXIX), 1988, pp.26-27; Id., 'Filosofia e teologia nella crisi del XIII secolo', Belfagor, XIX, 1964, p. 7; Id., 'Discussioni sulla "doppia verità"', *Cultura e scuola*, I, 1962, p. 101, where Gregory noted that Albert – as well as Siger – "was not interested in God's miracles while he was discussing natural objects within a naturalistic context".
22. TH, II, 559-60.
23. *Cf.* II.1, n.5 ff.
24. TH, II, pp. 708-709.
25. TH, II, pp. 529-530. *Cf.* now G. C. Anawati, 'Albert le Grand et l'Alchemie', in *Albert der Grosse, seine Zeit, sein Werk, seine Wirkung*, hrsg. v. A. Zimmermann, Berlin-New York 1981, pp. 126-133; P. Kibre, 'Albertus Magnus and Alchemy', in *Albertus Magnus and the Sciences*, ed. J. A. Weisheipl, Toronto 1980, pp. 187-202 (and *cf. ibid.* the papers by J. M. Riddle and J. A. Mulholland, pp. 203-204, and by M. G. George, pp. 235-260).
26. P. Kibre, 'The Alkimia minor ascribed to Albertus Magnus', *Isis*, XXXI, 1940, pp. 267-300; XXXIX, 1949, pp. 267-306; 'An alchemical Tract ascribed to Albertus Magnus', *ibid.*, XXXV, 1944, pp. 303-316; 'Alchemical Tracts Attributed to Albertus Magnus', *Speculum*, XVII, 1942, pp. 499-519; XXXIV, 1959, pp. 238-247; 'The *De occultis naturae* attributed to Albertus Magnus', *Osiris*, XI, 1954, p. 23. See also Albertus Magnus, *Libellus de alchimia*, transl., intr. and notes by V. Heines, Berkeley-Los Angeles 1959.
27. Cfr. TH, II, pp. 610-611, 676. The passages of *Speculum* (XII/28-36, III/4-8 and passim) concerning the problem of the animation of the stars will be examined in ch. 6 n. 15 and mainly in chapters 7 and 8.
28. M.-T. d'Alverny-F. Hudry, eds., 'Al-Kindi *De radiis*', *Archives d'histoire doctrinale et littéraire du Moyen Age*, XLI, 1974 (but 1975), pp. 169-170.
29. *Ibid.*, p. 173, 178; *cf.* H. Denifle and E. Chatelain eds., *Chartularium cit.*, I, pp. 486 art. 4; p. 543 ff. (Prologue).
30. *Ibid.*, p. 140 where Aquinas' *Summa contra gentiles*, III, ch. 104: "Quod opera magorum non sunt solum ex impressione caelestium corporum", in *Opera omnia iussu edita Leonis XIII P.M.*, XIV, Roma 1926, p.325, is quoted in extenso: "Fuerunt autem quidam dicentes quod huiusmodi opera nobis mirabilia, quae per artes magicas fiunt, non ab aliquibus spiritualibus substantiis fiunt, sed ex virtute caelestium corporum. Cuius signum videtur quod ab exercentibus huiusmodi opera stellarum certus situs consideratur. Adhibentur etiam quaedam herbarum et aliarum corporalium auxilia, quasi ad praeparandam inferiorem materiam ad suscipiendam influentiam virtutis caelestis". Aquinas' text corresponds litterary to al-Kindi's chapter "de virtute verborum" quoted in the following footnote. *Cf.* also *Summa contra Gentiles*, III, ch. 84: "Quod corpora caelestia non imprimant in intellectus nostros", *Ibid.* p. 248 ff., and ch. 105:"Unde magorum operationes efficaciam habeant", p. 330 ff.

31. M.-T. d'Alverny -F. Hudry, 'Al Kindi *De radiis*' *cit.* pp. 247-248, where al-Kindi introduces the sentences we have translated with a summary of the opposite thesis: "Non autem solummodo ad Deum diriguntur obsecrationes, sed etiam ad spiritus qui ab aliquibus hominibus esse creduntur, licet eorum existentia sensibus hominum non sit perceptibilis. Credunt enim plurimi angelos esse substantias incorporeas habentes potestatem faciendi motus in rebus elementatis. Credunt etiam homines corpore solutos spiritualem existentiam retinere et quandoque motus facere in hoc mundo, et ad hoc faciendum affectuosis precibus hominum induci. Sunt autem alii quorum scientia et fides a sensu tantum derivatur et ideo spirituum naturam esse non credunt in aliquo modo existendi qui ad humanam cognitionem possit pervenire. Quod enim motus et ymagines fiunt in aere vel alio elemento vel elementato, que per naturam vulgo notam fieri non solent, non est ex operatione spirituum, sed tantum ex condicione celestis armonie materiam aptante ad talis motus et talium ymaginum receptionem per actiones aliarum rerum corporearum eandem materiam moventium ad similitudinem armonie, ut sunt orationes et nomina et etiam aliqua alia, ut herbe et gemme".
32. A. Cortabarria Beitia, 'Al Kindi vu par Albert le Grand', *Revue des études islamiques*, 1977, pp. 117-146; Cortabarria considers however al-Kindi's *De diversitate adspectus lunae* which is not less astrological than the *De radiis*; see *ibid.* p. 121 ff. *Cf.* Cortabarria, 'Las obras y la filosofia de Alfarabi y Al Kindi en los escritos de S. Alberto Magno', *Estudios filosoficos*, 1951–52, pp; 191–209.
33. TH, II, pp. 702-703.
34. Roger Bacon, *Opus maius*, ed. Bridges, Oxford 1897-1900, I, p. 394, quoted in TH, II, p. 676.
35. TH, II, p. 661.
36. TH, II, p. 662.
37. TH, II, p. 660.
38. *Opus tertium*, ed. Brewer, in *Opera quaedam hactenus inedita*, London 1859, p. 44: "Et sicut logica docet proprietates sophistici argumenti *ut videntur*, sic haec scientia revolvit omnes artes magicas, ut doceat eas reprobare, ut reprobata omni falsitate, sola veritas artis et naturae teneatur. Sed haec non reprobat falsa quae ignorat, nec vera cum falsis, sicut faciunt fere omnes. Et Gratianus et multi minus bene dixerunt in hac parte, quia non omnia sunt magica quae ipsi docent reprobari et reprobant falsa quae ignorabant. Sed homo qui reprobat aliquid, debet scire eius conditiones, et sic falsum reprobare ut veritas semper maneat illaesa". Ibid., pp. 26-27: "Sed pro certo Sancti non reprobaverunt has scientias, de quibus loquor, licet maxima videatur hoc de mathematica, scilicet astronomica, propter judicia, et quia multi mathematici imposuerunt necessitatem libero arbitrio. Sed Sancti non reprobant mathematicam, quae est pars philosophiae, sed quae est pars artis magicae, ut manifestum est per Sanctos. Nam Isidorus ... dicit quod astronomia duplex est: una est naturalis et alia superstitiosa, et mathematica una derivatur a mathesis media correpta, et illa est pars philosophiae". Upon texts of this kind F. Palitzsch, *R. Bacons zweite Schrift über die kritischen Tage* (Dissertation), Borna-Leipzig 1919, pp. 12-15, based the thesis favoring Bacon's authorship of the *Speculum.* Palitzsch only examined the medical-pharmacological passages in chapters XIII and XV, indicating the election of the hours best fitted for the taking of drugs and therapeutics. Haskins, *Studien cit.*, p. 288 quoted Palitzsch, but he twisted or simply turned upside down the conclusions of this student.

39. TH, II, p. 666.
40. TH, II, p. 674.
41. TH, II, p. 674-675.
42. TH, II, p. 675.
43. TH, II, p. 675.
44. TH, II, p. 676-677.
45. TH, II, 705 n.; *cf.* p. 551: "He was really much greater as a natural scientist than as a theologian. But we have now to examine what grounds there are for calling him 'magnus in magia and in magicis expertus'. Magic is often mentioned by Albert, both in his Biblical and Aristotelian commentaries, both in his theological writings and his works on natural science".
46. Bonaventura da Bagnorea, *Commentaria in IV libros Sententiarum*, II, d. 14, pars 2, a. q. 2, in *Opera omnia*, II, Quaracchi 1882-1902, p. 360-61: "Luminaria caelestia impressionem habent super elementa et elementaria corpora; impressionem, inquam, non unicam tantum, sed multimodam". Further passages from this same quaestio are discussed by R. Jehl, *Melancholie und Acedia. Ein Beitrag zu Anthropologie und Ethik Bonaventuras*, Paderborn 1984, pp. 37-43, nn. 93 and 106 in particular; pp. 286-287 and n. 94.
47. G. Paré, *Les idées et les lettres au XIIIe siècle*, Montréal 1947, p. 228; see also p. 234, according to Paré astrological fatalism counted on supporters and sympathizers within the Faculty of Arts, and that the doctrine the *Roman de la Rose* deployed against them was the one commonly taught by scholastic theologians. The latter were used to maintain that the human body, together with all the bodies belonging to the sub-lunar world, is subjected to the influence of the stars; when the human body was born, celestial bodies inscribed in it good or bad dispositions, even though man's practical reason is capable of dominating those influences.
48. Bonaventura, *Collationes in Hexaëmeron*, in *Opera Omnia*, V, Quaracchi 1891, pp. 443-444: "sunt autem duae intersectiones in caelo super eclipticam, per quam transit luna, quae vocantur caput et cauda draconis; draco vocatur propter circulum, quasi tenens caudam in ore [...] Similiter vir contemplativus eclipsatur dupliciter et cadit turpiter et multum periculose". Texts from Bonaventura's *Collationes in Hexaemeron* are quoted in J. Goergen, *Des hl. Albertus Magnus Lehre von der göttlichen Vorsehung und dem Fatum*, Vechta i. Oldenburg 1932, pp. 101-102, 128 and by T. Crowley, *Roger Bacon. The Problem of the Soul in his Philosophical Commentaries*, Louvain-Dublin 1950, p. 56 n. 143 (and analysed once more by J. M. G. Hakkert, *The Meaning of Experimental Science (Scientia experimentalis) in the Philosophy of Roger Bacon*, Ph. D. Thesis, Toronto 1983, pp. 149, 151: on astrology Bonaventura writes *ibid.*, p. 351, that it deals "de influentia, et haec partim est secura et partim periculosa, et haec est astrologia. Periculosa est propter iudicia quae sequuntur; et ab hac fluit geomantia, vel nigromantia, et ceterae species divinationis". *Cf.* the different version ("reportatio") edited by F. Delorme, *Collationes in Hexaëmeron*, Quaracchi 1934, p. 56: "alia de influentia superiorum et regulatione inferiorum, et dicitur astrologia. Et haec in parte est vera et in parte periculosa propter deludia quae sequuntur, et ideo in plures haec dividitur, ut sunt necromantia, hydromantia, geomantia, pyromantia, ut sunt etiam auguria, divinationes, sortilegia et cetera huiusmodi", ibid., p. 58: "subdivisiones autem astrologiae, etsi sint aliquando opportunae ut sciantur, ut patet de suscitatione Samuelis, de miraculis magorum Pharaonis et miraculis Antichristi falsis, non multum tamen eis est insudandum".

A passage from Bonaventura's *Collationes in Hexaëmeron* is analysed in R. K. Emmerson and R. B. Herzman, 'Antichrist, Simon Magus and Inferno XIX', *Traditio*, XXXVI, 1980, p. 383. This relationship between Antichrist and necromancy is to be noted because both Roger Bacon and the *Speculum* insist on it.

49. Bonaventura, *Collationes de donis Spiritus Sancti*, in *Opera Omnia*, V, Quaracchi 1891, p. 498: "Secundus error est de necessitate fatali, sicut de constellationibus: si homo sit natus in tali constellatione, de necessitate erit latro, vel malus, vel bonus. Istud evacuat liberum arbitrium et meritum et praemium: quia, si homo facit ex necessitate quod facit, quid valet libertas arbitrii? Quid merebitur? Sequitur etiam, quod Deus sit origo omnium malorum. Verum est, quod aliqua dispositio relinquitur ex stellis; sed tamen solus Deus principiatur animae rationalis. Dicit Ieremias: Confundetur vehementer, quia non intellexerunt opprobrium sempiternum. Opprobriumm sempiternum habebunt qui sic errant". On this work *cf.* Hadrianus a Krizovlian, 'Controversia doctrinalis inter magistros franciscanos et Sugerium', *Collectanae franciscana*, XXVII, 1957, p. 131, n. 31.

50. Bonaventura, *Collationes de decem praeceptis*, in *Opera omnia*, V, pp. 514, 515. *Cf.* P. Robert, 'St. Bonaventure, Defender of Christian Wisdom', *Franciscan Studies*, III, 1943, p. 170: "That the principal errors denounced by Bonaventure in these two series [of *Collationes* preached in 1267-1268] were all included among the thirteen propositions condemned by Etienne Tempier is the first sign of the importance of the Minister General's intervention".

CHAPTER FIVE

1. To sum up the attitude of recent scholars, who – contrary to d'Alverny and Litt – when possible chose not to mention the *Speculum astronomiae*, nor to give their opinion on it, see A. Fries, *s.v.* 'Albertus Magnus', in Deutsches Literatur des Mittelalters. Verfasserlexicon, I, Berlin, 1977, col. 134: "nicht sicher unecht das gedruckte *Speculum astronomiae*, das der Astrologie im Abendland Jahrhunderte hindurch die theoretische Rechtfertigung geliefert hat". For the two recent syntheses here mentioned *cf.* ch. 4, n.1.

2. J. H. Sbaralea, *Supplementum et castigatio ad Scriptores trium Ordinum S. Francisci*, Roma 1806, p. 177; P. G. Golubovich, *Bibliotheca bio-bibliographica della Terra Santa e dell'Oriente francescano*, I, Quaracchi, 1906, pp. 223-224; from the *Chronica f. Salimbeni de Adam O. M.* (cfr. ed. O. Helder-Egger, M. G. H. SS, XXXII, Hannover-Leipzig 1905, p. 703), Golubovich took the information concerning the journey undertaken by the blessed Giovanni da Parma in 1249, when he led a party of twelve brothers to Constantinople, Nicomedia and Nicaea. As Brother Elia had already done, Giovanni returned from the journey he undertook with Bonaventura of Iseo bringing back "the great part of ... the experimental sciences to be found" in their alchemical codices. According to Salimbene "fuit autem frater Bonaventura antiquus tam in ordine quam in aetate, sapiens, industrius et sagacissimus, et homo honestae et sanctae vitae, et dilectus ab Icilino de Romano". Bonaventura of Iseo was minister in various provinces of his order, and in 1254, at the ecumenical Council of Lyon, he represented as socius the General Father Crescenzio da Iesi. At Lyons, Bonaventura of Iseo might have learned that the Pope had asked Albert to examine the "natural" and the occult writings; Bonaventura of Iseo talked of his friendship with Thomas and Albert in the only one of his works that has been

preserved, though Salimbene says that he had also written *Sermonum de festivitate et de tempore magnum volumen*. The extant work, compiled in Venice between 1256 and 1268, was a "liber medicinalis et alchimiae" entitled "*Liber Compostillae* multorum experimentorum veritatis ... ex dictis multorum philosophorum qui delectati sunt in scientiis secretis secretorum, experimentorum artis operis auri et argenti, que apud nos vocatur alchimia" (its description is given by Lopez, *Archivum franciscanum*, I, 1908, pp. 116-117; see also a note by A. Pattin, in *Bulletin de philosophie médiévale*, XIV, 1972, pp. 102-104, who described the ms. Riccardianus 119 (L. III.13), to which one must add CLM 23809). In the *Prohemium quarti operis* (f. 143va of the ms. Riccardianus) Bonaventura of Iseo named Albert and used the appellation common in the documents of the time: "fui amicus domesticus et familiaris f. Alberti Theutonici de O. P.: multa contulimus de scientiis et experimentis secretis secretorum, ut nigromancie, alchimie et cetera". In the ms. CLM 23809, f. 3v, the prologue published by Sbaralea and quoted by M. Grabmann in *Mittelalterliches Geistesleben*, II, München, 1936, pp. 385-396, offers a different *lectio* that includes Thomas: "fui amicus domesticus f. Alberti Theutonici et f. Thome de Aquino O. P., qui sic fuerunt probi viri et magni compositores scripture"; this sentence is followed by the passage quoted below, footnote 3. When Sbaralea was writing, there existed a further ms. in the Franciscan convent of Città della Pieve, from which, perhaps, derived the shortened text found in the fifteenth-century miscellany in the Riccardiana.

3. Quoted by Grabmann, *op. cit.*, p. 395: "Nam f. Albertus in diebus vitae suae habuit gratiam a domino papa propter eius famam sanctitatis et intellectus et prudentiae, et licite potuit addiscere, scire et examinare et probare omnes artes scientiarum boni et mali, laudando libros veritatis et damnando libros falsitatis et erroris. Inde multum laboravit in complendo inceptos libros Aristotelis et novas compilationes librorum fecit de multis artibus scientiarum, ut astrologiae, geomantiae, nigromantiae, lapidum pretiosorum et experimentorum alchimiae". Grabmann commented as follows: "wir haben hier auch eine zeitgenössische Zuteilung von Schriften an Albert, die ihm abgesprochen werden. Vor allem ist hier das sogenannte *Speculum astronomiae*, ein Gutachten über Schriften zur Astronomie und Nigromantie, das P. Mandonnet Roger Bacon Zuteilt, als Werk Alberts hingestellt, eine Zuteilung der auch P. Meersseman zuneigt. Desgleichen erscheint hier Albert auch als Verfasser eines Werkes über Alchemie. Bonaventura de Iseo bringt auch fol. 122v-125v Exzerpte aus den Büchern über Alchemie von Roger Bacon: "Incipiunt collecta et extracta de libro Rogeri et Alberti [...] videntur esse in concordia de istis receptis secundum quod est receptum in libris eorum, cum quilibet eorum composuit unum librum de arte alchimie multe veritatis experte". Ich konnte bisher diese Texte in keinem der mir bekannten gedruckten und ungedruckten Roger Bacon und Albert zugeteilten Werke über Alchemie feststellen. Ich kann hier noch nicht ausführlicher untersuchen, ob dieser Text des Bonaventura de Iseo, so wie es wörtlich lautet, authentisch ist". *Cf.* R. Lemay, *Abû Ma'shar cit.*, pp. XXII- XXIV n. and his paper 'Libri naturales et sciences de la nature dans la scolastique latine du XII siècle', *Proceedings of the International Congress of the History of Science*, Tokyo 1974, p. 64: "la publication du *Speculum Astronomiae* par Albert aura réussi à effectuer cette "épuration" des *libri naturales* promise par la papauté dès 1231 mais longtemps retardée. Muni d'une autorisation spéciale, vraisemblablement lorsqu'il assista au Concile de Lyon en 1245, Albert rédigea le *Speculum* dans le but, non pas d'expurger Aristote, qu'il avait déjà d'ailleurs commencé à commenter, mais de faire le partage entre la bonne et la mauvaise science de la nature. Il y passe en

revue à la lumière de l'orthodoxie à peu près tous les ouvrages de science naturelle alors connus et qui sont en immense majorité d'origine arabe. Il créait ainsi un guide officieux qui autorisât l'usage d'une grande partie de ces ouvrages de science, tout en rejetant dans l'hétérodoxie ceux qu'il qualifie de 'négromantiques' à cause de l'invocation des démons. La distinction et séparation des ouvrages d'Aristote d'avec les *libri naturales* était déjà chose accomplie dans l'esprit d'Albert comme chez certains de ses contemporains. Guillaume d'Auvergne en particulier avoue (*De Legibus*, cap. 25; ed. Lyon I 78) avoir lu dans sa jeunesse tous les *libri naturales* qu'il condamne maintenant presque sans réserve, tandis qu'il recourt sans trop de scrupules aux doctrines d'Aristote et d'Avicenne. Ainsi le maître anonyme du manuscrit de Ripoll mentionne formellement le fait que les *libri naturales* furent brûlés. Grabmann s'étonne sans raison de cette déclaration, puisqu'aussi bien les témoignages contemporains de Guillaume le Breton et de Césaire d'Heisterbach sont non moins explicites: 'jussi sunt omnes comburi – perpetuo damnati sunt et exusti'". As far as the anonymity of the work is concerned, it is useful to reproduce in this context the remarks Richard Lemay put forward in the above mentioned letter of November 30th, 1973, remarks which were the result of his own studies on this issue: "Dans le sillage de Thorndike, il m'a longtemps paru que le *Speculum astronomiae* était bien l'oeuvre d'Albert le Grand à cause de l'excellente connaissance des *libri naturales* du XIIIe siècle révélée par ce texte essentiellement bibliographique et canonique (visant à défendre l'orthodoxie). Seul un esprit averti, renseigné, dévoué à ces sciences et familiarisé avec toute leur bibliographie comme l'était Albert le Grand peut sérieusement être considéré comme l'auteur de ce catalogue critique et canonique. Ni Philippe de Thoiry, ni Roger Bacon lui-même ne rempliraient toutes les conditions psychologiques impliquées dans cet ouvrage. Les informations puisées par Grabmann chez Bonaventure de Yseo semblent concluantes. Albert a agi au nom de la Papauté, vraisemblement après le Concile de Lyon de 1245, où les *libri naturales* étaient encore prohibés, mais la promesse de les expurger, non remplie par la commission de 1231, fut renouvelée, et cette fois accomplie d'une façon plus spécifique et plus au point; mais la source des erreurs attribuées à Aristote en science naturelle depuis l'arrivée des traductions de l'arabe était maintenant perçue avec plus d'exactitude: c'étaient les *libri naturales*, non plus d'Aristote, ni exclusivement ni même principalement, mais bien tous les livres d'astrologie et de nécromancie etc. qui faisaient ample référence aux théories d'Aristote et de Ptolémée ainsi qu'aux Arabes. Entre 1248 et 1250 Albert eut de multiples occasions de rencontrer le Pape personnellement; il a pu dès lors recevoir directement de lui et de vive voix, mais d'une façon semi-officielle, cette mission dont parle Bonaventure de Yseo, bien placé pour avoir eu connaissance de cette mission. Compagnon de Frère Elie et de Jean de Parme, versé lui même dans l'alchimie et partisan du joachimisme, Bonaventure fut aussi grand voyageur en France, à la Cour pontificale et en Orient. Il fut dès lors en excellente posture pour saisir la véritable perspective de la tâche assumée par Albert le Grand et d'en connaître le résultat [...]. Il sait que la "compilation" des livres de sciences, produite par Albert ouvrait la porte de la légitimité pour la pratique des *libri naturales*, et cette compilation est bien le *Speculum Astronomiae*. Aucun autre ouvrage du XIIIe siècle ne correspond plus exactement à cette mission *accomplie* en sa totalité selon Bonaventure. En sa majeure partie, le *Speculum* est une revue critique de la bibliographie des *Libri Naturales* connus au XIIIe siècle. [...] C'est aussi un jugement critique de leur doctrines du point de vue de l'orthodoxie. Ces deux caractéristiques du *Speculum* montrent bien pourquoi Albert n'a pas cru nécessaire, ni justifié d'y apposer son nom, et pourquoi également certaines copies manuscrites de l'ouvrage portent un autre nom. L'oeuvre étant une liste purement bibliographique (les textes, titres, rubriques, doctrines rapportés dans le *Speculum* sont en majeure partie une simple compilation ou reproduction des titres et nombreuses rubriques des ouvrages considérés), les commentaires ou jugements de valeur considérant les diverses doctrines n'étant qu'une application des doctrines orthodoxes approuvées par l'Église; ... ainsi, spécialement eu égard à l'intervention directe de la Papauté, comme l'atteste Bonaventure de Yseo, le document dans son ensemble prenait la valeur d'un texte canonique et quasi officiel. Les rédacteurs de tels textes ne s'appropriaient pas en général ces textes en les signant de leur nom. Les textes canoniques ou législatifs du moyen âge sont en général anonymes du moins sous le rapport de l'auteur de leur rédaction. Le caractère canonique du *Speculum* explique encore pourquoi certaines personnalités officielles chargées du maintien de l'orthodoxie, comme le Chancellier Philippe, ont dû posséder ce document et même y apposer leur signature, qui est alors celle d'un officiel utilisant le texte pour fins jurisdictionnelles, et non pas pour affirmer leur paternité de l'ouvrage. En tout cas les docteurs du moyen âge dans leurs très grande majorité y ont reconnu le rôle d'Albert le Grand tel que décrit par Bonaventure de Yseo, et c'est ce qui donnait son autorité intellectuelle au catalogue". The documents concerning the condemnation are quoted and analyzed – with exclusive reference to Aristotle – by F. Van Steenberghen, *La philosophie au XIIIe siècle*, Louvain-Paris 1966, pp. 104-111.

4. *Cf. Speculum*, II/6; II/16-17; II/77-81; XI/38-44; XI/137-139, and XII/102 ff.
5. Among the first research in that field see B. Geyer, 'Zur Datierung des Aristotelesparaphrases des hl. Alberts des Grossen', *Zeitschrift für katholische Theologie*, LVI, 1932, pp. 432-436, developed interesting critical remarks concerning the characteristics of Albert's "paraphrases", and argued that they were composed between 1256 and 1275. This is a completely different stand from that taken by P. Mandonnet, 'Polémique averroiste de Siger de Brabant', *Revue thomiste*, V, 1897, pp. 95-105, who argued that all these comments were written between 1245 and 1256. Original is the contribution by Weisheipl, 'The Problemata' *cit.*, p. 313; he claimed that the *De animalibus* – a work composed before the *De causis* – could not have been written before 1268, and the *Problemata* confirmed that the works mentioned above, as well as the *Metaphysica*, were prior to 1271. In view of the fact the the *De causis* was explicitly the last Albertinian commentary to the Aristotelian corpus in its widest sense, it is clear that it must have been completed in or before 1271; yet many of its parts were written twenty years earlier. We will, however, know their dates precisely only when all the critical *editio coloniensis* will have been published. For instance, whereas the natural corpus is dated 1248-1260, when Albert lived in Cologne, the *Physica* was begun between 1251 and 1252, but was finished "paucis annis ante annum 1257" according to his editor P.Hossfeld. Previously this "first commentary" used to be dated back to the years 1245-1248 according to a commonly accepted chronology, which Weisheipl himself accepted, *s. v.* 'Albert', in *New Catholic Encyclopedia*, I, New York 1967, pp. 257-258 (see also his article 'The Life and the Works of St. Albert the Great', in *Albertus Magnus and the Sciences. Commemorative Essays*, ed. by J. A. Weisheipl, Toronto 1980), a scholar I am following for issues concerning chronology, unless otherwise stated.
6. *Speculum cit.*, XII/28-36: " Quod apud Albumasar [...] plenissime reprehensione dignius invenitur, est illud quod dicit [...] scilicet quod planetae sunt animati anima rationali; sed

quod dicit, dicere recitando videtur, cum dicat Aristotelem hoc dixisse, licet non inveniatur in universis libris Aristotelis quos habemus, et forte illud est in duodecimo aut decimotertio *Metaphysicae*, qui nondum sunt translati et loquuntur de intelligentiis, sicut ipse promittit"

7. Fundamental G. Vuillemin-Diem, *Praefatio* to her edition of *Aristoteles latinus. XXVI/2: Metaphysica: translatio anonyma sive media*, Leiden 1976, and bibliography there cited (see especially p. XIII).
8. G. Vuillemin-Diem, 'Die *Metaphysica media*. Übersetzungsmethode und Textverständnis', *Archives d'histoire doctrinale et littéraire du Moyen Agex*, XLII, 1976, p.7 ff., and especially pp.13-14 on Albert and Thomas; D. Salman, 'Saint Thomas et les traductions latines des *Métaphysiques* d'Aristote', *Archives d'histoire doctrinale et littéraire du Moyen Age*, VII, 1932 [but 1933], pp. 85-120; B. Geyer, 'Die Übersetzungen der Aristotelischen *Metaphysik* bei Albertus Magnus und Thomas von Aquin', *Philosophisches Jahrbuch*, XXX, 1917, pp. 392; the almost literal similarities between the passage in the *Speculum* quoted above and two Thomistic texts Geyer referred to should be emphasized: *De anima* (written in 1267-68; *cf. Sententia libri de anima*, ed. R.-A-Gauthier, in *Opera omnia iussu Leonis XIII P.M. edita*, T. XLIV Roma-Paris, Editori di S. Tommaso-Vrin, 1984, p. 234/312-319): "Haec enim quaestio hic determinari non potuit, quia nondum erat manifestum esse aliquas substantias separatas, nec quae, nec quales sint. Unde haec quaestio ad metaphysicum pertinet, non tamen invenitur ab Aristotele soluta, quia complementum eius scientiae nondum ad nos pervenit, vel quia nondum est totus liber translatus vel quia forte preoccupatus morte non complevit"; *De unitate intellectus* (written in 1270): "Huiusmodi autem quaestiones certissime colligi potest Aristotelem solvisse in his libris, quos patet eum scripsisse de substantiis separatis, ex his quae dicit in principio XII *Metaphysicae*, quos etiam libros vidimus numero 14, licet nondum translatos in nostram linguam" Geyer disagreed with the interpretation previously put forward by Grabmann, *Forschungen über die lateinischen Aristoteles Übersetzungen des XIII. Jahrhunderts*, München 1916 (= Beiträge zur Geschichte der Philosophie und Theologie des Mittelalters, XVII/ 5-6). Grabmann, who then dated the Albertinian commentary to 1256, believed that the reference to the translation by Moerbeke for books XI as well as XIII-XIV (*Metaphysica novae translationis*) was completed after 1260; for this Greek-Latin complete translation the dating admitted by Grabmann, Pelster, and Geyer himself, is between 1268 and 1273; more precisely according to G. Vuillemin-Diem, *Aristoteles latinus cit.*, p. XXXI: "paulo post 1262-1263". Geyer maintained however that the entire commentary was written by Albert after 1260, since it bespoke the use of the *translatio nova sive anonima* throughout. *Cf.* Geyer, 'Die von Albertus Magnus in *De anima* benutzte Aristotelesübersetzung und die Datierung dieser Schrift', *Recherches de théologie ancienne et médiévale*, XXII, 1955, pp. 322-326, where he referred to Pelster, Franceschini and Grabmann and remarked: "Man war nämlich allgemein der Ansicht, dass Albert im allgemeinen die Moerbeke-Uebersetzungen nicht gekannt oder wenigstens nicht benutzt habe", and confirmed this also in this specific instance. See also A. Mansion, 'Sur le texte de la version latine médiévale...', *Revue néoscolastique*, XXXIV, 1932, pp. 65-69; W. Kübel, 'Die Übersetzungen der Aristotelischen *Metaphysik* in den Frühwerken Alberts des Grossen', *Divus Thomas* (Freiburg), XI, 1933, pp. 241-268; F.Pelster, 'Die Übersetzungen der aristotelischen *Metaphysik* bei Albertus Magnus und Thomas von Aquin', *Gregorianum*, XVI, 1935, pp. 338-339; F. Ruggiero, 'Intorno all'influsso di Averroè su S. Alberto Magno', *Laurentianum*, IV, 1963, pp. 27-58; etc.

A further issue, discussed by A. Dondaine, *Secrétaires de Saint Thomas*, Roma 1956, p. 188 n., concerns the chronological relationships between the *De causis*, and the book *Lambda* of the *Metaphysics* (the *De causis* was considered complementary to it): "et haec quidem quando adiuncta fuerit undecimo *Primae philosophiae* opus perfectum erit" also about the *intelligentiae*. The issue bears heavily on the possible dating of the *Speculum* among Albert's works, in view of the vexed question of the lack of those books "qui nondum sunt translati": no difficulty is however implied if we abandon the connexion – hypothesized by Mandonnet – between the work and the condemnation of 1277, and we accept an earlier date of composition. In his *Prolegomena* to the critical edition of the *Metaphysica* in Albertus Magnus, *Opera omnia*, Münster 1960-1964, t. XVI, pp. 1-2, the eminent scholar has again gone over the entire issue, and has acknowledged that in his commentary Albert always used the *translatio media* – which he constantly followed between 1250 and 1270; for instance, in the Dionysian commentaries he employed the *vetus* (Greek-Latin, books I-X and XII) and the *nova translatio* (Arabic-Latin, books II-X e XI); in the *Commentary to the Sentences*, in the *De quatuor coaequaevis* and in the published as well as unpublished parts of the *Summa de creaturis*, he went back even to the *vetustissima* (Greek-Latin version of books I-IV) which he found useful in order to clarify a few obscure passages, thanks to its literal faithfulness to the original. The contemporary use of various translations, and the comparison between them, is typical of Albert, as the collations made by W. Kübel, 'Die Übersetzungen der Aristotelischen Metaphysik' *cit.*, have shown. The *translatio media*, Greek-Latin, books I-X, XII-XIV, reviewed and in part completed the more ancient versions, since it now included book N (XIV), and was used even by Thomas in his *Quaestiones de veritate* (which according to its editor A.Dondaine, *Opera omnia cit.*, XXII, Roma 1975, pp. 5*, 7* were written between 1256-59, and already quoted before 1264-65, in the *Speculum naturale* by Vincent of Beauvais), etc. According to Geyer, after 1270 the "translatio Moerbekana quasi universaliter divulgata et recepta est": the *media* was thus completely checked against the Greek text, was freed from the additions derived from Averroes, and contained for the first time the book Kappa (XII). Geyer is in any case convinced that Albert's commentary to the *Metaphysica* was likely composed in the years 1262-63, that is, at a time when Albert was probably aware of the Greek text Moerbeke worked on, but did not use it. The same hypothesis might hold true for the *Speculum*, if this work was written before 1270. Geyer himself points out that in the *Physica*, the first of the commentaries composed during his stay in Cologne, Albert laid down the program of following "eodem numero et nominibus" Aristotle's works, and of adding "etiam alicubi partes librorum imperfectas, et alicubi libros intermissos vel omissos, quos vel Aristoteles non fecit, et forte si fecit ad nos non pervenerunt". Thus, far from constituing a chronological impossibility (as Geyer claimed in 'Das *Speculum astronomiae' cit.*), the entire issue emphasizes a methodological procedure typical of Albert. This conclusion finds confirmation in the data collected by F. Pelster, 'Kritische Studien', *cit.*, and in the passage from the chronicle that he quoted on p. 146 n, concerning William of Moerbeke who "transtulit omnes libros Aristotelis... quibus nunc utimur in scholis ad instantiam fratres Thomae de Aquino. Nam temporibus domini Alberti translatione veteri omnes communiter utebantur". On this point, L. I. Bataillon, 'Status quaestionissur les instruments et techniques de travail de St. Thomas et St. Bonaventure', in *1274. Année charnière. Mutations et continuités*, Paris 1977, p. 650, expressed radical doubts. *Cf.* G. Vuillemin-Diem, '*Die Metaphysica media*, Übersetzungsmethode und Textverständnis', *Archives d'histoire doctrinale et littéraire du Moyen Age*, 42, 1975 [1976], pp. 7 ff.

9. *Cf.* the anonym "Quaestionensammlung" discovered by M. Grabmann, *Mittelalterliches Geistesleben*, II, München 1936, p.188: "Plures autem libros *Metaphysicae* non habemus translatos, quamvis in greco, ut dicitur, sint usque ad viginti duo".
10. Y. Congar, 'In dulcedine' *cit.*, pp. 47-57. *Cf.* L.-J. Bataillon, 'Status questionis' *cit.*, pp. 650-651, and especially p. 653: "Un Maitre médieval ne travaillait pas seul, mais était entouré d'assistants (*socii*), dont les plus avancés, les bachelliers, tenainet un rôle important dans les disputes universitaires. Il s'y joignaient éventuellement d'autres sécraitaires ou copistes". See also the mention of "socii nostri" as interlocutors quoted from Albert's *Physica* [l. II, tr. 2, c. 21; ed. Hossfeld, in *Opera omnia*, IV/1, Münster 1987, p. 129/25] by J. Goergen, *Des hl. Albertus Magnus Lehre von der göttlichen Vorsehung und dem Fatum*, Vechta i. Oldenburg 1932, p. 100, and the classic study by A. Dondaine, *Sécretaires de St. Thomas*, Roma 1956.
11. *Speculum*, Proem/10-11: "Vir zelator fidei et philosophiae, utriusque scilicet in ordine suo", and below, ch; 7.
12. *Speculum*, XII/107-109.
13. *Speculum*, II/17-20.
14. F.S. Benjamin and G.J. Toomer, *Campanus of Novara and Medieval Planetary Theory. 'Theorica Planetarum'*, ed. with Intr., Engl. Translation and Commentary, Madison 1971, p.19; the two editors did not find "assurance of Campanus's authorship". The authorship has now been maintained by M. Pereira, 'Campano da Novara autore dell'*Alma gestum Parvum*', *Studi medievali*, 19 (1978), pp. 769-776, and has been accepted by A. Paravicini Bagliani, 'La scienza araba nella Roma del Duecento: Prospettive di ricerca', in *La diffusione delle scienze islamiche nel Medio Evo Europeo. Convegno intern. promosso dall'Accademia Naz. dei Lincei, Fondazione L. Caetani,* e *Università di Roma 'La Sapienza'*, Rome 1987, p. 153.
15. A. Paravicini Bagliani, 'La scienza araba' *cit.*, p. 152; *id.* 'Un matematico nella corte papale del secolo XIII: Campano da Novara', *Rivista di storia della chiesa in Italia*, XXVIII, 1973, pp. 98-129; *id.* 'Nuovi documenti su Guglielmo da Moerbeke, *Archivum Fratrum Praedicatorum*, VII, 1982, pp. 135-143; *cf.* Benjamin and Toomer, *Campanus cit.*, p. 11, on Campanus's "membership in a 'remarkable scientific group', associated with the papal court during the third quarter of the 13th century, that included Moerbeke, Witelo and Johannes Gervasius and perhaps even Thomas Aquinas". *Cf.* also M. Grabmann, *Guglielmo di Moerbeke*, Roma 1946, pp. 56-62.
16. J.A. Weisheipl O.P., 'The Life and Works of St. Albert the Great' *cit.*, pp. 36, 38-39.
17. P. Simon O.P., 'Prolegomena', in Albertus Magnus, *De fato*, in *Opera Omnia*, XVII/1, p. xxxvi. Among the curial scholars also Witelo was strongly, and critically interested in the nature and influence of stars, cfr. his *De nature daemonum*, recently ed. by J. Burchardt, Wrocław 1978, and by E. Paschetto, in her *Demoni e prodigi*, Torino 1978.
18. It is well known that Albert was considered to be an "auctoritas" by his contemporaries; equally well known are the corresponding critiques by Roger Bacon. See J.M.G. Hackett, 'The Attitude of Roger Bacon to the Scientia of Albertus Magnus', in *Albertus Magnus and the Sciences*, *cit.*, pp. 63-64, and the bibliography listed by Hackett in this article, as well as in his *The Meaning of Experimental Science* (Scientia experimentalis) *in the Philosophy of Roger Bacon*, Ph.D. Thesis, Toronto 1983. It should also be pointed out that

Campanus was in his turn quoted by Bacon – this time without polemical allusions – among the illustrious mathematicians praised in the *Opus Tertium* (1267). *Cf.* Benjamin and Toomer, *Campanus cit.*, p. 7 and n. 20.

19. *Cf. Speculum*, XV/24-41, with Benjamin and Toomer, *Campanus cit.*, pp. 23-24 n. 87: "Tangere cum ferro membrum illud vulnerando est causativum doloris et dolor causat fleuma [reuma CLM] propter quod inquit in cirurgia cavendum est ab incisione in membro luna existente in signo significationem habente super illud membrum [...] Item narrat Campanus se vidisse hominem imperitum in astris qui in periculo squinantie minuerat sibi de brachio luna existente in geminis quod signum dominatur super brachia et absque ulla manifesta egritudine excepta modica brachii inflatione die septimo mortuus est. Novit etiam quendam ut asserit patientem fistulam in capite membris virilis et ipsum fuisse incisum Luna existente in Scorpione quod signum dominatur super partem illam corporis et eadem hora incisionis in manibus tenentium obiit nulla [*add.* CLM: alia] causa concurrente". Unfortunately, of this work by Campanus we only have fragments quoted by the dominican Nicholas of Lund (de Dacia) – not of Lynn as write Thorndike and Benjamin – in the canons of his calendar (end of fifteenth century). The discovery of the complete text of Campanus's *Canon pro minutionibus et purgationibus* would allow and indeed require a profitable comparison with the *Speculum astronomiae*. It is noted that the only sentence added by the *Speculum* is inserted between the two clinical cases ("Et audeo dicere me vidisse ex hoc quasi infinita accidentia accidisse"); this does not exclude that he was borrowing from more cases listed by Campanus in the text now lost. This was typical of Albert, when he was claiming to report personal observations, as was shown by P. Hossfeld, 'Die eigenen Beobachtungen *cit.*, pp. 170-171.
20. P. M. Tummers, 'Albertus Magnus' View on the Angle with Special Emphasis on His Geometry and Metaphysics', *Vivarium*, XXII, 1 (1984), p. 35. Cfr. Albertus [Magnus], '*Commentaar op Euclides' Elementen der Geometrie', Inleidende studie, analyse en uitgave van Boek I*, P. M. Tummers ed., Nijmegen 1984, 2 voll., where in *Proemium* (II, p. 1), talking of the uncertainty in knowledge of all things composed with matter, Albert cites "magnus in disciplinalibus Ptolomaeus". See also A. G. Molland, 'Mathematics in the Thought of Albertus Magnus', in *Albertus Magnus and the Sciences, cit.*, pp. 463-478; P. M. Tummers, 'The Commentary of Albert on Euclid's Elements of Geometry', *ibidem*, pp. 479-499; P. Hossfeld, 'Zum Euklidkommentar des Albertus Magnus', *Archivum Fratrum Praedicatorum*, 52, 1982, pp. 115-133.

CHAPTER SIX

1. Aside from Quétif and Echard, on Bernard de la Trille Nemausensis see Glorieux, *Répertoire cit.*, I, p. 155; T. Käppeli, *Scriptores Ordinis Praedicatorum*, I, Roma 1970, p. 234-237; P. Künzle, *s. v.*, *Enciclopedia filosofica*, I, Firenze 1967, 2nd ed., cols. 873-874; *Id.*, 'Notes sur les questions disputées '*De spiritualibus creaturis*' et '*De potentia Dei*' de Bernard de Trilia,O.P.', *Bulletin de philosophie médiévale*, VI, 1964, pp. 87-90; W. W. Wallace, *s. v.*, *Dictionary of Scientific Biography*, II, New York 1970, p. 20, pointed out that in his commentary on Sacrobosco's Sphere Bernard was more favorable to Ptolemy than to Aristotle-Alpetragius, and that he offered a combination of the theory of the precession of the equinoxes put forward by Hipparchus and the one of trepidation by Thebit,

following "lines suggested by Albertus Magnus, whom Bernard appears to have studied closely". The latter conclusion had already been advanced by P. Duhem, *Le système du monde*, Paris 1958, III, p. 326 ff., Ch. VI, "L'astronomie des Dominicains": Duhem reconstructed Bernard's descriptive astronomy and summarized parts of the commentary (cf. especially pp. 363-383). This chapter is also of interest for information on pupils of Albert such as Thomas and Ulrich von Strassburg, and critics of Albert within the Order, that is, according to Duhem, Dietrich von Freiberg. Pursuing his customary attitude, Duhem did not examine the astrological features of these Dominican texts; Thorndike did instead offer some data on them in his *The Sphere of Sacrobosco and his Commentators*, Chicago 1949, pp. 25-26, and in particular on Bernard (b. *ca.* 1240; d. 1292), who belonged to the Dominican Province of Provence and wrote in Nîmes and Avignon his commentary on Sacrobosco, still unpublished today. F.J. Roensch, *Early Thomistic School*, Dubuque, Iowa 1964, pp. 84-88, 289-296, gives a short (theological and biographical) notice of Bernard of Trilia, as well as of Giles of Lessines, pp. 89-92. Bernard studied in Paris (but between 1260 and 1265, therefore without being able to listen to Albert); he was in any case familiar with Albert's *De caelo*, and probably met Albert himself in the chapters of the Order. In the *Tabula Stams* (ed. Denifle, *Quellen zur Gelehrtengeschichte cit.*, p. 239) are recorded his "questiones super totam astrologiam" that can perhaps be identified with such commentary on Sacrobosco, together with numerous and better known theological works. In any case, we should consider Bernard as an author who autonomously developed his views, rather than as a witness of the activity that led Albert to write his *Speculum astronomiae*.

2. It is perhaps possible that Dietrich von Freiberg was not a direct pupil of Albert; he was in any case a careful reader and follower of his theories. W. A. Wallace, *The scientific Methodology of Theodoric of Freiberg*, Fribourg /Schw. 1959, p. 26, pointed out that Dietrich studied in Albert's province when the latter was still alive, and his paraphrases were used by all Dominican students. Thedoric quoted several authors Albert loved, and also referred to Albert's teaching in his *De miscibilibus in mixto* as well as in the *De intelligentiis et motoribus caelorum* (see, for example, some chapters of the latter: Cap. I. 2: "De differentia intelligentiarum et motorum caelestium corporum, quos animas caelorum vocant, cum ratione disserendi". Cap. II: "Quod tam intelligentiae quam motores corporum caelestium sint intellectus in actu per suam essenciam et quod secundum hoc sunt principia rerum causalia secundum philosophos et qualiter".). Born in *ca.* 1250, Dietrich was a lecturer in Freiberg, and it was only in 1276 that he went to Paris for his studies; after Albert's death, he became Prior of Würzburg and Provincial of Germany. He died after *ca.* 1310, according to data provided by Wallace, Flasch and Sturlese, modifying the previous conclusions of Glorieux, *Répertoire cit.*, I, pp. 162-65. See now Dietrich von Freiberg, Opera Omnia, ed. K. Flasch et al., Hamburg 1977-1985, part III, pp. xvii-xxxvii, and the 'Einleitung' by K. Flasch, pp. 1-46; *cf. Tractatus de animatione caeli*, a work which Loris Sturlese, its editor, *ibid.* p. 8, dates to the beginning of the 1280s and in any case before 1286; *cf.* Sturlese, *s.v.*, *Deutsche Literatur des Mittelalters: Verfasserlexikon*, II, Berlin 1979, pp. 127-137, and 'Il '*De animatione caeli*' di Teodorico di Freiberg', in *Xenia Medii Aevi Historiam illustrantia oblata T. Kaeppeli O.P.*, Roma 1978, pp. 175-247; at pp. 179-180, the author briefly refers to discussions by Ulrich von Strassburg, Berthold von Moosburg and Heinrich von Lubeck on the same theme.

3. In his *History* (II, 540 n.), Thorndike noted that Albert and his pupils were looked on by contemporaries as a team. *Cf.* Y. Congar, '"In dulcedine"' *cit.* above, and L. J. Bataillon, 'Status quaestionis' *cit.*, pp. 650-653 who has recently expanded upon the cooperation between Thomas Aquinas and Albert, and has studied the text of the *Commentum et quaestiones super Ethica* he was going to make use of in further studies, together with other "paraphrases d'Albert écrites par des scribes qui ont été au service de Thomas". In his 'Further consideration' *cit.*, p. 422, Thorndike discussed this problem when commenting on the pseudographical *Experimenta*, and pointed out that the allusions to "the brothers who have performed experiments" are a topos meant to imitate Albert. It is, however, difficult to suppose that writers of the caliber of Giles of Lessines or Theodoric chose the pseudoepigraphic form for their works. On the other hand, the *Speculum astronomiae* was an introductory work too perfect to be the product of a compiler of occult pseudoepigraphic works of the kind Thorndike studied: indeed, Thorndike knew them so well that he never equated the *Speculum* with their works.
4. Thorndike, ed., *Latin Treatises on Comets between 1238 and 1368 A. D.*, Chicago 1945, pp. 62 ff., 91, 185-187; Thorndike pointed out that Gerard's interpretation of the comet of 1264 was "largely indebted to Albertus Magnus, from whom passages ... of considerable length are embodied", and that Albert's commentary on the *Meteorologica* written before 1264 (between 1254-1257 according to Weisheipl, 'The Life and the Works' *cit.* p. 35) made reference to a comet that appeared in 1240; see also pp. 192, 194. *Cf.* P. Hossfeld, 'Die Lehre des Albertus Magnus von den Kometen', *Angelicum*, 57, 1980, pp. 533-541; *Id.*, 'Der Gebrauch der aristotelischen Übersetzung in den *Meteora* des Albertus Magnus', *Medieval Studies*, 42, 1980, pp. 395-406.
5. Thorndike, *Latin Treatises cit.*, p. 193 (Gerard's text) and also see p. 181 ff. (Giles's text) for the filling out of the quotation from Seneca: Thorndike (p. 1) was convinced that "Gerardus wrote later in the century than Aegidius".
6. Unfortunately, the biography of the Dominican from Feltre is largely unknown: his name has been given as Gerardus de Silcro or Silteo, as in Luíz de Valladolid, *Scriptores O. P.*, Roma, Archivio Generalizio dei Domenicani, ms. XIV lib. 99. p. 388; at my request, the latter text has been kindly examined by the archivist Father Emilio Panella, who has confirmed my hypothesis that in this fifteenth-century manuscript "the ductus of s and of f are identical, with the exception that the f bears a horizontal sign", and that "under the pen of a scribe, the transition to 'Siltro' would have been extremely easy". It is therefore clear that the better-knonw toponym "Feltre" found in many fourteenth-century codices of Gerard's works is largely to be preferred. Cf. Käppeli, *Scriptores O.P. cit.*, II, Roma 1975, pp. 34-35, only added the indication of two mss. (London, Wellcome Medical Library, 308; XV sec.: *Summa de astris* (Parts I-II only); Bamberg, Staatliche Bibliothek, ms. astron.-mathem. 4; XIV-XV sec., ff. 65^r-68^r: "Domenicani anonymi cuiusdam magistro Johanni [de Vercellis] O. P. dicatum... ad indagandam altitudinem cuiuslibet stellae novae... specialiter... de altitudine... stellae quae anno praeterito... 1264 apparuit") to those already known to Thorndike, *Latin Treatises cit.*, pp. 185-195, and Grabmann, *Mittelalterliches Geistesleben cit.*, II, p. 397; III, pp. 255-279. It is my intention to publish elsewhere the *Summa de astris*, the best manuscripts of which have already been noted by Grabmann (Bologna, Bibl. Archiginnasio, ms. A 539, and Milano, Bibl. Ambrosiana, ms. C 245 inf.) I will quote in the following pages which follow from ms. Bologna, Arch. A 539, giving the partition of the *Summa* there used, and compare with the Milanese ms.,

which I have not had the time necessary to study thoroughly. Its partitions do not correspond to the Bolognese ms. If that eventually would import that the Milanese ms. contains a different and earlier version, written before 1264, we should – assuming the *Summa de astris* as a *terminus post quem* – consider also the *Speculum astronomaiae* as possibly earlier, and perhaps date it but not the second residence of Albert at the papal court, to the first or immediately later, 1256-1258. The *Summa* – which cites from Thomas's *Quaestiones de veritate* dated 1256-1259– cannot be earlier than that.

7. Grabmann, 'Aegidius von Lessines', in his *Mittelalterisches Geistesleben cit.*, II, pp. 512-530; see especially p. 520 n.25 on Giles' method; p. 525 on the *De crepuscolis* (studies by P. Mandonnet, 'Giles de Lessines et son *Tractatus de crepusculis*', *Revue néoscolastique de philosophie*, 22, 1920, pp. 190-194); Grabmann quotes ibid., p. 514, Giles's words "Albertus quondam Ratisponensis episcopus, ob cuius reverentiam rationes predictam positionem confirmantes addidimus". The *Tabula Scriptorum O.P.* called *Tabula Stams*, cited by Grabmann, *ibid.*, p. 524, writes on Giles: "plura scripsit de astrologia". See also *Dictionary of Scientific Biography*, V, New York 1972, pp. 401-402 C. Vansteenkiste, *s.v.* 'Giles of Lessines', New Catholic Encyclopaedia, VI, New York 1967, p. 484: "His relations with Albert the Great suggests that he studied under this master probably in Cologne". Giles's first work, *De cometis*, shows "an interest for natural sciences not uncommon in the school of Albert". *Cf.* also Käppeli, *Scriptores O.P. cit.*, I, Roma 1975, pp.13-15. See also Giles's *Summa de temporibus*, Bk. III, *i.e*, the *Computus*, formerly attributed to Roger Bacon in his *Opera hactenus inedita*, ed. R. Steele, VI, Oxford 1926, p. 1, "Qualiter diversimode consideretur tempus ab astrologo, physico et medico". *Cf.* the very interesting observation on Giles's "use of past authors" and especially "of Arabic authors for the astrological significance of comets" in Thorndike, *Latin Treatises on Comets cit.*, p. 95 ff.

8. J.-P. Mothon, *Vita del b. Giovanni da Vercelli, sesto Maestro Generale dell'O.P.*, Vercelli 1903, p. 255 n.: "Il 5 Giugno 1267 in occasione della collocazione del corpo di S. Domenico nella tomba monumentale costruita a Bologna, secondo la *Chronica Ordinis* del fr. Sebastianus de Olmedo O.P. e secondo la Chronica Ordinis edita nel 1690 alla fine dei *Libri Contitutionum* 'quae translatio cum ageretur, apparuit cometa super templum nostrum Bononiensem, ibidem permansit donec cerimonia finita esset'". But as Gerardus de Feltre shows, by addressing his *Summa* and already in 1265 the Bamberg fragment on the comet of 1264 (*cf.* above n.6) to John of Vercelli, interest in comets among the Dominicans had begun some years previously; *cf.* Thorndike, *Latin Treatises on Comets cit.*, p. 193, where he cites from the *Summa de astris*: "Ego autem cum multis aliis anno ab incarnatione Domini 1264 in Lombardia vidi cometam". Earlier were the observations made by Giles of Lessines, starting with an eclipse observed in Paris on August 5, 1263. *Cf.* Thorndike, *ibid.*, and also 'Aegidius of Lessines on Comets', in *Studies and Essays in the History of Science and Learning offered to G. Sarton*, New York 1946, p. 413, where he mentions Giles's "own observations of the comets of 1264 [...] adduced only incidentally and briefly" in his *De essentia, motu et significatione cometarum.*

9. *Cf.* R. Creytens, 'Hugues de Castello astronome dominicain du XIVe siècle', *Archivum fratrum praedicatorum*, XI, 1941, pp. 95-96: "On a exagéré ou mal compris certaines ordonnances des chapitres généraux concernant l'étude des sciences naturelles. En dehors de l'alchimie, qui a toujours été prohibée sous peine graves, on ne retrouve pas d'ordonnances, sauf une seule au chapitre provincial de Viterbe 1258 (MOPH, XX, 22),contre

les études astronomiques". Hugo's text studied by Creytens was strictly astronomical, but the dominican scholar took the opportunity to list several dominican astrologers and to agree with Thorndike's statement that "hardly any class or group of men in the later Middle Ages were more given to astrology and occult arts and sciences than the friars" (p.91).

10. Grabmann, *Mittelalterliches Geistesleben cit.*, II, p. 271.
11. Gerardus a Feltre, *Summa de astris, Prologus II* (ms. Bologna, f. 2): "summa haec de astris compilata et conscripta est ex dictis Ptolemaei, Albumasar, Alfargani, Alchabitii, Omar, Çahelis, Messeala, qui fuerunt auctores magisterii astrorum".
12. *Speculum*, XII, 38-39 = *Summa*, P. III, d. iii, c.3; *Speculum*, XII, 66-73 = *Summa*, P. III, d. iv, c. 1; *Speculum*, XIV, 63-76 = *Summa*, P. III, d. iv, c. 3; etc.). Gerard expresses the awareness of the need of studying seriously the astrological discipline ("si quispiam adversus mathematicos velit scribere imperitus matheseos, risui pateat").
13. *Ibid.*, P. I d. vii, c.1 (ms. Bologna, f.13va): "frater Albertus ordinis nostri ... magnus philosophus".
14. Thomas Aquinas, *Quaestiones disputatae de veritate*, in *Opera omnia iussu Leonis XIII P.M. edita*, XXII, Roma 197O-1976, pp.161-171;Q. 5, art. 9-10: "Quaeritur utrum per corpora caelestia disponat divina providentia inferiora corpora [et humani actus]".
15. *Summa cit.*, P. III, d. iv, c. 1 (ms. Bologna, f.68v): "Dicunt enim astrologi blasphemando quod omnes actus humani et mores, omnia quorum bona et mala, imo ipsa electio animae rationaliter eveniunt de necessitate, secundum dispositionem superiorum corporum, ad quod probandum introducam famosiores auctores ipsorum. Albumasar [...] Ptolemeus".
16. *Ibid.*, P. III, d. iv, c. 1 (ms. Bologna, f.68rb): "amplius manifestabimus eorum stultitias: et enim insaniunt dicentes".
17. *Ibid.*, P. III, d. viii, c. 1 (ms. Bologna, f.73va-74ra): "probabimus per eorum auctores quos vocant philosophos, cum non fuerint philosophi, sed contemptibiles ut ribaldi".
18. *Ibid.*, P. III, d. viii, c. 1 (ms. Bologna, f.74rb): "Patet hoc est contra Sacram Paginam".
19. *Ibid.*, P. III, d. xi (ms. Bologna, f. 33rb): "igitur adversarii christianae fidei obmutescant".
20. *Ibid.*, P. III, d. xii, c.1 (ms. Bologna, f.79vb): "reprobatis astrorum iudiciis tanquam infidelibus et blasphemis".
21. *Ibid.*, P. III, d. iv, c. 2 (ms. Bologna, f.68vb): "auditis blasphemiis".
22. *Ibid.*, P. III, d. viii, c. 1 (ms. Bologna, f.73vb): "apparet infidelitas eorum qui iudicia tradiderunt".
23. *Ibid.*, P. III, d. xi (ms. Bologna, f.78rb): "hec pestis heretica que enim de se multos fidei articulos subruit: ideoque tales *a consorcio fidelium sunt eliminandi*".
24. *Ibid.*, P. III, d. x. (ms. Bologna, f.77vb): "Amplius numquid credendum est adversariis fidei christianae, qui iudicia tradiderunt in his quae contra pietatem christianam non sunt? Lego in libris eorum ritum paganorum, ritum etiam saracenorum, sed de commendatione religionis christianae nullam percipio fieri mencionem ab his qui post incarnacionem Domini nostri sua confecere scripta. Ipsi etiam a Machometo suorum annorum ducunt principium: Albumasar *Introductio in scientiam iudiciorum astrorum* et Messehala *De receptione*, ut de caeteris sileam, non in nostra lingua sunt compositi, sed a Ioanne Hyspalensi translati sunt ex arabico in latinum".
25. *Ibid.*, P. III, d. vii (ms. Bologna, f.73rb): "astrologi non sunt dii, sed inimici dei".
26. *Ibid.*, P. III, d. i, c. 2 (ms. Bologna, f.62ra), on Augustine's *De doctrina christiana*, "ubi agitur de artibus magorum, aruspicum, augurum, incantatorum, sociat eis astrologos".

27. *Speculum*, Proemium 6/2-3: "verae sapientiae inimici, hoc est domini nostri Iesu Christi".
28. *Speculum*, XVII/9-10: "quae ad verum non merentur dici scientiae, sed garamantiae".
29. *Summa*, P. III, d. ii, c. 3 (ms. Bologna, f.65va): "anima etenim non ex necessitate sequitur complexionem corporis, sed ex voluntate [...], ideoque actus humani dependent a causa voluntaria et non a positione syderum".
30. *Ibid.*, P. III, d. ii, c. 1 (ms. Bologna, f.64rb): "vel iussu Dei aut nutu Dei"; *cf.* P. III, d. vi, c. 1: "si Deus voluerit immutare, sub quibus verbis ... latet venenum ad interficiendos simplices".
31. *Speculum*, XVI/9 and *passim*.
32. *Summa.*, P. III, d. iii, c. 1 (ms. Bologna, f.66rb): "blasphemat deum, quia corpora superiora cogunt hominem peccare et beneficere".
33. *Ibid.*, P. III, d. iii, c. 1 (ms. Bologna, f.66va): "ad ministerium, non ad dominium creaverit deus stellas".
34. *Ibid.*, P. III, d. iii, c. 2 (ms. Bologna, f.66va): " non materialiter, nec formaliter seu finaliter: igitur caelum horum esset causa efficiens, et non causa agens per electionem. [...] sed per naturam et ideo causa efficiens et naturaliter efficiens".
35. *Ibid.*, P. III, d. iii, c. 2 (ms. Bologna, f.66va.b): "Prima causa non aufert operationem suam a secunda causa, sed fortificat eam, ut patet per hoc quod in libro *De causis* dicitur: ergo si stellae faciunt hominem homicidam et latronem, multo magis prima causa idest deus, quod nephas est dicere". Cf. the article 167 condemned in 1277, *Chartularium cit.* cit. above ch. 2, n. 6.
36. *Ibid.*, P. III, d. viii, c. 1 (ms. Bologna, f.73vb ff.).
37. *Ibid.*, P. III, d. v, c. 1 (ms. Bologna, f.71va): "Avicenna ponit quod sicut corpora nostra mutantur a corporibus caelestibus, ita voluntates nostrae immutantur a voluntate animarum caelestium, quod tamen est omnino hereticum".
38. *Ibid.*, P. III, d. v, c. 1 (ms. Bologna, f.71va): "Dignum etenim est, ut qui in sordibus est sordescat adhuc, et caeca mente de uno errore in alio labatur. Et in hoc errore fuit Albumasar, quem iudices astrorum multum venerantur".
39. *Ibid.*, P. III, d. iii, c. 3 (ms. Bologna, f.68ra): " cum Albumasar dicat quod planetae sunt animalia rationalia, [...] tamen non habent electionem".
40. *Speculum*, XII/38-39.
41. *Speculum*, XII/38-39: "cum dicat Aristotelem hoc dixisse, licet non inveniatur in universis libris Aristotelis quos habemus".
42. *Summa*, P. III, d. iv, c. 3 (ms. Bologna, f.70rb): "de virtute magnetis vel aliorum lapidum, de iudiciis medicorum et de medicinis, de diversitate aeris et complexionum et de aliis effectibus naturalibus".
43. *Ibid.*, P. III, d. vii (ms. Bologna, f.73va.b): "in causis suis, sicut cognoscitur frigus futurum in signis et dispositionibus stellarum".
44. *Ibid.*, P. III, d. ii, c. 2 (ms. Bologna, f.65rb): "Simili quoque modo complexio alicuius hominis non solum est ex positione siderum, immo contrahitur a parentum natura, ab alimentis, ab exercitationibus, ab aeris qualitatibus et huiusmodi".
45. *Ibid.*, P. III, d. ii, c. 2 (ms. Bologna, f.65ra): "Sed si sidera essent causae sanitatis et aegritudinis, essent quidem causae remotae".
46. *Ibid.*, P. III, d. VII, c. unicum (ms.Bologna, f.73ra-b).
47. *Ibid.*, P. III, d. ii, c. 3 (ms. Bologna, f.65va): "scientiae coniecturae, ut communiter in iudicia medicorum et astrologorum apparet".

48. *Ibid.*, P. III, d. ii, c. 2 (ms. Bologna, f.65ra): "Eorum causae magis se habent ad unam partem quam ad alteram, et ista sunt contingentia quae ut in pluribus habent causam determinatam, ut accidentia corporum naturalium inferiorum, quare causae naturales quamvis sint determinatae ad unum tamen recipiunt impedimenta: et huiusmodi effectus in causis suis non possunt cognosci infallibiliter, sed cum quadam certitudine coniecturae, ut naturales eventus in inferioribus, sicut pluviarum et huiusmodi [...]. Unde si homines cognoscerent omnes causas naturales, quod in vita praesenti contingere non potest, quaedam quae contingentia videntur, ut de pluvia et aliis accidentibus aeris, aliquibus causis pensatis cognoscerent ut necessaria. Dum omnes causas cognoscerunt, nam praeter motum superiorum corporum concurrit dispositio materiae inferiori, quae contingentiae subiacet, et ideo contingens est effectus".
49. *Cf.* P. III, distinctio IV: "Utrum omnia de necessitate contingant" (ms. Bologna, f.68rb-71rb); *Speculum*, XIV/48-74.
50. *Speculum*, XII/2-3.

CHAPTER SEVEN

1. *De quatuor coaequaevis cit.*, tr. III, q. xvi, a. 2; p. 73: "Ista omnia diximus secundum philosophos, qui non contradicunt quibusdam Sanctis negantibus caelum animam habere, nisi in nomine solo, qui abhorrent nomen animae et tamen bene concedunt quod intelligentiae quaedam sive angeli movent caelum iussu Dei." *Cf. Speculum cit.*, Proem/10-11.
2. *Metaphysica cit.*, L, XI, tr. ii, c. 21; p. 509/87-94: "Hoc autem ideo non est dictum quod aliqua sit influentia superioris super inferiorem, nisi per motum corporis animati, secundum Peripateticos. Sed quia animatum primum indeterminatas et universalissimas influit et exquiritur formas, et illae determinantur ad materiam plus et plus, secundum quod magis et magis descendunt ad materiam hanc et illam."
3. *Metaphysica cit.*, L. XI, tr. ii, c. 21; p. 509/94-510/5: "Si autem est aliqua alia irradiatio superioris super inferiora, sicut dixit Plato et sicut dicunt theologi, illa per rationem investigari non potest, sed oportet quod ad illam investigandam ponantur alia principia ex revelatione spiritus et fide religionis: et de hac non est loquendum in philosophia peripateticorum, quia cum eis ista scientia non communicat in principiis." See below n. 16 the passage mentioned from the *De generatione*.
4. *Metaphysica cit.*, L. XI, tr. ii, c. 26: p. 516/81-88: "ita intelligentia agens quae movet orbem et stellam vel stellas, luminari invehit formam, et per lumen luminaris traducit eam in materiam, quam movet, et hoc sic tangens materiam educit eam de potentia ad actum. Et huius signum est, quod sapientes astronomi per haec principia, quae sunt loca stellarum, pronosticantur de effectibus, qui luminibus stellarum inducuntur rebus inferioribus."
5. *Metaphysica cit.*, L. XI, tr. ii, c. 21; p. 510/21-24: "Quod Deus deorum pater dixit superioribus caelestibus diis, quod ipse esset qui sementem generationis faceret et ipsis traderet sementem illum ad ex equendum."
6. *Metaphysica cit.*, L. XI, tr. iii, c. 2; pp. 535/91-536/1: "ille [motus artis] non est a natura, sed a principio extrinseco et est cum violentia aliqua, nisi quando artifex est naturae minister, sicut est medicus et alchimicus aliquando." See also De mineralibus, L. II, tr. iii, c. 1 and c. 5, ed. Jammy, II, pp. 238, 241-242.

7. *Metaphysica cit.*, L. XI, tr. ii, c. 26; p. 516/85-88: "sapientes astronomi per haec principia quae sunt loca stellarum pronosticantur de effectibus, qui luminibus stellarum inducuntur rebus inferioribus."
8. *De causis proprietatum elementorum*, L. I, tr. ii, c. 9, in *Opera omnia cit.*, V/2, p. 78/35-37: "dicit Philosophus quod astronomia est altera pars physicae et Ptolomaeus dicit quod iudex, elector et observator astrorum errabit, si non sit physicus".
9. *Metaphysica cit.*, L. XI, 2, c. 22; p. 510/41-48: "Numerum autem et pluralitatem ferentiarum sive circulationum ex quibus cognoscitur numerus motorum, oportet cognoscere ex illa philosophia de numero mathematicarum scientiarum, quae maxime est propria talium motuum investigationi: haec autem est astrologia, quae tales motus ex tribus investigat, ex visu videlicet et ratione et instrumentis"; *ibid.*, p. 511/3-4, where he lists "instrumenta diversa, de quibus hic agere non est conveniens, sicut est armilla, et instrumentum aequinoctii, et instrumentum latitudinis stellarum, et astrolabium". J. M. Schneider, 'Aus Astronomie und Geologie des hl. Alberts des Grossen', *Divus Thomas* (Freiburg), X, 1932, pp. 52-54, has written on Albert's specifically astronomical knowledge, and on his use of sources such as the Almagestum and Abû Ma'shar; *cf.* B. Barker Price, 'The Physical Astronomy and Astrology of Albertus Magnus', in *Albertus Magnus and the Sciences cit.*, pp. 155-186.
10. J. A. Weisheipl, 'Albertus Magnus and the Oxford Platonists', *Proceedings of the American Catholic Association*, XXXII, 1958, pp. 124-139: *Id.*, *The Problemata determinata cit.*, p. 129, n. 15 ff. See also D. C. Lindberg, 'On the Applicability of Mathematics to Nature: R. Bacon and his Predecessors', *British Journal for History of Science*, 1982, pp. 3 25, especially p. 25: "Judged by 13th-century standards, rather than perceived as progress toward modern science, it is Albert the Great who was the innovator. It is he who transformed the debate, by abandoning the 'Platonism' of Grosseteste and the 12th-century and reviving a relatively pure version of the Aristotelian doctrine. Bacon, as viewed by the new Aristotelians of the second half of the 13th-century, must have seemed quite old fashioned".
11. Pangerl, *Studien cit.*, pp. 339-341, and the passages there cited.
12. *Metaphysica cit.*, L. XI, tr. 2, c. 22; pp. 511/6-14, 31-33: "naturalibus et doctrinalibus iam quantum licuit elucidatis"; "secundum quod divinum quoddam existit in nobis, sicut scientia naturalis perficit eundem [intellectum], prout est cum tempore et quemadmodum perfectus est a doctrinalibus in quantum ad continuum"; "illa namque est de substantia sensibili incorruptibili sicut de caelo et motibus eius: de his enim facit theoricam speculationem. Aliae vero mathematicae de nulla sunt substantia, sicut arithmetica, quae est circa numeros, et geometria, quae est de quantitate immobili, et musica, quae est circa numeros harmonicos in cantu modulato. Quia igitur *astrologia sola considerat motus orbium caelestium, sumamus ex ipsa quaecumque in ea sapientius dicta esse videntur*... non investigantes veritatem de his, secundum nostram sententiam, sed *recitando ea quae dicunt quidam probabiliores mathematicorum.*" (Italics mine) *Cf. Speculum*, XII/6-8.
13. *De fato*, ed. P.Simon, in *Opera omnia*, XVII, 1, Münster 1975, p.68/42-56, "Fluit enim a multis stellis et sitibus et spatiis et imaginibus et radiationibus et coniunctionibus et praeventionibus et multiplicibus angulis, qui describuntur ex intersecationibus radiorum caelestium corporum, et productione radiorum super centrum, in quo solo, sicut dicit Ptolomaeus, omnes virtutes eorum quae sunt in caelesti circulo, congregantur et adunantur. Haec autem talis forma media est inter necessarium et possibile; necessarium enim

est, quicquid est in motu caelestis circuli, possibile autem et mutabile, quicquid est in materia generabilium et corruptibilium. Forma autem ista causata ex caelesti circulo et inhaerens generabilibus et corruptibilibus, media est inter utrumque."

14. *Ibid.*, p. 68, n.50 where is quoted *Centiloquium*, *verbum* 1: "Per materiam habemus de re cognitionem dubiam, per formam vero certam, et haec iudicia quae trado tibi sunt media inter necessarium et possibile".
15. *Ibid.*, p.69/30-43: "licet sit ex necessario tamen est mutabilis et contingens. Cuius causam optime assignat Ptolomaeus in *Quadripartito*, dicens quod virtutes stellarum per aliud et per accidens fiunt in inferioribus: 'per aliud quidem quia per sphaeram activorum et passivorum, per quorum qualitates activas et passivas inhaerent inferioribus; per accidens autem, quia cum haec forma effluat a causa necessaria et immutabili, accidit ei habere esse in rebus contingentibus et mutabilibus. Ex duobus ergo habet mutabilitatem, scilicet ex qualitatibus elementorum, per quos defertur ad generata, et ex esse generatorum, in quo est sicut in subiecto. Hoc igitur est fatum."
16. *De generatione et corruptione*, L. II, Tr. III, c. 5: "Et est digressio diclarans ea quae dicta sunt de periodo"; ed. Jammy, II, p. 67b: "Potest tamen hoc impediri per accidens, per cibum malum vel mortem violentam vel alio quocumque modo: et hoc vocat Aristoteles *materiae inaequalitatem*: quia scilicet per accidentia multa aliter disponitur quam moveantur [sic] a circulo, et ideo diversimode moriuntur homines citius et tardius quam per naturam sint, et similiter etiam alia animalia." On Albert's use of the idea of materiae inaequalitas see passages from his *De divinis nominibus*, c. 4 nr. 103, already quoted by J. Goergen, *Des hl. Albertus Magnus Lehre von der göttlichen Vorsehung und dem Fatum*, Vechta i. Oldenburg 1932, p. 126. This book compares Albert's theories on fate and foreknowledge with the *Summa halensis*, (see also his 'Untersuchungen und Erläuterungen zu den *Quaestiones de fato, de divinatione, de sortibus* des Mag. Alexander', *Franziskanische Studien*, 19, 1932, pp. 13-38) and gives the richest treatment to date of Albert's astrological theory. Unfortunatelly it has been seldom read and considered. *Cf.* the recent, but shorter article by B. Barker Price, 'The Physical Astronomy and Astrology of Albertus Magnus', *cit.*
17. *De generatione et corruptione cit.*, p. 67a: "Et hoc etiam modo innotescit, quoniam qui sciret vires signorum et stellarum in ipsis positarum in circulo periodali dum nascitur res aliqua, ipse quantum est de influentia caelesti praenosticari posset de tota vita rei generatae: sed tamen hoc necessitatem non poneret, quia posset impediri per accidens, ut dictum est." (Cf. *Summa theologiae*, P. I, tr. XVII, q. 68: "De fato"; ed. Jammy, XVII, p. 380b).
18. *De intellectu et intelligibili*, tr. I, c. 4; ed. Jammy, V, p. 241a: "si enim cognitivum mortalium flueret et regeretur ab intelligentia ultimi orbis vel alicuius aliorum orbium vel omnium, tunc ipsa in suis motibus et operibus cognitionis et affectus necessario subderetur motibus astrorum, eo quod omne effluens ab aliquo continetur et restringitur ab illo in potentiis operationum".
19. *De quatuor coaequevis*, tr. III, q. xviii, a. 1; ed. Jammy, XIX p. 76a: "Fatum autem quod imponat necessitatem liberi arbitrii etiam contra philosophiam est ponere, nisi per hunc modum quo dicimus animam hominis inclinari et mutari".
20. *De intellectu et intelligibili cit.*, p.67a: "Quod autem anima praecipe sub motibus astrorum restringitur contra omnes est peripateticos et contra Ptolemaeum. Ipsa enim et superiora in sphaeris apprehendit et ab his ad quae motus astrorum inclinat, libere avertitur et alia

advertit per sapientiam et intellectum, sicut testatur Ptolemaeus."

21. *Summa theologiae*, P. I, tr. XVIII, q. 68; ed. Jammy, XVIII, p. 381a: "Talis enim stellarum qualitas trahere potest corpora et mutare animos etiam plantarum et brutorum, sed animam et voluntatem hominis, quae ad imaginem Dei in libertate sui constituta est, domina est suorum actuum et suarum electionum nec mutare nec trahere potest coactiva coactione, licet forte eatenus qua anima inclinatur ad corpus secundum potentias quae affiguntur organis (sicut sunt potentiae animae sensibilis et animae vegetabilis) *anima humana inclinative, non coactive* a tali qualitate trahi possit." (Italics mine).
22. *Ibid.*: "Et hoc est quod dicit Augustinus quinto libro *de Civitate Dei*, capite 6 sic: 'Cum igitur non usquequaque absurde dici possit ad solas corporum differentias afflatus quosdam valere sidereos, sicut in solaribus accessibus et recessibus videmus etiam ipsius anni tempora variari et lunaribus incrementis atque decrementis augeri et minui quaedam genera rerum: sicut echinos et conchas et mirabiles aestus Oceani: non autem animi voluntates passionibus sideribus subdi. Hoc igitur modo dicere fatum haereticum est. Primo autem modo dicere fatum non est haereticum".
23. *De quatuor coaequaevis cit.*, tr. III, q. 18, a. 1; ed. Jammy XIX, p. 75a: "astra habent virtutem in transmutatione elementorum et in mutatione complexionum et in motibus hominum et insuper etiam in habitibus inclinantibus ad opera et etiam in eventibus praeliorum." Immediately after Albert quotes "philosophos praenosticantes [sic!] in astris sicut Ptolemaeus docet in Tetrascum, et Albumasar, et Alcabitius [ed.: Acabir], et Messahallach, et Aristoteles et Gergis [ed.: Gorgis] et multi alii".
24. *Ibid.*: "cum ergo opera et praelia sint de his quae subsunt libero arbitrio, [astra] videntur habere potestatem super liberum arbitrium".
25. *Ibid.*, p. 75b: "Nostrorum actuum sydera nequaquam sunt causa, nos enim liberi arbitrii a Conditore facti domini nostrorum actum existimus."
26. *Super Ethica*, ed. W. Kübel, in *Opera omnia cit.*, XIV/1, Münster 1968-1972, p. 174/65-66: "quod etiam operationes nostrae sint necessariae, sicut innati sumus ex superioribus motibus."
27. *Ibid.*, p. 174/75-80: "homo habet principium suae generationis in motibus caelestibus, sicut dicitur in XVIII De *animalibus* [777 b 16], quod omnes diversitates, quae contingunt in conceptis, reducuntur ad principia quaedam caelestia: ergo et operationes humanae habent principium in illis."
28. *Ibid.*, p. 174/81-87: "Praeterea, ex quocumque habentur principia cognoscendi opera alicuius, ex illo etiam sunt principia illorum operum, quia eadem sunt principia essendi et cognoscendi; sed in astrologia docetur quomodo operationes humanae praecognoscantur ex motibus caelestibus; ergo habent principia effectiva in illis."
29. *Ibid.*, p. 175/7-8: "Si dicatur quod motus corporales sequuntur motus superiores, sed non operationes animae".
30. *Ibid.*, p. 175/11-12: "sicut anima imprimit in corpus, ita intelligentia in animam".
31. *Ibid.*, p. 175/47-49: "Ex hoc habetur, quod etiam in corporalibus non est necessitas ex principiis caelestibus".
32. *Ibid.*, p. 175/50-73: "Ptolemaeus dicit in scientia *De iudiciis* capitulo III: 'Non est existimandum, quod superiora super sua significata procedant inevitabiliter, velut ea quae divina dispositione contingunt...' Item quod terrestrium variatio naturali calle procedit, quae variatur, et primas rerum superiorum actiones *accidentaliter suscipiunt* et quod quaedam accidentium generali nocumento non ex alia rei proprietate contingunt hominibus,

velut in grandi commutatione aeris, a qua vix possumus cavere nobis, multorum eveniunt hominum exitus'... Ex hoc iterum videtur, quod *superiora non inducant necessitatem inferioribus* et quia multa eveniunt in hominibus ex propria complexione vel aliis causis quam de effectibus stellarum." (Italics mine).

33. *Ibid.*, p. 176/30-38: "fata sive inflexiones primarum causarum non trahunt necessitatem inducendo voluntati, sed tantum inclinando et quasi disponendo ad aliquid... huiusmodi dispositiones non possunt ex toto tolli a natura hominis, quin sit semper pronus ad iram; possunt tamen in anima fieri habitus contrarii talibus dispositionibus, si velit niti in contrarium."
34. *Ibid.*, p. 176/15-21: "aut secundum quod est actus corporis, quantum ad vires affixas corpori, et sic *per accidens* imprimitur in illam ex motibus caeli, inquantum sequitur passiones corporis; aut secundum quod nullius corporis est actus quantum ad potentias rationalis animae, et sic nulla impressio fit in ipsam ex motibus caeli."
35. *Ibid.*, p. 176/21-29: "Voluntas autem, quae est principium nostrorum operum, quibus sumus boni vel mali, est potentia animae rationalis, et sic patet quod non necessario sumus boni vel mali ex dispositione nativitatis secundum effectus stellarum, sed quod tantum relinquuntur ex eis dispositiones in natura corporis, quibus est *habilitas ad iram vel concupiscentiam*, sed *anima non necessario sequitur has*." (Italics mine).
36. *De mineralibus cit.*, l. II, tr. iii, c. 3; ed. Jammy, II, p. 240a: "Et enim in homine duplex principium operum, natura scilicet et voluntas: et natura quidem regitur sideribus, voluntas quidem libera est: sed nisi renitatur, trahitur a natura et induratur, et cum natura moveatur motibus siderum, incipit voluntas tunc ad motus siderum et figuras inclinari. Probat hoc Plato ex operibus puerorum qui libertate voluntatis non adhuc renituntur naturae et siderum inclinationi."
37. *De quatuor coaequevis cit.*, tr. III, q. 18, a. 1; ed. Jammy, XIX, p. 75b: "Astra habent vim et rationem signi super ea quae sunt in materia transmutabili et etiam super illa quae sunt obligata ei".
38. *Ibid.*: "Et dico illa obligata materiae, quae de necessitate sequuntur transmutationes materiae, sicut est anima vegetabilis et sensibilis".
39. *Ibid.*: "Quaedam sunt quae dependentiam habent ad materiam et obligationem secundum quid et non simpliciter, sicut est animus hominis. Unde dicimus sanguinem accensum circa cor inclinare ad iram animum hominis, et tamen non de necessitate irascitur; et secundum quod animus inclinatur ad materiam et complexionem, sic etiam habet vim constellatio secundum quid et non simpliciter. Aliter enim periret casus secundum liberum arbitrium et consilium, si nihil esset contingens ad utrumlibet dici de futuro, sicut optime disputat Philosophus."
40. See *Summa theologiae cit.*, P. I, tr. xvii q. 68, m. 1; ed. Jammy, XVII, p. 381b: "dicendum quod fatalia divinationes non sunt nec auguria, sed fatalia quaedam prognosticationes sumptae sunt a signis probabilibis non necessariis; sicut Hippocrates docet in libro *De prognosticis*. Sic enim ordo in stellis est in gradu primo et dispositio sive qualitas adhaerens generatis est qualitas influxa a stellis et est in gradu secundo. Et ideo a Ptolomaeo in *Centilogio* tales dispositiones vocantur stellae secundae"; see *ibid.*, p. 389a "tales dispositiones Ptolemaeus vocat stellas secundas", as well as the earlier text of the *De fato cit.*, a. 1, p. 66/50-56, and the even earlier one (1245-48) of the *Physica cit.*, l.II, tr. ii, c. 21; p. 129/70-71: "augures in illis [imaginibus animae] tanquam in secundis stellis, ut dicit Ptolomaeus, ponunt considerationes".

41. *De quatuor coaequevis cit.*, tr.III, q.18, a.1; p.75b: "Et quando dicitur quod stellae habent vim in inferioribus, intelligitur quod habent vim sicut causae primae universales moventes causas immediatas et propter quid: et ideo non semper sequitur de necessitate effectus ad constellationem".
42. *Ibid.*: "Signa autem sunt magna imbrium et aeris transmutationis. Fortassis utique quis dixerit, quoniam et praeliorum non sunt causa, sed signa; sed et qualitas aeris a Sole et Luna et astris alio et alio modo facta diversas complexiones et habitus et dispositiones constituit. Intendit enim Damascenus quod signum minus dicit quam causa: causa enim, ut dicit Boetius in *Topicis*, est quam de necessitate sequitur effectus. Signum autem est causa remota et non de necessitate causans sine coniunctione aliarum causarum".
43. *Ibid.*: "motus animae...non reducitur at motum caeli, sed tantum motus corporales, in quibus etiam non est necessitas ex motibus superioribus, quin possint aliter evenire, quia res non recipiunt necessitatem a primis causis, sed a causis proximis, alioquin omnia essent necessaria, cum necessaria sit causa eorum quae sunt frequenter et eorum quae sunt raro. Causae autem proximae sunt variabiles et ideo motus. Unde non est ex eorum virtute, quod superiora non consequuntur necessitatem, sed ex defectu mutationis, quae est in eis."
44. *Super Ethica cit.*, L. III, 1. vii; p. 176/56-64: "Astrologi prognosticantur futuros eventus ex causis primis, quae non inducunt necessitatem, et ideo non est iudicium ipsorum necessarium, sed *coniecturale*, sicut Ptolomaeus dicit, propter quod variantur in causis secundis; et ideo dicit etiam quod certior esset prognosticatio si possemus scire virtutes superiorum causarum secundum quod sunt incorporatae causis secundis, et hoc vocat stellas secundas."
45. *Ibid.*, p. 176/77-80: "quia in motibus caeli sunt principia omnia contrariarum dispositionum, omnes habent principia in motibus caelestibus inferiores dispositiones corporum, tamen nulla necessario contingit."
46. *De fato cit.*, p. 69/65-69: "istas qualitates per contrarietatem inventam in materia et diversitatem dispositionum materiae saepe excluduntur effectus motus caelestis. Propter quod Ptolemaeus dicit quod sapiens homo dominatur astris."; see also *De natura locorum* [written before 1259], p. 277b: "sicut dicit Ptolemaeus in *Quadripartito* effectus stellarum impediri possunt per sapientiam peritorum virorum in astris"; lastly, see *Summa theologiae cit.*, P. I, tr. xvii, q. 68, m.1; ed.Jammy, XVII, p. 381a: "nulla coactiva inhaeret rebus qualitas, quia etiam corpora non cogit, ut dicit Ptolemaeus in *Centilogio*. Dicit enim quod sapiens homo dominatur astris."
47. *Speculum*, XIII/54-59: "Quod si propter hoc condemnetur ista scientia... certe eadem ratione non stabit magisterium medicinae... Quod si magisterium medicinae destruatur, multum erit utilitati reipublicae derogatum", and in general see chapter XV.
48. *De fato cit.*, pp. 69/69-70/4: "ubi dicit Commentator [Haly] quod si effectus circuli caelestis minorando humores corpora disponit ad quartanam, sapiens medicus hoc praevidens per calida et humida corpora disponit ad sanguinem et tunc excluso effectu caelestis quartanam non inducitur".
49. *De causis proprietatum elementorum* cit., L.I, tr. 2, ch. 4; p. 67/55-61: "numerantur propter hoc cretici dies secundum Lunam, et vocatur regina caeli, quia regit humiditates corporum inferiorum [...] animalium membra, praecipue oculi in quorum compositione abundat aquae natura, recipiunt maximas alterationes et augmenta et diminutiones secundum Lunam."

50. *Summa theologiae* cit., P. I, tr. XVII, q. 68, m. 4; ed. Jammy, XVII, p. 386ff.; especially p. 390b, where in the conclusions, Albert conceded that the argument of the twins was not sufficient to refute the thesis of the physical conditioning at the time of conception or birth: "Primum ergo dictum de geminis sic calumniantur, quod licet in uno utero concipiantur, et ex uno concubitu, tamen non concipiuntur nisi ex diviso semine ad diversa loca matricis, et sic diversa sunt centra sive corda conceptorum. Diversitas autem centri totum variat circulum nativitatis."
51. *De quatuor coaequaevis cit.*, tr. III, q. 15, a. 3; p. 67b; "Dicuntur enim quaedam stellae calidae et siccae sicut Sol; et quaedam frigidae et siccae, sicut Saturnus; quaedam frigidae et humidae, sicut Luna; quaedam calidae et humidae, sicut Iupiter. Et videtur hoc multipliceter inconveniens. Primo: quia tales complexiones qualitatum activarum et passivarum non sunt nisi in materia generabili et corruptibili... Ad hoc sunt tres responsiones scriptae in libro de substantia orbis: prima, quod istae qualitates aequivoce sunt in stellis et elementis... sed ista solutio non videtur stare... Alia solutio est quod stellae habent istas qualitates et tamen non afficiuntur eis". See also *Metaphysica cit.*, L. XI, tr. ii, c. 25; p. 515/44ff: "omne autem quod est determinatum mixtione et complexione primarum qualitatum, est attributum septem sphaeris planetarum." In these texts Albert expressed awareness of the difficulty to integrate the astrological concepts he was discussing with the Aristotelian idea of the "fifth element", and yet he did feel he had solved this contradiction, or decided not to criticize astrology; in the *Speculum* (V/16-17: "naturae planetarum in semetipisi, secundum quas dicuntur calidi, frigidi, humidi, sicci") this classification is taken as a matter of fact.
52. See above ch. 7 n.8 and 9; and *cf. Speculum*, II/74 ff., IV/31-33, XIII/106-109.
53. Bacon, *Opus maius*, ed. Bridges *cit.*, p. 254ff. ("Et per hanc viam mathematicae non solum certificamur de professione nostra, sed praemuninur contra sectam Antichristi"); Id., *Un fragment inédit de l'"Opus tertium"*, éd. P. Duhem, Quaracchi 1909, p. 157: "Et hac scientia mirabili utetur Antichristus et longe potentius quam Aristoteles, et ideo dividet mundum gratuito, ut dicit Scriptura. Nam omnem regionem et civitatem infortunabit et reddet imbellem et capiet omnes sicut aves inviscatas". *Cf.* D. Bigalli, *I tartari e l'Apocalissi cit.* above ch. 4 n.34, in particular its chap. XII; and E. R. Daniel, 'Roger Bacon and the *De seminibus Scripturarum*', *Mediaeval Studies*, XXXIV, 1972, pp. 462-467. See also the *De novissimorum temporum periculis*, ed. in Guilelmus de Sancto Amore, *Opera*, Constantiae 1632, pp. 17-72, but now attributed to Nicholas of Lisieux, one of his followers (see M. M. Dufeil, *Guillaume de Sainct-Amour et la Polémique universitaire parisienne 1250-59*, Paris 1972).
54. *Summa theologiae cit.*, P. II, tr. viii, q. 30; ed. Jammy, XVIII, p. 177b: "Tempore Antichristi maior erit potestas daemonum quam nunc;... sicut dicitur secunda ad Thessalonicenses, 2, adventus Antichristi erit secundum operationem Satanae in omni virtute et prodigiis et signis mendacibus".

CHAPTER EIGHT

1. Albertus Magnus, *Problemata determinata*, ed. J. A. Weisheipl (and P. Simon) in *Opera omnia* VII/1, Münster 1975, p. 54: "Decima septima quaestio de phantasia fatuitatis procedit" and *cf. ibid. passim. Cf.* J. A. Weisheipl, 'The Celestial Movers in Mediaeval

Physics', *The Thomist*, XXIV, 1961, p. 287: "To the casual reader these questions, too, might appear to be useless in this age of scientific progress. Angels, it is frequently thought, have no place in a discussion of scientific questions"; [see this article also printed in *The Dignity of Science. Studies in the Philosophy of Science Presented to W. H. Kane*, ed. by J. A. Weisheipl, Washington 1961, pp. 150-190].

2. H. C. Wolfson, 'Immovable Movers in Aristotle and Averroes', *Harvard Studies in Classical Philology*, LXIII, 1958, pp. 249-251 (especially p. 243). Wolfson, p. 244, observes that the idea of a relation that is causally emanative was brought in only by Avicenna, who, together with Averroes, introduced the name of intelligences for Aristotle's motors. Unfortunately, "the problem of the souls of the spheres is not dealt with in this paper"(p. 251 n.1).
3. P. Mandonnet – reviewing J. Destrez, 'La lettre de St. Thomas d'Aquin dite lettre au Lecteur de Venise' and M.-D. Chenu, 'Les réponses de St. Thomas et de Kilwardby à la consultation de Jean de Verceil' (both in *Mélanges Mandonnet*, Paris, 1930) in *Bulletin Thomiste*, VII, 1930, p. 135 – was correcting, rightly in my opinion, Chenu's observation that the list of 43 questions represented "les résidus d'une dispute quodlibétique (*Mélanges cit.*, p. 211). However, Chenu must be given credit for linking the *Responsio de 43 articulis* with the two redactions of Aquinas's Letter to the Venetian Reader (*ibid.*, pp. 211, 191) and with some of the propositions condemned by Tempier in 1277 (p. 214): "Sans doute il n'y a pas trace, dans notre questionnaire, de déterminisme astral; ni de nécessité des intermédiaires cosmiques pour l'agir divin; et cela suffit à séparer complètement son cas des tendances suspectes et des erreurs dénoncés par le document épiscopal. Mais ce sont le même préoccupations cosmologiques qui apparaissent".
4. Chenu, 'Les réponses' *cit.*, p. 211: "le plus gros bloc des questions a manifestement trait à l'action des corps célestes sur les phénomènes terrestres, en particulier à l'influence des anges dans cette action des corps célestes"
5. J. A. Weisheipl also remarks on the rapid and wide circulation of Thomas's *Responsio* to the 43 problems or articles ('The Celestial Movers' *cit.*, p. 286), which was inserted into a rather numerous series of *Responsiones* (see *Opera Omnia*, XVII, ed. H.-F. Dondaine, Roma 1979, pp. 300ff.) In the Introduction to his own critical edition of the *Responsio de 43 articulis* and of Thomas's *Responsione* to Bassiano of Lodi (*Opera omnia*, XLII = *Opuscula* III, Roma 1979, p. 265), H.-F. Dondaine notes an analogous question-and-answer exchange between the General and Thomas Aquinas in the *De forma absolutionis* (*ibid.*, XL, pp. C5-C6). I have already noted (ch. 2, n. 1 and *passim*) the wide circulation and the importance of the literary genre of "consultationes".
Before the critical edition by H.-F. Dondaine ('Robert Kilwardby, *De 43 questionibus*', *Archivum Fratrum Praedicatorum*, XLVII, 1977, pp.5-50, the Father Provincial's document had already been discovered and most of it published in 1930 by Chenu (see n. 2 above). Albert's text was only identified in 1960 by the librarian N. R. Ker, who, before publishing it in his *Mediaeval Manuscripts in British Libraries, I: London*, Oxford, 1969, vol. 1, p.249, communicated it to D. A. Callus, 'Une oeuvre récemment découverte de St. Albert le Grand', *Revue des sciences philosophiques et théologiques*, 44, 1960, pp. 243-261, see especially J. A. Weisheipl, "The *Problemata determinata XLIII* ascribed to Albertus Magnus (1271)', *Mediaeval Studies*, XXII, 1960, pp. 303-354.
6. Chenu, "Les réponses" *cit.*, pp. 212-213).

7. *Ibid.*, p. 222. The italics are mine and put there to show that even Chenu had come to terms with the atemporal nature attributed to Aquinas's thought.
8. T. Litt's monograph, *Les corps célestes dans l'univers de saint Thomas d'Aquin*, Louvain-Paris 1963, has prompted lively discussion. Among various other articles and reviews, see B. Montagnes, 'Bulletin de philosophie: Anthropologie', *Revue des sciences philosophiques et théologiques*, XLVII, 1963, pp. 703-704; XLIX, 1965, p. 116. He praises Litt's "rigorously historical study", which allows us to read "les textes de St. Thomas sans commettre d'anachronisme". However, he observes that "s'il y a un domaine où Saint Thomas suit Aristote sans discussion, ni hésitation, c'est celui de la representation physique du cosmos, dont la pièce maîtresse est constituée par la théorie des corps célestes ... Or cette hiérarchie physique est-elle purement et simplement confondue par Saint Thomas avec les degrés d'être? ... même pour Saint Thomas la superposition physique des corps célestes incorruptibles aux corps sublunaires corruptibles n'est pas identique à la hiérarchie métaphysique de substances materielles et des substances séparées, entre lesquelles l'homme tient une place originale". Cfr. J.L. Russell, 'St. Thomas and the heavenly bodies', *Heytrop Journal*, VIII, 1967, pp. 27-39, whose point of departure is the (super-rogatory) statement that such a theory "is now completely outdated and has disappeared from scholastic philosophy" (p. 27) and that "Thomas's theory may seem strange and implausible to the modern reader" (p. 33). But he concludes: "An understanding of mediaeval ideas on celestial causality will throw light on several problems in the history of philosophy. It explains, for instance, why natural science never developed to any great extent during the Middle Ages ... when in the seventeenth century, the mediaeval theory of celestial causality collapsed, philosophers found themselves with no theory of physical causality wich could stand up to scientific scrutiny and, still worse, with an inherited climate of opinion which took it for granted that physical substances cannot act for themselves.... The vacuum created by the failure of mediaeval cosmology remained unfilled".
9. F. Van Steenberghen, 'Deux monographies sur la synthèse philosophique de Saint Thomas', *Revue philosophique de Louvain*, LXI, 1963, pp. 90-91. Although he praises the Cistercian Litt's historical research and repeats his fundamental query "à quel point St. Thomas était étranger au souci caractéristique de l'esprit scientifique moderne", which prompted "des salutaires réflexions sur les conditions de succès de la renaissance thomiste", Van Steenberghen concludes that it is still "possible et légitime de reprendre les thèses fondamentales de l'ontologie du maître: mais sa métaphysique spéciale du monde corporel ou sa cosmologie doit être repensée de fond en comble, en tenant compte de ce que la science contemporaine nous apprend sur la nature et les propriétés des corps".
10. Mandonnet, review-article *cit.*, p. 137.
11. M.-D. Chenu, 'Aux origines de la "science moderne", *Revue des sciences philosophiques et théologiques*, XXIX, 1940, p. 209. The Duhem reference is to "Les précurseurs parisiens de Galilée", in *Etudes sur Léonard de Vinci*, Paris 1913, pp. 34-53.
12. Mandonnet, review-article *cit.*, p. 138.
13. M.-D. Chenu, 'Aux origines' *cit.*, pp. 212-213.
14. M.-D. Chenu, 'Aux origines' *cit.*, pp. 208-209. *Cf.* Duhem, *Études sur Léonard cit.*, pp. 34-53, and especiallly pp. vi-xi, cited by Chenu in order to exemplify Duhem's theses "en passe de devenir classiques": "Buridan proposes a formula of the law of projectile

movement which is so precise that we can recognize the role which Galileo will attribute to *impetus* or *momentum*, Descartes to *quantity of movement*, and Leibniz to *live force*".

15. Chenu, 'Les réponses' *cit.*, pp. 221-222.
16. Chenu, 'Les réponses' *cit.*, pp. 218-219.
17. Chenu, 'Les réponses' *cit.*, pp. 219. Chenu continues on Kilwardby (pp. 219-221), who supports an opinion which is "rationabilis nec philosophica carens ratione", according to which celestial bodies, heavy or light like the ones compounded by elements, are moved by their nature and by their natural inclinations ("Rationabiliter ponitur quod non moventur illa corpora a spiritibus, sed instinctu propriorum ponderum", q. 3; "Unicuique enim stellae vel orbi indidit Deus inclinationem quasi proprii ponderis ad motum quem peragit. Ut ex multorum motuum correlata proportione una fiat sufficiens conservatio generis et generabilium usque ad tempus perinfinitum", q. 2). Only in exceptional and "reserved" cases do angels remain "rectores et gubernatores ..., quorum ministerio, nutu Dei, stetit sol contra Gabon et retrocessit sol in horologio Achaz" (q. 2). Thomas knew this point of view and rejected it with equal energy in his q. 5: "Quod autem corpora celestia a sola natura sua moveantur sicut corpora gravia et levia est omnino impossibile; unde nisi moveantur a Deo immediate, consequens est quod vel sint animata celestia corpora et moveantur a propriis animabus, vel quod moveantur ab angelis, quod melius dicitur". Chenu's comment recalls the importance given by Thomas to secondary causes. "As a good Aristotelian, he grants natures, the internal principles of movement, their full and autonomous efficiency". This is indeed one of the fundamental aspects of his philosophy of the natural world which he opposed to the Augustinianism and to the hylomorphism of Avicebron's *Fons vitae*. However, "Thomas's 'naturalism' is surrounded by a metaphysics of being and causality which incapsulates it without breaking its design or detracting from its efficiency". From the *De veritate*, q. 5, a. 8 and the *Summa theologiae*, I Pars, q. 110, a. 1, Chenu adduces that in the *Responsio de XLIII articulis* and parallel texts, "we find ourselves on a metaphysical level of generally Platonic inspiration, although it has been radically modified by the substitution of a theory of causality for the theory of participation. Kilwardby, on the other hand, stays on the physical level, on the level of *motus*; one might almost say on the level of experience. The angel has nothing left to do. This level is only inhabited by the intrinsic nature of the heavenly bodies – as it was then conceived – and it is to this alone that he recurs in order to explain the celestial movements".
18. On Grosseteste see R. C. Dales, 'Mediaeval De-animation of the Heavens', *Journal of the History of Ideas*, XLI, 1980, pp. 540-542, where starting from Weisheipl's studies the author briefly reconstructs the question's antecedents, observing that "heavens were the last parts of the cosmos to lose their souls", p. 531. On Giles of Lessines, see Weisheipl, '*The Problemata*' *cit.*, p. 308 and 'The Celestial Movers' *cit.*, p. 307, who cites the *De unitate formae*, P. II, c. 5, ed M. De Wulff, Louvain 1902 (= Les Philosophes Belges, I), p. 38 which, in differentiating between angels and intelligences, asserts: "Haec est positio multorum magnorum et praecise domini Alberti quondam Ratiponensis episcopi".
19. Dietrich von Freiberg, *Tractatus de animatione caeli*, hg. v. L. Sturlese, in *Opera omnia. III: Schriften zur Naturphilosophie und Metaphysik*, mit einer Einleitung von K. Flasch, Hamburg, F. Meiner Verlag, 1983, pp. 11, 46; L. Sturlese, 'Il *De animatione caeli* di Teodorico di Freiberg', in *Xenia Medii Aevi Historiam illustrantia oblata T. Kaeppeli O.P. cit.*, pp. 175-247. See too Weisheipl, 'The Celestial Movers'*cit.*, pp. 307-308, n. 61, who

quotes from the *De intellectu et intelligibili*: "tenendum quod dicti philosophi, loquentes de intelligentiis, non loquebantur de angelis, de quibus scriptura sacra loquitur, quae loquitur mysteria abscondita a sapientibus et prudentibus et revelat ea parvulis".

20. See S. Donati, 'La dottrina di Egidio Romano sulla materia dei corpi celesti', *Medioevo*, XII, 1986, pp. 229-280, who says that the supporters of the negative theory were Averroes and, after him, Siger, Geoffroy de Fontaines, Petrus de Alvernia, and the Dominican Durandus de Sancto Portiano. Thomas Aquinas was the first person responsible for the thesis of the hylemophic composition of celestial bodies, which was also accepted by Hervé Nédellec, Jacques de Metz and Henri de Gand. Giles of Rome, finally, maintained the thesis of the identify of "matter" in celestial and corruptible bodies, which was then accepted by Jacobus of Viterbo, Agostino Trionfo, and Ockham himself (pp. 231-233).
21. This characteristic remark is in 'Aux origines de la "science moderne"' *cit.*, p. 210. Contrary to what Weisheipl wrote ('The Celestial Movers' *cit.*, p. 288: "Mandonnet was quick to point out the modernity of Kilwardby's universal mechanics. This suggestion was developed at some length by F. Chenu"), here Chenu explicitly takes his distance from Mandonnet and says that he must therefore "take full responsibility for his own interpretations" even though he is grateful to Mandonnet for "information and reflections", which have helped him to evaluate the implications of Kilwardby's answer ('Aux origines' *cit.*, p. 207, n. 2). If, as we shall see, the differences between Chenu and Duhem concern the thirteenth- or fourteenth-century, Aristotelian-Averroist or nominalist origins of the *vis motiva* theory, those between Chenu and Mandonnet have some bearing on the timeliness and validity of Thomas's "concordisme" and his synthesis. On p. 208, Chenu observes: "Les théologiens mediévaux ... identifièrent aux intelligences motrices des philosophes les anges de la révélation chrétienne. De ce concordisme et des problèmes qu'il introduit, on peut suivre les détours chez Saint Thomas par example *Quest. de potentia*, qu. 6, art. 6, ou dans le *de substantiis separatis*".
22. *Opera omnia cit.*, XLII, Roma 1979, pp. 354-356, where Mandonnet's dating is used without bringing in any new elements, and no other biographical data is available on Gerard. But, due to the ingenuousness of his questions, there is no temptation to identify him with Gerard of Feltre, even though we also know too little about the latter to exclude that he was a reader at Besançon.
23. *Opera omnia cit.*, XLII, Roma 1979, pp. 355-356. With the exception of the last article which dealt with sin and confession, Thomas criticized all of the others ("Nec tamen huiusmodi sunt extendenda, nec populo praedicanda", p. 355/40-42): "Primus igitur articulus est quod stella quae Magis apparuit figuram habebat Crucis; secundus articulus est quod habebat figuram hominis; tertiam quod habebat figuram Crucifixi... Quartus articulus est quod parvule manus pueri Ihesu nati creaverunt stellas" (p. 355/9-12, 28-29).
24. *Opera omnia cit.*, XLIII, Roma 1976, pp. 207-208 where Dondaine proposes this date and confirms that the dedicatee was a papal chaplain in Orvieto between August and December 1263, while Thomas was there as the Dominican reader in theology. The two Italians' friendship continued, and the episode, in which Jacobus questioned Aquinas on the subject of *sortes*, is a curious one, because Thomas was consulted in order to resolve the long contest between his friend and another candidate for the position of bishop of Vercelli after 1268. The two candidates had an equal number of votes from the local canons, and since there was no pope then (*sede vacante)* the contest was undecided. See

ibid., p. 237, cap. V/ 125 ff., where Thomas does not allow for this type of drawing by lots: "si id quod est per divinam ispirationem faciendum aliquid forte velit sorti committere, sicut ad ecclesiasticas dignitates sunt homines promovendi per concordiam electionis quam Spiritus Sanctus facit". Dondaine points out the originality of this work, which corresponds to q. 95, "De superstitione divinativa" in Thomas' *Summa Theologiae IIa IIae*. A precedent for these texts is in the *Summa halensis*, since in Thomas's commentary on the *Sententiae* there are only the most cursory of references to magic and "divinatio per daemones" and no mention at all of the various types of "sortes", "spatulomantia" (chapt. 3/65), geomantic procedures (chapt. 3/69-173, chapt. 4/44-47), and "aruspicium" (chapt. 3/115-116). According to the *Speculum astronomiae*, all of the latter were divinatory practices "quae non merentur dici scientiae, sed garamantiae" (XVII/9-10 and *passim*). About these practices in the whole Middle Ages cf. D. Harmening, *Superstitio. Überlieferungs- und theoriegeschichtliche Untersuchungen zur kirchlich-theologischen Aberglaubenliteratur des Mittelaters*, Berlin, E.Schmidt, 1979.

25. In his critical edition Dondaine uses Walz's dating of post-1260 (*Opera omnia cit.*, XLIII, Roma 1976, pp. 189-190). Another series of 108 articles on the divine attributes could be added to these writings by Thomas, all of which were promoted by Italian correspondents or dedicatees. They were criticized in the catechism of Peter of Tarentasia, Regent Master at the University of Paris and a Dominican. Thomas was examining questions posed by the Dominican General John of Vercelli. However, like John's other consultation with Aquinas of 1269 (*De forma absolutionis*), their content is not pertinent to this study. The *Responsio de 108 articulis*'s authenticity was recently established by H.-F. Dondaine (*Opera omnia cit.*, XLII, Roma 1979, pp. 264-266) and dated between 1264 and 1268, and it does have one article which is worth citing: "Quod vero XCVIII proponitur: 'Sol est agens improportionatum, effectum communem in inferioribus facit', non est verum si simpliciter accipiatur, sed solum secundum respectum: sicut dicitur quod terra ad primum celum optinet locum puncti et non habet proportionem ad ipsum, scilicet secundum aspectum nostrum".
26. *Opera omnia cit.*, XLII, pp. 163 ff. where Dondaine accepts Mandonnet's dating of Aquinas's second stay in Paris (1269-1272). Mandonnet had pointed out parallel passages ("échos abrégés") in the *Summa Theologiae, IIa IIae*, q. 96 and in *Quodlibet XIII*, art.12. It can be inferred from the date that the "Miles ultamontanus" to whom the piece is dedicated is an Italian. This dedicatee, together with Jacob of Tonengo and Bassiano of Lodi, confirms the impression that Central and Northern Italy were particularly prone to those interests and discussions which lay behind Gerard of Feltre's *Summa de astris* and the *Speculum astronomiae*.
27. Mandonnet, review-article *cit.*, p. 136: "Robert Kilwardby ... fut prié de fournir une qualification des articles. Il le fit, à mon avis, indubitablement après avoir lu le referé de Thomas d'Aquin, et il n'est pas difficile de voir qu'il vise beaucoup plus à repousser les doctrines de cet adversaire qu'à donner son jugement sur le propositions en discussion". It is already well known and confirmed by Dondaine ('Robert Kilwardby' *cit.*, p. 6) that Thomas received the request on 1 April 1271, and threfore it should not be dated in connection with the General Chapter held at Montpellier during the following May. Mandonnet had proposed that dating because Kilwardby participated in the General Chapter as English Provincial. The several week difference in dating is slight, but it forces us to partly modify Mandonnet's thesis on Kilwardby, who "n'avait pas attendu le chapître de

Montpellier pour connaître les positions de son célèbre adversaire [Thomas d'Aquin]; mais il trouva là l'occasion de rompre directement une lance avec lui, puisque les circonstances les plaçaient l'un en face de l'autre" (review-article *cit.*, p. 139).

28. P. Duhem, *Système du monde*, Paris 1958, V, pp. 440-465 and *passim*.
29. J. A. Weisheipl, 'The Celestial Movers', *cit.*, p. 312: "Once Albert has established in his reply to the Master General that angels are not the same as intelligences discovered by philosophers, he can easily dismiss the first five questions as fatuous: the existence of angels, the messengers of God, cannot be proved in philosophy. They have nothing to do with the problems of natural science; and even if God were not the first mover of the Heavens – which He really is – the existence of angels would still not be demonstrated".
30. C. Vansteenkiste, 'Il quinto volume del nuovo Alberto Magno', *Angelicum*, XXXIX, 1962, pp. 205-220. His advice to compare the *Problemata determinata XLIII* with the work being reviewed – none other than the *Metaphysica* – is worthy of note.
31. K. Flasch, "Einleitung" in Dietrich von Freiberg, *Opera omnia cit.*, III.

CHAPTER NINE

1. *Speculum*, XI/4: "Est unus modus abominabils, qui suffumigationem et invocationem exigit" and *cf. passim*; XVI/2-3: "Partem vero quae est de imaginibus astronomicis propter vicinitatem quam habent ad necromanticas, non defendo".
2. D. Pingree ed., *Picatrix.The latin Version*, London, The Warburg Institute, 1986 (= Studies of the Warburg Inst., XXXIX).
3. L. Thorndike, "Traditional Mediaeval tracts concerning engraved astrological Images", *Mélanges Pelzer*, Louvain 1947.
4. *Cf.* above ch. 4 *passim*.
5. O. Lottin, 'Problèmes concernants la *Summa de creaturis* et le *Commentaire des Sentences* de saint Albert le Grand', *Recherches de théologie ancienne et médiévale*, XVII, 1950, pp. 319-328, who summarized his own former researches and those by Mandonnet, Pelster, Grabmann, Chenu, Doucet, etc., and proposed a chronology for the *Summa de creaturis*, placing it earlier than the commentary to the *Sentences* (whose second book is dated 1246), indeed, even earlier than 1243; on the other hand Weisheipl in *Albertus Magnus and the Sciences cit.*, dates the *Summa* 1245-1250. Cf. also the 'Prolegomena' to Albertus Magnus, *De bono*, ed. by H. Kühle, B. Geyer, C. Feckes, W. Kübel, in *Opera omnia*, XXXVIII, Münster 1921, p. XI ff., who consider this book of the *Summa de creaturis* "arctissime secundum tempus...coniunctus cum Scriptis super Sententias". They and Kübel, editor of the other unedited sections of the *Summa de creaturis (De sacramentis, De incarnatione, De resurrectione) Opera omnia*, XXVI, 1958, all agree with Lottin in placing the *De quatuor coaequevis*, earlier than the other sections and of the Commentary *In secundum Sententiarum*.
6. *In Secundum Sententiarum*, L.II, d.XIV, C.a.6, in *Opera omnia*, ed. Jammy, XV, p.147B: "omnes philosophi arabum dixerunt et probaverunt multipliciter, quod caelum movetur ab anima coniuncta sibi: et hoc dicit Aristoteles, et Avicenna, et Averroes, et Algazel, et Alpharabius, et Maurus Albumasar, et Rabbi Moyses, et quod habet motorem triplicem, scilicet causam primam, quae est desideratum primae intelligentiae, quae est plena formis explicabilibus per motum sui orbis: sed quia intelligentia simplex est, ideo non potest

intendere motum particularem in hoc vel in illo situ, et ideo tertius motor coniunctus coelo est anima secundum ipsos, et natura caeli est dispositio ad motum: quia naturaliter circulariter fertur et in compositione non est contrarietas".

7. *De quatuor coaequaevis cit.*, tr. III, q. xvi, a. 2; ed. Jammy, XIX, p. 73b: "Ista omnia diximus secundum philosophos, qui non contradicunt quibusdam Sanctis negantibus caelum animam habere, nisi in nomine solo, qui abhorrent nomen animae et tamen bene concedunt quod intelligentiae quaedam sive angeli movent caelum iussu Dei."
8. *Metaphysica cit.*, L. XI, tr. ii, c. 21; p. 509/87-94: "Hoc autem ideo non est dictum quod aliqua sit influentia superioris super inferiorem nisi per motum corporis animati secundum Peripateticos. Sed quia animatum primum indeterminatas et universalissimas influit et exquiritur formas, et illae determinantur ad materiam plus et plus, secundum quod magis et magis descendunt ad materiam hanc et illam." *Cf. Speculum*, XIII/25 ss.: "Ego autem dico, quod omnis operatio causae agentis supra rem aliquam est secundum proportionem materiae recipientis ipsam operationem, ut unus idemque ignis operatur in luto arefactionem atque liquefactionem in cera".
9. *Metaphysica cit.*, L. XI, tr. ii, c. 21; p. 509/94-510/5: "Si autem est aliqua alia irradiatio superioris super inferiora, sicut dixit Plato et sicut dicunt theologi, illa per rationem investigari non potest, sed oportet quod ad illam investigandam ponantur alia principia ex revelatione spiritus et fide religionis: et de hac non est loquendum in philosophia peripateticorum, quia cum eis ista scientia non communicat in principiis."
10. *Metaphysica cit.*, L. XI, tr. ii, c. 26; p. 516/81-88: "ita intelligentia agens quae movet orbem et stellam vel stellas, luminari invehit formam, et per lumen luminaris traducit eam in materiam, quam movet, et hoc sic tangens materiam educit eam de potentia ad actum. Et huius signum est, quod sapientes astronomi per haec principia quae sunt loca stellarum, pronosticantur de effectibus, qui luminibus stellarum inducuntur rebus inferioribus."
11. *Metaphysica cit.*, L. XI, tr. ii, c. 21; p. 510/21-24: "Deus deorum pater dixit superioribus caelestibus diis, quod ipse esset qui sementem generationis faceret et ipsis traderet sementem illum ad exsequendum."
12. *In Secundum Sententiarum cit.*, II, d. III, A, a.3, in *Opera omnia*, ed. Jammy, XV, p.36B: "An theologi vocant angelos illas substantias separatas quas philosophi vocant intelligentias? ...Videtur quod sic: quia (1) Ita dicit Avicenna, quod intelligentiae sunt quas populus et loquentes in leges angelos vocant. (2) Item, hoc idem dicit Algazel ante finem *Metaphysicae* suae. (3) Item Rabbi Moyses dicit hoc in secunda collectione *Ducis Neutrorum*: ergo videtur quod ipsi hoc intendunt".
13. Albertus Magnus, *Problemata determinata*, ed. J.A. Weisheipl [and P.Simon], in *Opera omnia*, XVII/1, Münster 1975, p. 50/55-62: "Apud nullos enim infallibiliter probatum est angelos esse motores corporum caelestium. Sed quidam arabi et quidam iudaei dicunt quod apud vulgus angeli sunt intelligentiae, nec illi probaverunt hoc esse verum, nec dictum vulgi approbaverunt: quin immo sicuti divina scriptura loquitur de angelis et philosophia de intelligentiis, intelligentiae non sunt angeli".
14. *Ibid.*, p. 50/30 ff.
15. *Ibid.*, p. 50/15-23.
16. *Ibid.*, p. 48/42-54: "Quia si dicitur quod angelus aliquando missus ad Abraham venit, dicit Rabi Moyses, quod fuit propheta vel bonus homo, quia intelligentia nec venit nec recedit, quae omnia de angelis. Et si aliquando hoc non est, dicunt quod est virtus quaedam caeli movens homines instictu naturae ad aliquid [...] Ex hoc ergo patet, quod

de angelis non loquuntur, quemadmodum scriptura loquitur".

17. *De quatuor coaequevis cit.*, p. 71b: "motus processivus est ad omnem partem, scilicet ante, retro, dextrum, sinistrum, sursum, et deorsum: principium autem motus coeli non movet nisi in uno circa medium". *Cf. De caelo cit.*, p. 153/22-25: "totum orbem philosophi assimilaverunt uni animali, in quo principale membrum et locum cordis habens sit hoc, propter quod etiam solis orbis in medio orbium positus est a natura, sicut cor in animali"; however, *cf. ibid.*, p. 28/42-43 where Albert, following Averroes, *De caelo*, bk. 2 text. com. 42, and Averroes' *Destructio* (*Tahafut al-Tahafut* <*The Incoherence of the Incoherence*>, Transl. from the Arabic, Introd. and Notes by S. van den Bergh, London 1954, I, pp. 285, 310: "XIV Discussion: to refute their proof that heaven is an animal moving in a circle in obedience to God") discussed the "sacerdotes Aegypti, qui primum in scholam ad inquirendum de caeli natura intraverunt. Dixerunt enim illi quod caelum animal est, quod nutritur umore oceani... Contra quos Avicenna in *Sufficientia de libro caeli et mundi* procedit ex diffinitione et proprietatibus nutrientis et nutriti et ex diffinitione augmenti"; he was even more severe (pp. 135/55-61) against the thesis – that reached Leonardo and Galilei – of the nutrition of a planet: "Aegyptii autem alias causas assignaverunt ex gravitate et levitate astrorum sumentes causas harum diversitatum, quia videbatur eis, quod astra attracto humore oceani nutrirentur et tunc essent graviora et descenderent, et digesto illo essent leviora et ascenderent. Quorum sermo fabulosus est et contemptibilis..."; *Ibid.*, L. II, tr. 3, c. 9: "Quod stellae non moventur secundum motum processivum"; p. 161/35-46: "Possent enim aliquis dicere, quod quia caeli habent motores, qui sunt intellectus, et habent aliquem actum animae in corpora illa, eo quod movent ea, quod illi intellectus movent ea, sicut corpora animalium moventur ab anima, quae habent vires movendi localiter, quibus viribus respondent organa motus, sicut pedes et alas in animalibus, et in isto motu diceret forte aliquis stellas per se moveri. Sed quod hoc omnino sit incoveniens, manifestum est ex hoc quod nos nullo modo videmus instrumenta motus in stellis, quia nec habent pedes neque alas. Quae enim habent huiusmodi organa, non sunt rotunda omnino". *Cf.* Denifle-Chatelain, *Chartularium cit.*, art.102: "quod anima caeli est intelligentia, et orbes caelestes non sunt instrumenta intelligentiarum, sed organa, sicut auris et oculus sunt organa virtutis sensitivae"; on it see also Hissette, *Enquete sur les 219 articles cit.*, p.139: "La proposition paraît expliciter la thèse avicennienne selon laquelle l'âme du ciel est la forme de son mobile. Guillaume de la Mare, dans ses *Declarationes*, l'attribue à tort à Thomas d'Aquin".

18. *Problemata determinata cit.*, p. 49/61-64: "non est dubium quod corpora caelestia non movent angeli. Angeli enim habent aliam distinctionem ad actus virtutis assistricis et ministrativae, sicut tradunt Dionysius et Gregorius".

19. *Ibid.*, p. 50/1-20: in this context Albert quotes Gregorius Magnus, *Homelia 34 in Evangelia* /PL 76, 1254 ff. 9; *ibid.*, p. 49/81-50/5, he cites extensively the definition of angel from chapter 9 of Ps.Dionysius' *De caelesti hierarchia*: "caelestes animi et divini intellectus theophaniis et theoriis a deo in ipsos descendentibus illuminati et purgati et perfecti per conversionem ad fontem illuminationis primae, qui recipiendo per modum passionis perficiuntur".

20. *Ibid.*, p. 50/15-23. *Cf.* very similar observation by Albert in his almost contemporary commentary to *De causis et processu universitatis*, L. II, tr. ii, c. 35, in *Opera omnia*, ed. Jammy, V, p.612B: "Prima enim potestas intelligentiae est producendi formas ex seipsa. Proxima autem potestas sub illa est formas intellectuales intellectualiter recipere, et ad

quiditates rerum determinare. Perceptio autem illa duobus modis est, scilicet per applicationem quae vocatur *influentia*, et per quandam generationem quae vocatur *exitus de potentia ad actum* et motus quidem vel mutatio. Cuius exemplum est quod dicunt astronomi, quod inferior stella applicatur superiori, et non econverso, sicut Jupiter Saturno et non Saturnus Jovi. Superior enim potestas semper eminet et non restringitur ab inferiori. Inferius autem lumen, vel inferior potestas format et determinat et distinguit superiorem, sicut lumen Saturni se habet ad lumen Jovis, et sicut lumen Jovis se habet ad lumen Martis, et sic de aliis".

21. *Ibid.*, p. 49/59-62: "Quod enim quaeritur, an omnia quae moventur naturaliter, moveantur ministerio angelorum movente corpora caelestia, non est dubium quod corpora caelestia non movent angeli". *Cf.* p. 50/35-38: "Et si sic est, quod certissime probatum est, tunc angeli per ministerium non movent corpora caelestia, et sic ulterius sequitur, quod nec alia inferiora corpora moventur ab ipsis'
22. *Ibid.*, p.48/55 ff. "quidam aliam defendunt posicionem dicentes virtutem intelligentiae alicuius orbis sive caeli influxam inferioribus vocari angelum".
23. *Ibid.*, p. 50/39 ff.: "Si quis autem dicat quod deo imperante movent sphaeras caelestes, ille motus erit motus oboedientiae et non naturalis. Et de hoc nihil secundum philosophiam determinari potest, quia principia philosophiae, quae sunt dignitates per se notae, non sufficiunt ad hoc. Et ideo sic dicens, quia non est tenens principia philosophiae, nihil debet loqui cum philosopho; dicit enim Aristoteles, quod non est sermo geometrae cum non geometra."
24. *Ibid. cit.*, L. XI, tr. ii, c. 10; p. 495/53-56. "quod caelestes quidem circuli habent animas, sed praeter animas sunt intelligentiae separatae operativae, praesidentes eis, et has intelligentias secundum vulgus angelos vocant."
25. Ibid. cit., L. XI, tr. ii, c. 10; p. 495/5-10: "animas has intellectu et imaginatione et desiderio sive appetitu distingui... oportet eas esse intellectuales intellectu activo quia formas per motum sui orbis efficiunt sicut artifex explicat formam artis per artis instrumenta."
26. *Ibid.*, L. XI, tr. ii, c. 1; p. 535/38-40: "ad hoc enim est caelum, ut sit instrumentum intelligentiae, sicut manus est instrumentum architectonici."
27. *Ibid. cit.*, L. XI, tr. ii, c. 10; p. 495/24-26: "ideo dederunt illi animae etiam desiderium, sensum autem negaverunt inesse ideo, quia caelestis circulus nullius est sensibilis receptivus."
28. *Ibid.*, p. 495/77-79: "ipsas animas de virtutibus animae nihil habere dixerunt nisi agentem universaliter intellectum et desiderium sive appetitum."
29. *Ibid.*, p. 496/49-54: "ideo animae caelorum superflue haberent tales virtutes, cum virtutes corporis caeli ad hoc sufficiunt...: nunquam fit inoboedientia inter motorem et id quod movetur".
30. *Ibid.*, p. 496/65-68: "quia non considerant motum caelestium secundum principia motus, sed potius secundum numerum et mensuram quantitatis suae."
31. *De quatuor coaequaevis cit.*, tr. III, q. xvi, a. 2; p. 73: "non est contrarium fidei quosdam angelos miraculosa facere et legibus naturae concurrere: ita non est contrarium fidei quosdam angelos iuvare naturam in movendo et gubernando sphaeras caelorum, quos angelos moventes sive intelligentias philosophi dicunt animas. Sancti vero timentes ne forte dicere cogantur caelos esse animalia, si concedunt ipsos habere animas, negant motores caelorum esse animas. Et ita patet quod non est contradictio inter eos: antiqui enim deos et angelos dicebant animas mundi."

32. *De motibus animalium*, L. I, cap. 3, ed. Jammy, V, p. 112: "si enim haec essent ex solo animae imperio vel motorum caelestium sine motu orbis, non esset scientia per artem inventa, unde prognosticarentur talium [monstrorum] nativitates. Habemus autem scientiam nobis a multis rusticis traditam, ex qua talia prognosticamus ex situ stellarum et motu orbis."
33. *Ibid.*, p. 147b: "est sine labore et poena, ut dicunt, et non inducit lassitudinem, ut motus animae quo movet corpus nostrum".
34. *De quatuor coaequevis cit.*, p. 69b: "Hoc etiam patet ex dicto Averrois in libro *De substantia orbis*, qui dicit, quod coelum non lassatur in motu: eo quod motor movet ipsum secundum convenientiam formae mobilis ipsius, qualiter non movet anima corpus animatum elementatum: et ideo animalia in motu corporis lassantur".
35. *De quatuor coaequevis cit.*, p. 70b: "Augustinus *Super Genesim ad Litteram*,... reliquit pro dubio, utrum stellae sint animalia, vel non. Nam si esset haeresis, videtur quod ipse determinare deberet".
36. *In Secundum Sententiarum cit.*, p. 147b:"rationes supradictae non probant nisi quod non moveantur a natura quae sit forma corporis movens: et hoc dicunt etiam alii philosophi, sicut astronomi et Ptolemaeus, et Albategni, et Albumasar, et Geber, et alii quamplures."
37. *De quatuor coaequevis cit.*, Tract. III, q. 16, a. I: "Utrum motor primus sit Deus", p. 68b: "Cum igitur primus motor sit, qui non alia causa motus, immobilis per se et per accidens, ut habitum est, movet caelum et mundum, primus motor est Deus". *Cf.* p. 74: "primus motor movet non motus, et movet inferiores, ut desideratum movet desiderium." For sake of space I cannot analyse here this important Aristotelian idea, according to which not only God, but the Heavens act as final causes. It has had enormous resonance in the Middle Ages, and even in Dante.
38. *Ibid.*, p. 69b: "Solutio: dicendum quod, si velimus loqui secundum Philosophos, ponemus in coelo triplicem motorem, scilicet Deum qui est motor extra, non proportionatum mobili: et hoc attendit Ptolemaeum, qui dicit, quod nihil movet caelum nisi solus Deus, et Rabbi Moyses... et de hoc etiam intelligitur dictum Boetii in libro *De consolatione philosophiae* [*cf.* p. 69a where is cited the whole *Metrum nonum* of book. III]. Secundus motor est forma coniuncta coelo non divisibilis divisione coeli: et isti motori convenit, quod sit mobilis per accidens, quemadmodum probat Rabbi Moysis, et convenit ei, quod virtus sua proportionatur mobili; et hoc dicit Commentator super librum *De coelo et mundo*...Et hoc tangit Avenalpetras in *Astrologia* sua... Tertius motor est forma materialis divisibilis secundum divisionem coeli. Sicut enim est grave in terra et leve in igne, quae sunt potentiae ad motum sursum et ad motum deorsum, ita est quaedam forma in coelo que est potentia ad motum circularem". At p. 70a Albert prevents some objections with the statement "prima causa extra genus naturae est Deus, cuius causata sunt omnia creata".
39. *Ibid.*, p. 70a: "Articulus II: Utrum motor ille, qui est infra [ed.: intra] et non est divisibilis secundum quantitatem mobilis, sit anima mundi vel non?"; "videtur quod sic: 1)...prout videtur in secundo *De caelo et mundo*, ubi dicit Philosophus, quod si coelum habet animam, et est in ipso principium motus, tunc procul dubio sunt ei sursum et deorsum, dextrum et sinistrum. Ex quo patet quod Philosophus ponit pro causa dextri et sinistri caelum habere animam". He goes on: "Vita est in coelo fixa et sempiterna in saecula saeculorum, quae non finitur neque deficit et est melior vita", and Albert notices that some readers had concluded: "ergo anima caeli est motor caeli".

40. *Ibid.*, p. 70a: "coelum habet intellectum, qui est forma indivisibilis secundum quantitatem coeli, et ille est anima"; "corpus coeleste movetur ex se et anima".

41. *Ibid.*, p.70b: "corpus coeleste non est necessarium in suo esse, sicut est dispositio in corporibus animalium quae sunt hic:... apparet quod animae eorum sunt necessariae in esse corporum suorum, et quod non salventur nisi per sensibilem animam et imaginativam. Corpus autem coeleste, quia est simplex et intrasmutabile ab aliquo extrinseco, non indiget in suo esse anima sensibili, nec imaginativa, sed tantum indiget anima spiritum movente in coelo, et virtute quae non sit corpus neque sit in corpore secundum divisionem ipsius ad largiendum ipsi permanentiam aeternam et motum aeternum, qui non habet principium neque finem." Thus heaven's soul "non est sensibilis, sed motiva secundum locum, quae largitur ei permanentiam et motum".

42. *Ibid.*, p. 70b: "Caeli sunt animalia rationalia, scilicet apprehensores creatoris. Haec autem est veritas probata ex parte Legis, et non secundum corpora motiva sicut ignis et terra, sicut putant insipientes,sed sicut dixerunt Philosophi, animalia obedientia creatori, et laudant ipsum et cantant ei canticum sublime".

43. *Ibid.*, p. 70b: "expresse dicit quod caelum habet animam et phantasiam, cui obedit universa materia mundi, sicut corpus animalis obedit animae animalis: et sicut corpus animalis immutatur secundum imagines delectabilis vel tristis apprehensas ab anima animalis, ita materia elementorum mutatur ad imaginationem motorum coeli", and he mention a surprising example: "et ideo fiunt quandoque terraemotus et scissurae terrae in inferioribus".

44. *Ibid.*, q. XVI, a. II, p. 72b: "quod ipse vocat phantasiam et imaginationem applicationem intellectus ad particularia naturae".

45. *Ibid.*, p.70b: "Nullus corpus simplex potest esse animatum, ...coelum est corpus simplex, ergo non potest esse animatum"; *ibid.*, p. 71a: "non videmus ibi organa sensuum: ergo non habet animam sensibilem. Si forte dicatur quod habet animam intellectualem praeter sensibilem et vegetabilem, et illa non indiget organis, neque in se neque in suis operationibus. Contra: Intellectus non efficitur in actu nisi per abstractionem a phantasmatibus: si igitur habet intellectum, aut ille intellectus nunquam erit in actu et sic erit sicut dormiens, aut oportet quod habeat phantasiam et sensibilem animam, et hoc absurdum est ponere in coelo".

46. *Ibid.*, p. 72a.

47. *Ibid.*, p. 72a: "Nos cum Sanctis confitemur coelos non habere animas, nec esse animalia, si anima secundum propriam rationem sumatur. Sed si vellemus philosophos ad idem reducere cum Sanctis, dicemus quod quaedam intelligentiae sunt in orbibus deservientes primo in motu orbium, et intelligentiae illae dicuntur *animae orbium*, et non univoce cum intelligentiis hominum, eo quod non egrediuntur in actum per abstractionem a phantasmatibus, sed ipsae revertuntur super essentiam suam, et per essentiam super aliud reditione completa." Here he mentions the definition of such an intelligence given in the pseudo-aristotelian *Liber de causis*, a text which shortly before 1271 was the subject of his last commentary, a very important one also for the present discussion.

48. *Ibid.*, p. 72a-b: "illae intelligentiae non habent nisi duas potentias, scilicet intellectum et appetitum moventem secundum locum; nec habent comparationem ad orbes secundum istam rationem animae, qua dicitur quod anima est entelechia corporis organici physici potentiam vitam habentis...Operatur autem ad corpus ut nauta [ed. erroneamente: natura] ad navem, hoc est secundum rationem movendi ipsum et regendi, sicut dicitur in libro *De*

anima" as well as "in secunda propositione *De causis*".

49. *Ibid.*, pp. 72b-73a: "dicimus quod motores sphaerarum per motum causarum causant omnem diversitatem quae est in inferioribus secundum naturam: et ideo cognoscentes se in quantum causae sunt, cognoscunt naturalia omnia. Similiter inferiores motores sphaerarum cognoscunt superiores per hoc quod sunt moti ab eis, ut desideratum movet desiderium. Et ita patet quod haec scientia non est in universali, nec in particulari: per hoc enim quod cognoscunt se, cognoscunt universalia et particularia causata a motoribus suis."

50. *Summa de creaturis cit.*, p. 73[b]: "Ista omnia diximus secundum Philosophos, qui non contradicunt quibusdam sanctis negantibus coelum animam habere, nisi in nomine solo, qui abhorrent nomine animae, et tamen bene concedunt, quod intelligentiae quaedam sive angeli movent coelum iussu Dei. Sicut ponimus secundum catholicam fidem quosdam angelos miraculosa facere et legibus naturae concurrere: ita non est contrarium fidei quosdam angelos iuvare naturam in movendo et gubernando sphaeras coelorum, quos angelos moventes sive intelligentias philosophi dicunt animas. Sancti vero timentes ne forte dicere cogantur coelos esse animalia, si concedunt ipsos habere animas, negant motores coelorum esse animas. Et ita patet quod non est contradictio inter eos: antiqui enim Deos et angelos dicebant animas mundi."

51. *De caelo cit.*, L. II, tr. 3, cap. 5; pp. 150-153. *Cf. ibid.*, p. 152/25-41: "Oportet absque dubio quod stellae, quae sunt quasi membra quaedam caeli, sint primi motores, ad quos reducuntur omnes alterationes et augmentationes et generationes materiae universalis generatorum et corruptorum. Habent enim stellae virtutem in se intellectuum moventium, qui sunt intellectus operativi formales, sicut est intellectus artificis formalis ad opus, quod producit; et actiones stellarum informantur ex illis, quemadmodum informatur calor complexionalis a virtute animae. Et ideo influunt per motum suum illas formas, sicut calor naturalis in cibum et corpus inducit formam carnis et sanguinis, quando informatus est a virtute animae. Et haec est causa etiam, quod quando sciuntur virtutes stellarum ex sitibus et motibus eius, tunc coniecturatio habetur verisimiliter de productione generandorum et duratione et omni formatione eorum"; see also *ibid.*, p. 152/81-92: "egregie dixit Aristoteles in *Primae philosophiae* l. VII [cap. 7, 1032 b/11-12] quod sicut sanitas est ex sanitate et domus ex domo in operibus artis, eo quod sanitas in corpore est ex sanitate quae est in anima medici, et domus quae est ex lignis et lapidibus est ex domo quae est in anima aedificantis domum, ita est in formis naturalibus. Et haec fuit causa inducens Platonem quod dixit omnes formas esse a Datore formarum et non esse in materia, cum tamen hoc non sit verum quia isti motores educunt eas de materia in quae sunt potentialiter et non secundum actum." *Cf.* p. 138/67 ff., where Albert pointed out that it was not a living and emotional being – easy to influence – but "intelligentia sive caelestis animus sive mens caelestis vel quocumque alio nomine motor primi caeli potest convenienter nominari"; he also added that "intellectus formarum moventium nihil extra se habet, quemadmodum habet intellectus hominis variabilis", and concluded: "hic motor non alio actu movet et alio causat naturalia inferiora in causis suis distincta, sed eodem sicut artifex eodem actu quo movet instrumentum inducit formam artis in materia artificiati". See also the other Albertinian texts touching upon the theme of the heavens seen as instruments quoted by TH, II, 581-582. The analogy between the intelligence of the heavens and those of the artist and the craftsman survived for a long time in the Renaissance, from Ficino's *De amore* (a commentary to the *Symposion*) to Della Porta and Bacon, through the Aristo-

telian Pomponazzi (*De incantationibus*, in *Opera*, Basileae 1567, Hildesheim 1970, p. 170: "quandoquidem ars imitatur natura"; p. 229: "sicut natura adiuvat artem, sic et ars naturam"), and more specifically in his unpublished *Quaestio de alchimia* "pro complemento tertii *Metheorologicorum*", written three years after the clandestine *De incantationibus*, ms. Ambros. R. 76, f. 110r, a work I am going to analyze and publish elsewhere); *cf.* also the distinctions I deployed in my 'Il problema della magia naturale nel Rinascimento', *Rivista critica di storia della filosofia*, XXV, 1973, pp. 292-293. It is useful to compare the first text here quoted from the *De caelo* with the thesis of the *Speculum* we have quoted on the issue of the "heavens not alive", but seen as a mere instrument of God, *cf.*. *De mineralibus cit.*, III, ii, c. 3; p. 240; *Metaphysica cit.*, p. l XI, tr. iii, c. 2; p. 535/93-94: "quando artifex est naturae minister, sicut est medicus et alchimicus aliquando."

52. *De caelo cit.*, p. 152/76-81: "Statuarius enim per se et essentialiter facit statuam, non tamen materialiter disponitur secundum formam statuae. Et sic stellae essentialiter agunt formas, sed habent eas spiritualiter et intellectualiter [non per essentiam materialem et corporalem], secundum quod sunt instrumenta intellectuum moventium."
53. *Speculum*, III/4-8: "sic ordinavit Deus altissimus sua summa sapientia mundum istum, ut ipse qui est Deus vivus, Deus caeli non vivi, velit operari in rebus creatis, quae inveniuntur in his quatuor elementis inferioribus, per stellas surdas et mutas sicut per instrumenta". He continues by pointing out that nothing will be more desirable for the preacher than to have a "scientia media" ("ligamentum naturalis philosophiae et mathematicae") which shows him how the changing of the heavenly bodies influences that of the terrestrial ones.

CHAPTER TEN

1. TH, II, p. 701.
2. Albumasar, *Introductorium maius*, transl. by John of Seville, tr. 1, diff. V, cap. 'De secta tertia'; ms. Laur. Plut. XXIX, 12, f.14v: "Contradixerunt scientiae astrorum et dixerunt quod planetis non sit significatio supra res de his quae fiunt in hoc mundo. Et hac ratione usi sunt et dixerunt quod stellae non significarent id quod possibile est, sed tantum necessarium et impossibile."
3. Lemay, *Abû Ma'shar cit.*, pp. 112-130 devoted the concluding section of the first part of his book to discussing the standpoints concerning free will defended in the *Introductorium*. Lemay quoted in his footnotes the translation by John of Seville that we have checked in the ms. Laur. Plut. XXIX, 12.
4. *Speculum*, XIV/68-71: "Non enim idem est, esse necessario quando est, et simpliciter esse ex necessitate. Antequam ergo sit, potest non esse, et tamen erit, quia non est necesse illam potentiam ad actum reduci."
5. *Speculum*, XIV/71-77: "Similiter de eo de quo significatum est, quoniam non erit in tempore determinato, et de quo verum est dicere quoniam non erit tunc, nihilominus semper ante hoc potest esse, et tandem revertitur ad naturam impossibilis. Et haec est sententia Albumasaris, a qua tamen famosus Aristoteles in aliquo declinare videtur, cum non concedat quod prius sit verum dicere. Me autem nihilominus sic dixisse non piget...". *Cf.* Pangerl, *Studien über Albert cit.*, p. 785, who used the *Speculum* as an authentic work and interpreted the discussion of the eternity of the world as a clear, and by no means the only, instance of Albert's independence from Aristotle: "Zur Ergänzung sei bemerkt,

dass Albert noch an anderen Stellen Ansichten des Aristoteles als unrichtig zurückweist (*Opera*, ed. Borgnet, III, 200; IV, 108, 523, 679; XI, 587, 627; XII, 424; X, 27). Wenn Albert im *Speculum astronomiae* (*Opera*, X, p. 643 = XII/20) in Bezug auf die Ewigkeit der Welt sagt, 'in quo solo ipse Aristoteles invenitur errasse', so wird man verstehen müssen: In welchem Punkte allein ein besonders gewichtiger Irrtum des Stagirites vorliegt." Without declaring the *Speculum* an authentical work Hossfeld compare it (ch. X and XV) to Albert's *De fato* in his critical edition *cit.*, pp. 67/1, 150/54, 154/88.

6. *Speculum*, XII/18-21: "ex praecepto suo [dei] stabit motus, sicut et coepit ex ipsius praecepto (in quo solo ille utilis Aristoteles invenitur errasse; nihilominus, regratiandus est in mille millium aliorum)."
7. *Speculum*, XII/60-61: "elegentius scilicet testimonium fidei et vitae aeternae."
8. *Speculum*, XIV/84-92: "nam in his quae operatur dominus per coelum, nihil aliud est caeli significatio quam divina providentia. In his vero quorum nos sumus principium, nihil prohibet etiam caelo non causam, sed significationem inesse: duarum enim partium contradictionis quarum alterutram potest homo eligere, sciebat deus ab aeterno quam illarum eligeret. Unde etiam in libro universitatis, quod est caeli pellis... potuit figurare, si voluit, quod sciebat; quod si fecit, tunc eadem est determinatio de compossibilitate liberi arbitrii cum divina providentia et cum interrogationis significatione". The image "caelum sicut pellis" had been borrowed from *Psalm* CIII,2 already by Peter Abelard in a famous context of his *Expositio in Hexaëmeron*, PL 178, cols. 744-745 dealing with "aquae supercaelestes".
9. *Speculum*, XIV/97-98: "quecumque non latent divina providentia sint etiam cognita apud caelum".
10. *Speculum*, XIV/100-101: "consilium magisterii astrorum est supersedere, quia dominus voluit celare a nobis".
11. M.-T. d'Alverny, 'Un temoin muet des luttes doctrinales du XIIIème siècle', *Archives d'histoire doctrinale et littéraire du Moyen Age*, XXIV, 1949, pp. 223-248, *cf.* pp. 228-230 and n. 1 p. 230. Of this pseudo-Aristotelian commentary, the author emphasized Albert's typical attitude "vis-à-vis des notions qui lui paraissaient scientifiques, et qu'il est soucieux d'accorder avec sa foi chrétienne". *Cf.* M.-T. d'Alverny – F. Hudry, 'Al Kindi *De Radiis*', *Archives d'histoire doctrinale et littéraire du Moyen Age*, 41, 1974 (but 1975), pp. 139-259.
12. Abû Ma'shar, *Introductorium maius*, tr. I, d. 5, transl. Joannes Hispalensis: ms. Laur. Plut. XXIX. 12; cf. *Speculum*, XIV/48 ff.
13. Historians have often emphasized Albert's willingness to add to Aristotle. This attitude is clear to those who pay attention to the structure of his course of philosophy. As far as his scientific works are concerned – besides the case of the addition of the *De vegetabilibus* and of the *De mineralibus* – of particular relevance are the corrections and integrations to Aristotle Albert had introduced within the theory of heaven. See, among other cases, *De caelo cit.*, p. 162/73-84: "Aristoteles... in secundo libro *Caeli et mundi* se excusat [quia nulla rationum istarum de motu processivo stellarum habet vim demonstrationis], dicens quod debent sufficere solutiones topicae et parvae in his quae sunt de caelo quaesita, eo quod ad ipsa cognoscenda perfecte non sufficimus. Nos tamen domino concedente collationem faciemus in Scientia astrologiae inter viam, quam invenit Alpetraz Abuysac, et viam quam secutus est Ptolemeus accipiens eam a Babiloniis et Aegyptiis, quorum scientiam se verificasse dicit Aristoteles in libro *Caeli et mundi*, ex quo videtur innuere quod et ipse consensit opinionibus eorum". At p.132/48-58, Albert had already insisted on the

insufficiency and fallacy of the instruments employed by astronomers: "licet in aliquo defectum sensus suppleat rectitudo intellectus". When he was examining the contrasts between Aristotle, Ptolemy and Alpetragius concerning descriptive astronomy (pp. 168-69) he was particularly explicit and independent, showing that he was not giving advantage to the Aristotelian auctoritas: "Nos autem magis consentimus Ptolemaeo Pheludensi", and called Ptolemy by the geographical name he was keen to get right in the *Speculum* (II/7), where he also declared that (II/75) "perspectiva enim Aristotelis ad supra dicta non descendit". On the theme of the relative speed of middle and inferior heavens, in *De caelo cit.*, p.169/4, 13-14 Albert stated: "dicimus generaliter non esse verum quod dicit Aristoteles... Et ideo, sicut dicit Maurus Abonycer [Abû Bakr], si viveret Aristoteles, oporteret vel ista improbare quae comperta sunt de motibus astrorum, vel oporteret eum suum dictum revocare". This criticism was softened, but not canceled, when a few pages later Albert claimed that, thanks to his own observations, he had found the way "salvare Aristotelem et veritatem, quam invenimus diligenti astrorum inspectione".

14. Albert, *De causis proprietatum elementorum*, in *Opera omnia*, V/2, Münster 1980, pp. 76-79, 78/86 ff. in particular: "causa universalis", "causa minus universalis, in qua quaedam caelestium conveniunt et quaedam terrestrium" and later "causa vero particularis... in qua conveniunt aut quaedam caelestia sola, aut quaedam terrestria sola" [...] "quorum autem Arabum sententia: ... huiusmodi prodigia in terra fieri ab imaginatione intelligentiae quae movet sphaeram lunae".
15. *De causis proprietatum elementorum cit.*, p. 78: "significat illa coniunctio magna accidentia et prodigia magna et mutationes generalis status elementorum et mundi: cuius causam debet dicere naturalis secundum ipsum quia scit astronomus". *Cf.* above II/3, n. 18.
16. *De caelo cit.*, p. 129/58-59: "stellas generantes et moventes materiam generatorum".
17. *De caelo cit.*, p. 131-132/60: "sive simplicia, sive composita non exprimitur totus decor corporis coelestis".
18. *De caelo cit.*, p. 131-132/60: "inquisitio difficilium est aliquando vituperabilis, ita aliquando est laudabilis". (Italics mine)
19. *Cf. Problemata cit.*, p. 321; *De causis proprietatum*, in *Opera omnia*, V/2, pp. 76/75-77/2: "Sunt autem quidam qui omnia haec divinae dispositioni tantum attribuunt et aiunt non debere nos de huiusmodi quaerere aliam causam nisi voluntatem Dei. Quibus nos in parte consentimus, quia dicimus haec nutu Dei mundum gubernantis fieri ad vindictam maleficii hominum. Sed tamen dicimus haec Deum facere propter causam naturalem, cuius primus motor est ipse, qui cuncta dat moveri. Causas autem suae voluntatis non quaerimus nos: sed quaerimus causas naturales, quae sunt sicut instrumenta quaedam per quae sua voluntas in talibus producitur ad effectum." *Cf. De fato cit.*, p. 78/4 with chap. 4 of the Hermetic Asclepius quoted in the commentary to the edition, as well as the passage quoted above from the *De quatuor coaequevis*. See also Albert's *De causis et processu universitatis cit.*, I, tr. 4, c. 6, in *Opera omnia*, ed. Borgnet, X, pp. 421-423; in our edition of the *Speculum*, V/18 "iussu Dei" replaces "nutu Dei".
20. *De caelo cit.*, p. 150/49 ss.: "De effectibus autem stellarum diversis duo in philosophia quaerantur, quis videlicet et quando et ubi sit effectus cuiuslibet stellae. Et hoc inquirere est electoris et divinantis per astra, cuius est eligere et scire horas, secundum quas ad figuras astrorum referentur ea quae fiunt in inferioribus. Et hoc oportet relinquere scientiae electorum, qui alio nomine vocantur geneatici [or better *genetliaci*, correcting *geomantici* found in ms. A and the eds.; *cf.* the same spelling (on which see below p. 281) in *De caelo*

below n. 23 and *Summa theologiae*, I, 17, 68 (ed. Borgnet, X, pp. 633-34)], eo quod principalius, quod inquirunt per stellarum figuras et effectus, sunt nativitates eorum quae generantur."

21. *De caelo cit.*, p. 150/58 ss: "Accidentia autem magna sunt sicut mutationes regnorum de gente in gentem et translationes sectarum et doctrinae novarum religionum, et huiusmodi". This passage was quoted, but not clarified, by TH, II, 586, n. 2. The second "volumen", quoted below, is without doubt the *Centiloquium* attributed to Ptolemy. As far as the first volume is concerned, the description corresponds almost exactly to the contents of the *Quadripartitum*; there is however, no correspondence between the list of contents Albert talks about, and the actual structure of the Ptolemaic work. For instance, Albert described it as a work "habens octo distinctiones", instead of the famous four. In any case, the "distinctiones" or "differentiae" were a typically Arabic way of subdividing texts; the *De magnis coniunctionibus* of Abû Ma'shar, for instance, a work dealing with exactly the same topics, contained eight "distinctions". It is reasonable to doubt that when he was writing the *De caelo*, Albert was greatly interested in descriptive astronomy –a field where he used al-Bitrûjî's terminology– whereas he was less taken by judicial astrology, and had not yet gathered the bibliographical data he was going to use in the *Speculum astronomiae*. Is it then possible that Albert confused Abû Ma'shar with Ptolemy? On pp. 170/70-73 of the *De caelo*, Albert unequivocal refers to the methodological approach preferred in the *Quadripartitum*, a work he repeatedly referred to in the *Super Ethica*, I, tr. 7, ch. 6 and *passim*; *cf.* above at nn. 40. ff., as well as in the late *Problemata determinata XLIII cit.*, pp. 330-331, 350 ("Alarba seu Quadripartitum": *cf. Speculum*, VI/3-5: "qui dicitur... arabice Alharbe, latine Quadripartitum"); see also analogous expressions by Albert in *Summa theologiae*, P. I, tr. XVII, q. 68, m. 1; ed. Jammy, XVII, p. 381b ("arabice Alarba, latine Quadripartitum"); *De mineralibus cit.*, IV, 3; *De XV Problematibus cit.*, pp. 326, 339, 353 and passim for more quotations from Ptolemy and other astrological authors.
22. *De caelo cit.*, p. 150/65-67: "de accidentibus parvis particularibus, sicut sunt eventus unius hominis nati in hac constellatione vel illa".
23. *De caelo cit.*, p. 150/67-71: "Secundum autem quod quaeritur de effectibus stellarum, est naturalis causa, propter quam stella dicitur habere hunc vel illum effectum, et hoc hic determinandum est et a geneaticis [sic] sive electoribus supponendum." This unusual latin word ("astrologos, qui et *geneatici* dicuntur") is to be found also in Thomas Aquina's *De iudiciis astrorum* and *Summa theologiae cit.*, IIa IIae, p. 95, a. 3, p. 453a. *Cf.* above, p. 281.
24. *De caelo* cit., p. 151/23-31: "Quod autem magis est difficile scire, est, secundum quam naturam sidera habeant facere fortunas et infortunia et vires ministrent non tantum exortis per naturam, sed aliquando factis per artem, sicut imaginibus vel vestibus incisis de novo vel aedificiis de novo factis vel huiusmodi. Haec enim omnia a causis mutabilibus sunt, possunt esse et non esse. Et ideo videtur quod regimen eorum non dependeat ab aliqua natura vel virtute stellarum."
25. *Speculum*, XI/103-106, 123-124: "tertius modus imaginum astronomicarum, qui... virtutem nanciscitur solummodo a figura caelesti... et habebit effectum iussu Dei a virtute caelesti".
26. *Cf.* n. 13 above, where we have provided the full text of this announcement in the *De caelo cit.*, p. 162/77-78.
27. *De caelo cit.*, p. 157/54-61: "patet quod astrologia, quam dicit se fecisse Alpetruauz Abuysac, secundum Aristotelis intellectum falsa est... De his tamen in Astrologia erit

inquirendum"; *ibid.*, 167/83-85: "Nos autem collationem faciemus in Scientia astrologiae [...] hae res omnes dicendae sunt in Astrologia et determinandae sufficienter per principia mathematica."

28. *Cf.* Pangerl, *Studien cit.*, pp. 339-341; Grabmann, 'Studien über Albert den Grossen', *Zeitschrift für katholische Theologie*, XXXVI, 1912, p. 339n; *Id.*, 'Der Einfluss Alberts' cit., *ibid.*, 1928, p. 169; Meerssemann, *Introductio cit.*, p. 61; Pelster, *Kritische Studien cit.*, p. 139, have only sketched an examination of the various manuscripts attributed to Albert. Besides a *Perspectiva*, that in fact looks like an excerpt from Roger Bacon, it is noted that in the ms. Escorial, III.&.8, ff. 293r-296v there is a *Questio Alberti de speculis* (inc. Queritur de forma resultante seu resiliente in speculo, que nec lumen, nec calor esse videtur. Queritur primo utrum sit vel non, quare dicit autor sex principiorum... exp. Et sic est dictum de hac questione. Explicit questio de speculo edita a gravissimo domino Alberto Magno). Still to be studied, are the mss. Wien, lat. 5309, ff. 127r-155v, XV century, *Albert Magni Summa astrologiae*, inc.: In hoc tractatu brevi...; Wien, lat. 5292, ff. 1-65v, CLM 56, ff. 1-122; Innsbruck 2511, ff. 1-15, containing an *Epitome in Almagesti Claudi Ptolemei* attributed to Albert (but entitled *Almagesti abbreviatum per magistrum Thomam de Aquino* in the CLM 56 of the A.D. 1434-36); these texts should, in part at least, be compared with the *Almagesum parvum* quoted in the *Speculum*, II/17-20 and studied by A. Birkenmajer, *Études*, Wrocław-Warszawa-Kraków 1950, pp. 142-47).
29. *De caelo* cit., p. 170/25-26: "Amplius in scientia astrologiae iam diximus, quod si Sol non esset orbicularis..."; ibid. p. 154/87-89 "in Astronomia enim et in Scientia electionum, deo favente, loquemur adhuc de stellis et determinabimus ea quae hic relinquuntur." The first passage (as well as *De caelo* cit., p. 132/87-88: "de quantitatibus et motibus superiorum in astronomia explicabitur") has been identified as a quotation from Averroes's commentary. *Cf.* P. Hossfeld, 'Die Arbeitsweise des Albertus Magnus in seinen naturphilosophischen Schriften', in *Albertus Magnus Doctor Universalis*, ed. G. Meyer and A. Zimmermann, Mainz 1980, p. 201. But the passage mentioning the *Scientia electionum* is original.
30. When we consider that the more ancient manuscripts had all the features of a textus (for instance, *cf.* the ms. Laur. Plut. XXIX. 12), it is highly probable that the *Speculum* was designed to be a bibliographical instrument and a theoretical propedeutic tool for the Faculty of Arts: in other words, the treatise might well belong to the genre of introductory handbooks F. Van Steenberghen has shown in *La philosophie au XIIIe siècle cit.*, pp. 121-132 to have been very popular: "Cette littérature d'introduction, dans laquelle les problèmes de classification jouent un rôle très important, est née de circonstances diverses: besoin de coordonner, d'expliquer et de vulgariser en vue de l'enseignement les écrits scientifiques ou philosophiques des grands penseurs; naissance de la bibliographie et de la bibliothéconomie..., soucis d'ordre pédagogique visant les méthodes à employer dans l'enseignement des différentes sciences et la succession chronologique des branches mise au programme des écoles; le progrès scientifique même...". This hypothesis finds support in the insertion of the *Speculum astronomiae* in the series of the Aristotelian and Albertinian *Parva naturalia*; before Jammy and Borgnet, see several mss. and the edition produced in Venice in 1517 by M. A. Zimara that amounted to the extension of the *curriculum studiorum* from the *Physica* and *Metaphysica* to all Aristotle's texts – including the *Historiae* – and to other works designed to fill gaps and omissions in the Aristotelian corpus. As far as Albert was concerned, we know that in the *De mineralibus*, after having

tried to find an analogous Aristotelian treatise, he decided to reconstruct the theories of the Greek philosopher from a few hints in the *Liber IV Meteorologicorum*, and then rather cavalierly proceeded to integrate them with theses found in Arabic sources, with medieval texts on stones, and with his own observations. *Cf.* D. Wyckoff, 'Albertus Magnus on Ore Deposits', *Isis*, 49, 1958, pp. 109 ff., and Albertus Magnus, *Book of Minerals*, transl. by D. Wyckoff, Oxford 1967, where Wyckoff strongly supported the Albertinian authenticity of the *Speculum* and the exact correspondence of the sources for chapter XI of this treatise with treatises II and III of the *De mineralibus*. Though Albert and the *Speculum* showed great care for philological precision, it was equally important to achieve encyclopedic thoroughness.

CHAPTER ELEVEN

1. It is inevitable to recall in this case the beginning of the *Quadripartitum*. *Cf.* Abû Ma'Shar, *Introductorium maius*, trans. Johannes Hispalensis, ms. Laurent. Plut. XXIX.12, f. 2v.
2. *Speculum*, III/31-33: "quasi omnes libros laudabiles, quos de ea pauper latinitas ab aliarum linguarum divitiis per interpretes mendicavit." *Cf.* G. R. Evans, '"Inopes verborum sunt latini". Technical language and technical terms in the writings of St. Anselm and some commentators of the mid-twelfth Century', *Archives d'histoire doctrinale et littéraire du Moyen Age*, XLIII, 1976, pp. 113-134.
3. *Speculum*, XII/34-35: "forte... nondum sunt translati".
4. *Speculum*, XII/106-109: "Si sunt in textu eius nomina ignotae linguae, statim subduntur in littera interpretationes eorum: quod si forte aliquorum interpretationes defuerint, paratus est vir earum copia exhibere."
5. *De quatuor coaequaevis cit.*, tr. III, q. xv, a. 1, ed. Jammy, XIX, p. 66b, where he employs the term "assub"; *De fato cit.*, a. 2, p. 70/23-26: "dicit Massehallach, quod caelestis effectus, quem ille alatir vocat iuvatur a sapiente astronomo, sicut in producendis terraenascentibus iuvatur aratione et seminatione"; *cf. De caelo cit.*, l, II, tr. 3, c. 15, p. 378/59-60: "alatyr hoc est circulum effectivum" and *Summa theologiae cit.*, P. I, tr. XVII, q. 68, m. 2; ed. Jammy, XVII, p. 384a: "Messeallach praecipuus in astris dicit quod Alkir [sic!] hoc est circulus celestis studio periti viri iuvatur ad effectum"; *De causis proprietatum cit.*, L. I, tr. 2, c. 4 and 9, in *Opera omnia*, V/2, pp. 67/81-82: "a plenilunio, quod interlunium a quibusdam vocatur, quod Arabes vocant almuhac, usque ad perfectum lunae defectum recedit"; "axem qui meguar sphaerae dicitur".
6. *Opus tertium cit.*, p. 90 (and *cf.* all chap. XXV): "vocabula infinita ponuntur in textibus theologiae et philosophiae de alienis linguis, quae non possunt scribi, nec proferri, nec intelligi, nisi per eos qui linguas sciunt. Et necesse fuit hoc fieri propter hoc quod scientiae fuerunt compositae in lingua propria et translatores non invenerunt in lingua latina vocabula sufficientia."
7. *Ibid.*, p. 91, on Gerard of Cremona, Michael Scotus, Alfred of Sarashel, Herman of Carinthia; *Compendium studii philosophiae*, in *Opera quaedam hactenus inedita*, ed. Brewer *cit.*, pp. 471-72 where Roger criticized the translators: "Unde cum per Gerardum Cremonensem, et Michaelem Scotum, et Alvredum Anglicum, et Hermannum Alemannum, et Willielmum Flemingum data sit nobis copia translationum de omni scientia, accidit tanta falsitas in eorum operibus quod nullus sufficit admirari."

8. M. Bouyges, 'Roger Bacon a-t-il lu des textes arabes?', *Archives d'histoire doctrinale et littéraire du Moyen Age*, V, 1930, pp. 311-315; G. Théry, 'Note sur l'aventure bélénienne de Roger Bacon', *ibid.*, XVIII, 1950-51; p. 129 ff. (and see p. 141 n. on Albert who "ne laisse passer aucun terme étranger sans essayer d'en retrouver l'origine").
9. *Cf.* A. Birkenmajer, 'La bibliothèque de Richard de Fournival, poète et érudit français du XIIIe siècle' [1922], now in his *Études d'histoire des sciences et de la philosophie du Moyen Age*, Wrocław-Warszawa-Kraków 1970, pp. 117-210; Birkenmajer reconstructed the reference in the *Biblionomia* to several scientific mss. in the Sorbonne, against the thesis defended by L. Delisle that "tous ces volumes [n']ont jamais été réunis que dans l'imagination de Richard de Fournival". P. Klopsch, *Pseudo-Ovidius de vetula. Untersuchungen und Text*, Leiden und Köln 1967, p. 90 n., considers the hypothesis that "Richard habe den Katalog seiner existierenden Bibliothek zugleich als den einer Normbibliothek darstellen wollen", to conclude that the thesis is highly improbable. Besides Birkenmajer, B. L. Ullmann has identified more than one hundred mss. that belonged to Fournival, in 'The Sorbonne Library and the Italian Renaissance', in his *Studies in the Italian Renaissance*, Roma 1955, pp. 41-53; Id., 'The Library of the Sorbonne in the XIVth Century', in *The Septicentennial Celebration of the Funding of the Sorbonne*, Chapel Hill 1963, pp. 33-41; M.-T. d'Alverny, 'Avicenna latinus. II', *Archives d'histoire doctrinale et littéraire du Moyen Age*, XXXVII, 1962, pp. 227-233; E. Seidler, 'Die Medizin in der Biblionomia des Richard de Fournival', *Südhoff's Archiv*, LI, 1967, pp. 44-54; M. Mabille, 'Pierre de Limoges copiste de manuscripts', *Scriptorium*, 1970, pp. 46-47; R. H. Rouse, 'Manuscripts Belonging to Richard de Fournival', *Revue d'histoire des textes*, III, 1973, pp. 253-69; P. Glorieux, 'Bibliothèques des maîtres parisiens. Gérard d'Abbeville', *Recherches de théologie ancienne et médiévale*, XXXVI, 1969, pp. 148-183; Id., 'Étude sur la *Biblionomia* de Richard de Fornival', *Recherches de théologie ancienne et médiévale*, XXX, 1963, pp. 206-231 (Glorieux did not consider the *Biblionomia* a hypothetical library, but a true encyclopedic collection characterized by Richard's strong interest in astrology. Glorieux deemed however that the collection reflected the cultural situation preceding the diffusion of Averroes, an author not included in the library, whose ideas, according to Glorieux, became known after 1230, even though there were translations of his works available from 1220).
10. P. Klopsch, *op. cit.*, pp. 78-79 and the bibliography he lists. *Cf.* D. M. Robothan ed., *The pseudo-ovidian 'De vetula', Text with introd. and notes*, Amsterdam 1968, pp. 1-14 on the diffusion of the *De vetula* in medieval libraries and among authors like Roger Bacon, Petrarch, Bradwardine, Pierre d'Ailly etc.
11. *Ibid.*, p. 79 ff. (*Opus maius R. Baconis*, ed. Bridges, I, p. 254); more quotations – not identifying the author of the *De vetula* – are given in the *Lamentationes Mathaei* [1298], in Walter Burleigh, Richard Bury, Thomas Bradwardine, Robert Holkot; these quotations are taken from representatives of the same British milieu from where the author of the *De vetula* absorbed the central cosmological theme of the metaphysics of light, clearly derived from Grosseteste. On the latter topic, see the critical remarks by Birkenmajer, 'Robert Grosseteste and Richard Fournival' [1948], now in *Études cit.*, p. 216, and by Klopsch, who emphasized that the authentic writings by Fournival did not dwell upon that theme.
12. Birkenmajer, 'Pierre de Limoges, commentateur de Richard de Fournival' [1949], now in *Études cit.*, pp. 222-35 (see the ms. Regin. lat. 1261, ff. 59r-60v). Thanks to the courtesy

of the late Dr. Alexandra Birkenmajer, I was able to consult the transcription by her father of this "genitura" full of biographical information.

13. Edited from the ms. Paris, BN, Fond Universitaire 636, by L. Delisle, *Le cabinet des manuscrits de la Bibliothèque Nationale*, II, Paris 1874, pp. 518-536, and pp. 527-528 in particular. Glorieux, 'Études sur la *Biblionomia*' *cit.*, has compared it with the *Laborinthus* by Everardus the German (vv. 599-686 in the Faral edition), with the list of Alexander Neckam (*Sacerdos ad altarem accessurus*, ed. Haskins, *Studies cit.*, pp. 356-376, who was the first to emphasize its shallowness in astrology), with the "guide" compiled in Paris between 1230 and 1240 (*Recipiendarius' Guide*), and lastly with the *curriculum studiorum* for the Parisian Faculty of Arts established by the decree of 1255 (*Chartularium cit.*, I, n. 246). By comparison with all these documents Glorieux finds the *Biblionomia* very thorough, especially as far as astronomy was concerned.
14. *Études cit.*, pp. 155-210. Without discussing the data offered by Birkenmajer, F. Carmody, *Astronomical and Astrological Science in Latin Translation*, Berkeley and Los Angeles 1956, p. 163 insisted on attributing the work to Geber.
15. *Speculum*, II/17-20, already discussed above, ch. 6, n. 8 ff. In the *Biblionomia* (ed. Delisle cit., pp. 527-554): "Liber extractionis elementorum astrologiae ex libro Almagesti Ptolomaei per Galterum de Insula usque ad finem sexti libri ex eo". It is worth noting that Fournival does not mention the second source (Albategni) underlined by the *Speculum*. The authorship of Gautier de Chatillon is rightly excluded by M. Pereira, 'Campano da Novara autore dell'Almagestum parvum', *Studi medievali*, 1978, p. 770, who concludes, p. 776; " Il passaggio [nell'*Almagestum parvum*] dalla stretta dipendenza dagli autori classici all'accettazione del ricco apporto fornito dall'astronomia araba [...] si accorda anche con l'ipotesi di una composizione dell'opera in due tempi successivi". This hypothesis would explain the possibility of so early a mention of the *Almagestum parvum* in the *Biblionomia*, probably written in *ca.* 1243. I am very grateful to Prof. A. Paravicini Bagliani whom I consulted on several points concerning Campanus: given the lack of documents on Campanus's early decades of work, prof. Paravicini Bagliani does not exclude the possibility that the composition of the *Almagestum parvum* could also be prior to that of the *Biblionomia*.
16. *Biblionomia cit.*, n. 53: "Mercurii Trismegisti liber de motu spere celi inclinati, qui intitulatur Nemroth ad Joanton"; *Speculum*, II/2-6: "primus tempore compositionis est liber quem edidit Nemroth gigas ad Johanton discipulum suum, qui sic incipit: Sphaera caeli etc., in quo est parum proficui et falsitates nonnullae, sed nihil est ibi contra fidem, quod sciam."
17. C. H. Haskins, *Studies cit.*, cap. XVI: 'Nimrod the Astronomer', pp. 336-354; Th, I, 415; A. Van de Vyver, 'Les plus anciennes traductions médiévales', *Osiris*, I, 1936, pp. 684-687; A. R. Nykl, 'Dante, Inferno XXXI/67', in *Estudios dedicatos a Menéndez Pidal*, Madrid 1952, III, pp. 321-24; R. Lemay, 'Le Nemrot de l'*Enfer* de Dante', *Studi danteschi*, XL, 1963, pp. 57-128; B. Nardi, 'Discussioni dantesche: II. Intorno al Nemrot dantesco e ad alcune opinioni di R. Lemay', *L'Alighieri. Rassegna di bibliografia dantesca*, VI, 1965, pp. 42-55, and the bibliography there listed; R. Lemay, 'Mythologie païenne éclarent la mythologie chrétienne chez Dante: le cas des Géants', *Revue des études italiennes*, XI, 1965, (= *Dante et les mythes*), pp. 236-279; S. J. Livesey-R. R. Rouse, 'Nimrod the Astronomer', *Traditio*, 37 (1981), pp. 203-266. The latter scholars re-examine the correspondence between Fournival's *Biblionomia* and the first bibliographical item in the

Speculum astronomiae. This corresponence has been underlined for the first time in my paper 'Da Aristotele a Albumasar' *cit.* – which I am reproducing now in these pages – at the *International Congress for Medieval Philosophy* (Madrid 1972), published for the first time in *Physis*, XV, 1974, pp. 375-398, and later in *Actas del 5° Congreso internacional de Filosofia Medieval* [1972], Madrid 1979, II, pp. 1377-1391, both printed before their paper: it is surprising that they do not acknowledge it.

18. *Biblionomia cit.*, n. 56: "Geber Hispalensis liber in scientia forme motuum superiorum corporum et cognitionibus orbium eorum et in evasione a quibusdam erroribus inventis in libris Claudii Ptolemai Phudensis (sic!), qui dicitur Elmegisti vel Megasinthasis, quem quidem corrupte nominant Almagesti".
19. *Speculum*, Proem/12-14: "exponens numerum, titulos, initia et continentias singulorum in generali et qui fuerunt eorumdem auctores."
20. *Speculum*, II/7-9: "quod de hac scientia utilius invenitur est liber Ptolemaei Pheludensis, qui dicitur grece Megasti, arabice Almagesti, latine Maior perfectus, qui sic incipit: Bonum fuit scire etc.". Albert often uses the name Pheludensis in the *De fato cit.*, p. 66/52 ff. and in the *De animalibus*, l. I, tr. ii, c. 2, p. 47/20 ss: "Sapiens Ptolemeus Pheludensis dixit quod divinans melius et verius pronuntiat accipiens iudicium a stellis secundis". This precision and the thoroughness exhibited in the *De caelo* to provide the content of the Ptolemaic works finds its explanation in the novelty of these texts within the schools, as is confirmed by the ms. Barcellona, Ripoll 109, also called *the Recipiendarius'Guide*. This guide was discovered by Grabmann and was studied by Van Steenberghen, *La philosophie au XIIIe siècle cit.*, pp. 119, 121-132. "Haec scientia traditur secundum unam sui partem in Ptolomeo, secundum autem aliam partem traditur in Almagesto, et isti libri combusti sunt". Apart from the interpretative hypotheses put forward by Van Steenberghen, I would like to suggest that in this enigmatic passage the first part relates to judicial astrology, and therefore that Ptolemy is mentioned as author of the *Quadripartitum*. I am at a loss as to how to interpret the expression "combusti sunt", if not with the hypothesis that there had been a prohibition followed by a burning at the stake, an episode which would have been recorded only in this Guide, composed in Paris in 1230. *Cf.* R. Lemay, 'Libri naturales' *cit.* above at ch. 3, n. 16. Fundamental for the translations, mss. and the historical relevance of the *Almagest* is P. Kunitzsch, *Der Almagest. Die Syntaxis Mathematica des Claudius Ptolemäus in arabisch-lateinischer Übersetzung*, Wiesbaden 1974, pp. 83-111, § B 'Die lateinische Übersetzung aus dem Arabisch' by Gerard of Cremona, and especially the description of ms. p. 91 ff., where one finds in ms. Paris lat. 14738, and in the printed ed. Venice 1515 (p. 95) as in the cit. passage of the *Speculum* the wrong incipit "Bonum fuit scire" taken from the *Sayings of Ptolemy*' a collection which in many mss. precedes the *Almagest* itself.
21. *Cf.* D. Pingree, 'The diffusion of Arabic Magical Texts in Western Europe', in *La diffusione delle scienze islamiche nel Medioevo europeo, Convegno intern. Accademia Lincei*, Roma 1987, pp. 81-83: "the *Speculum astronomiae* [...] probably was written in the late 1260's. If Albert was not its author, the only other candidate that can be seriously considered is Roger Bacon, who was in Paris from 1257 till his death in about 1292; but we shall see that the magical texts named by him are different from those known to the author of the *Speculum*. [...] If indeed his elaborate presentations of the incipits of many works [in Chapter 11 of the *Speculum*] were due solely to memory, he did truly possess a most remarkable faculty; notes on the manuscripts, presumably made in or near Paris, but

used in Cologne or in Italy, furnish a more plausible explanation of the fact that he no longer had access to all the texts that he had seen [...] the ***De ymaginibus*** by Thabit ibn Qurra and the ***Opus ymaginum*** ascribed to Ptolemy [...] are also found, one after the other, in this same order [in the ***Speculum***] on pp. 534-543 or Paris, BN lat. 16204 [...] which was almost certainly, since it was copied by a scribe, whose hand is identical or at least very similar to that employed by Richard of Fournival, in the room of "tractatus secreti" in his library at Amiens. I would even argue that the author of the ***Speculum***, whom I believe to be Albert, saw these books and most of the others that he describes in chapter 6 to 11 in that very room. This argument is strengthened when, following the splendid lead of Prof. Zambelli, one compares chapter 2 of the Speculum, on the astronomical books of the ancients, with the ***Biblionomia***". Pingree, ***Ibid.*** pp. 99-100, has drawn up a list of the items present in both the ***Speculum*** and the ***Biblionomia***, and it is worth reproducing here:

SPECULUM	BIBLIONOMIA
1. liber quem edidit Nemroth gigas ad Iohanton.	53. Mercurii Trismegisti liber de motu spere celi inclinati, qui intitulatur Nemroth ad Ioanton.
2. liber Ptolemaei Pheludensis, qui dicitur graece Megasti, arabice Almagesti, latine Maior perfectus	
2b. in commento Geber super Almagesti.	See 56 opposite 7 below.
3. in libro Messehalla *De scientia motus orbis*.	
4. ab Azerbeel hispano, qui dictus est Albategni, in libro suo.	55. Machometi Albateigny... Acharram liber.
5. ex his quoque duobus libris collegit quidam vir librum secundum stilum Euclidis, cuius commentarium continet sententiam utriusque, Ptolemaei scilicet atque Albategni.	54. Liber extractionis elementorum astrologie ex libro Almagesti Ptolomei per Galterum de Insula usque ad finem sexti libri ex eo.
6. apud Thebit motus sphaerae stellarum fixarum in libro.	58c. Thesbich filii Chore ... liber de motu accessionis capitum Arietis et Libre.
7. apud Ioannem vel Gebum Hispalensem motus Veneris et Mercurii in libro quem nominavit *Flores suos*.	56. Geber Hyspalensis liber in scientia forme motuum superiorum corporum et cognitionibus orbium eorum et in evasione a quibusdam erroribus inventis in libro Claudii Ptolomei Phudensis, qui dicitur *Elmegesti* vel *Megasinthasis*.
8. apud alium quendam ... super figura kata coniuncta atque disiuncta in libello.	58b. Thesbich filii Chore ... liber super figura alkara.
9. Alpetragius corrigere principia et suppositiones Ptolemaei.	57b. Avenalpetraugy liber de astrologia possibili et radicibus probabilibus loco earum Ptolomei.
10. liber eiusdem Ptolemaei, qui dictus est arabice *Walzagora*, latine *Planisphaerium*.	59a. Claudii Ptolomei Pheludensis liber *Walzagore*, id est plane spere.

11. apud Alfraganum Tiberiadem eaedem conclusiones, quae in *Almagesti* demonstratae sunt.	57a. Ameti filii Ameti, qui dictus est Alphraganus, liber de aggregationibus scientie stellarum et principiis celestium motuum per viam narrationis super conclusionibus Ptholomei.
12. in libro Thebit *De definitionibus*.	58a. Thesbich filii Chore liber *de diffinitionibus*.
13. *Liber canonum* Ptolemaei.	
14. Canones Machometus Alchoarithmi.	
15. librum Auxigeg, hoc est cursuum, Humenid magister filiae regis Ptolemaei, quem vocavit Almanach.	
16. Azarchel Hispanus in libro suo.	60. Alzerkel Hyspani liber tabularum.
17. demonstrationem planisphaerii [...] quem transtulit Ioannes Hispalensis.	59c. Iohannis Hyspalensis atque Linensis liber de opere astrolabii secundum Mascelamach.
18. alius Hermanni.	59e. Hermanni Secundi de compositione astrolabii.
19. alius Messehalla.	
20. alius secundum Ioannem Hispalensem de utilitatibus et opere astrolabii.	

22. We should not fail to mention a difficulty arising from not admitting a direct consultation of the manuscripts, and from refusing to date the composition of the *Speculum astronomiae* before the controversy at the University of Paris between the regulars and the seculars. At the time of the dispute, Albert was not living in Paris, but after teaching his Dionysian and philosophical courses when in Cologne, he came in 1264 to the papal court in Anagni where, besides discussing the unity of the intellect and of fate, and contributing to the defense of the mendicant orders, he could have met Campanus and other scientists. On the other hand, Gérard d'Abbeville, the heir of Fournival's manuscripts, was, with Guillaume de Saint Amour, among the strongest supporters of the seculars, and kept his stand up to the latest phases of their polemic; when he died, he gave his manuscripts to the Sorbonne on condition that the regulars could never have access to them. *Cf.* A. Teetaert, 'Deux questions inédites de Gérard d'Abbeville en faveur du clergé séculier' [1266-1271], in *Mélanges A. Pelzer cit.*, Louvain 1947, pp. 347-388. The author of the *Speculum* made use of several mss. not included in the library, and we cannot exclude that the *Biblionomia* today preserved in the single copy of the original catalog could have been loaned or given to be copied to some authoritative contemporary, who did not visit the library itself. It is, however, less difficult to suppose that Albert consulted both the manuscripts and their catalog soon after the death of Fournival, during his first stay in Paris (that is, after 1243 and before 1248), when Gérard did not have any reason, yet, to keep the regulars out of his library or in 1256-1257 or 1264 at the papal court, where Fournival had lived from 1239 on as 'familiaris' of Cardinal Robert of Sommercotes (*cf.* A. Paravicini Bagliani, *Cardinali di Curia*, Padova 1972, pp. 138-140) and perhaps made a copy of the *Biblionomia*: it is thus possible that Albert took from the ordering of Richard's astronomical books the first idea and the first notes for the *Speculum*. This text however is much richer than the *Biblionomia* in the field of astrology.

23. Pingree, 'The diffusion' *cit.*, pp. 84-88, 100-102 (where Appendix B compares the *Speculum* with the Paris and Oxford ms. cit. in this note): examining manuscripts which could have belonged to Richard or anyway been used for the *Speculum* Pingree has discovered that "70 % of Albert's catalogue of astrological books is based on what is now found in one manuscript – Paris, BN lat. 16204". This manuscript, in fact, originally included "twenty-four separate items", all mentioned – except for two – in the *Speculum*, and "it seems to have been copied by one of Fournival's scribes". Pingree concludes that "everything points to the conclusion that the *Speculum astronomiae* depends almost exclusively on the notes that Albert took of the manuscripts in the library of Richard de Fournival at Amiens". There is another codex (Oxford, Corpus Christi 248, 13th Century) which was already studied by Thorndike, 'John of Seville', *Speculum*, 34, 1959, pp. 37-38 and 'Notes on Manuscripts of the BN', *Journal of the Warburg and Courtauld Institutes*, XX, 1957, pp. 150-151: Thorndike observed "a strikingly close resemblance between portions of BN 16204" and this Corpus Christi 248, exactly with f. 82r, which contains a list (it looks to me like an offer for further copies of astrological treatises, like that of Albumasar's *Introductorium* which precedes fols. 3r-81r). Pingree has examined deeply this list of "books of Arabic authors which have perhaps been translated into Latin by John of Seville", and which, according to Thorndike, "represent copies made from the same exemplar of a collection of Arabic astrology in Latin translation" as BN 16204 or from this very codex. Pingree observed that to the eleven items listed in both manuscripts, the list in Corpus Christi 248 adds two works not contained in BN 16204, but present in the *Speculum* concerning Abu Ma'shar's *De revolutione annorum nativitatum*, which was item 5 of the Parisinus ms., one reads in ms. Corpus Christi 248: "Sequitur quod non habeo de sine [?] revolutione annorum ex libro Albumasar in revolutione nativitatis extracte". This note corresponds to the fact that "the entire subject of anniversary horoscopes" is omitted in the *Speculum*. Pingree observes that this Corpus Christi manuscript is not only based directly on Paris, BN lat. 16204 before it was deprived of *Aomar and Abu Ali, but written by the author of the *Speculum astronomiae* who notes correctly that he has not included in the *Speculum* an item of the Parisinus; so Pingree concludes that "if our hypothesis is correct, the Corpus Christi catalogue was written by Albertus Magnus himself or by a close associate".
24. Pingree, 'The diffusion' *cit.*, pp. 86-87.
25. R. Lemay, 'De la Scholastique à l'Histoire par le truchement de la Philologie', in *La diffusione delle scienze islamiche cit.*, p. 487: "Soit dit en passant, le *Speculum* n'est pas expressément reclamé comme son oeuvre propre par Albert, qui semble plutôt l'avoir composé sur l'injonction du Pape et en vue de procurer un guide 'orthodoxe' des *libri naturales*, qui permettrait d'éliminer comme 'par la bande' et sans un décret spécial les fameuses condamnations des *libri naturales* d'Aristote portées en 1210 et restées formellement en vigueur jusque là". Prof. Lemay concludes here: "Un moment capital de ce désengagement de la philosophie naturelle d'Aristote d'avec les ouvrages arabes d'astrologie fut la composition certainement par Albert le Grand et aux environs de 1250, du *Speculum astronomiae*". I am unable to check all the suggestions given by Prof. Lemay ("sur l'injonction du Pape", relationship with Aristotle's condemnations etc.) and I am proposing to date the *Speculum* several years later, but I consider Prof. Lemay's interpretation extremely valuable and interesting. See also his 'The Teaching of Astronomy in mediaeval Universities, principally at Paris in the 14th Century', *Manuscripta*, XX, 1976,

pp. 197-217 and his articles cit. above, n. 3 § II/1. He has now consecrated to the *Speculum* Appendix VI of 'De la Scholastique' *cit.*, p. 526 ff.: "S'il y a encore des médiévistes pour soutenir que le *Speculum astronomiae* n'est pas d'Albert, ce ne peut être que par suite de l'influence délétère des préjugés d'un Mandonnet, mais aussi par un manque de familiarité avec les réalités de l'astronomie-astrologie médiévale [... pour] s'ériger sans motif sérieux contre l'opinion universelle des savants latins du moyen age".

CHAPTER TWELVE

1. R. Sabbadini, *Le scoperte dei codici latini e greci ne' secoli XIV e XV*, Firenze 1967, 2nd ed., I, cap. I: 'Gli scopritori veronesi', pp. 4-20. Pastrengo's work was edited by Michele Biondo, Venice 1547, as the first of a series of analogous publications. Biondo introduced some corrections, particularly as far as the list of works attributed to Aristotle was concerned, which he and other humanists considered to be unacceptable and uncritical. On the interesting activity of Michele Biondo (1500-1565?) *cf.* the excellent entry by Giorgio Stabile in *Dizionario biografico degli italiani*, X, Roma 1968, pp. 560-63.
2. Sabbadini, *op. cit.*, I, p. 7 and 22. This conclusion is accepted by A. Avena, 'Guglielmo da Pastrengo e gli inizi dell'umanesimo a Verona', in *Atti dell'Accademia di agricoltura, scienze, lettere, arte e commercio di Verona*, LXXXII (S. IV, vii), Verona 1907, pp. 229-85, see p. 279 in particular; Avena painstakingly reconstructed the biography of Guglielmo and published documents relating to Guglielmo's contacts with Petrarch. Avena also pointed out the contrast between "la sua predilezione per l'astrologia" and "l'odio accanito ch'ebbe invece per i cultori di essa il Petrarca". The latter had trusted his library to Guglielmo during his journey to Rome on the occasion of the jubilee of 1350: according to Avena, this date represents the end of the period during which it is possible to argue that Guglielmo composed his *De originibus*. The work, it is claimed, was posterior to 1337, the year of the death of some of the figures quoted in a work explicitly excluding all reference to living contemporaries, and was also posterior to 1346, when a Veronese inscription allegedly by Livius, and quoted in the work, was discovered. Among the philosophical works quoted by Guglielmo, Avena indicated the Boetian translations of Aristotle, and the *Timaeus* commented on by Calcidius, as well as more modern works by Alain of Lille, Goffredo da Viterbo, Alexander Neckam, Vincent of Beauvais, Walter Burleigh, Ricoldo da Montecroce and Uguccione da Pisa. Thorndike too mentioned Guglielmo's strong "interest in Arabic astrologers" (TH, III, 592), yet, when in the article 'Traditional Medieval Tracts'*cit.* he made use of the *De originibus* to highlight some of the texts on images quoted in chap. XI of the *Speculum*, he was not aware that he was dealing with the very source of the work.
3. *De originibus cit.*, p. 13.
4. C. Cipolla, 'Attorno a Giovanni Mansionario e a Guglielmo da Pastrengo', in *Miscellanea Ceriani*, Milano 1910, pp. 743-88; Cipolla took those passages from the ms. Vat. lat. 5271, ff. 2-5r, and above all from the ms. Ottobonianus lat. 92, ff. 1v-3r, which are to be preferred to the ms. Marcianus lat. X, 51 employed by the sixteenth-century editor. A critical edition of this text though promised by the late Roberto Weiss is still lacking. The edition by Biondo prints only the last entries of the very long list of works attributed to Aristotle, but these entries are sufficient to betray one of Guglielmo's main sources (be-

sides Walter Burleigh): "Scribit Laercius, libro de vita phylosophorum, Aristotelis opera ad tercentorum voluminum summam accedere; alibi legitur quod ad mille". Yet, Guglielmo had been able to quote 146 titles only, among which the one quoted in the text (pp. 775-76) and the "De iudiciis in astrologia, qui incipit: Signorum alia" are indebted to two passages of the *Speculum astronomiae*, VI/21-24 (where, in any case, the vulgate has "Haly" and not "Aristoteles", the name found in the mss. and the edition by Cumont) and XI/25-33. In order to identify the second treatise, "Item de ymaginibus, qui omnium est pessimus qui loquuntur de ymaginibus, hunc ad Alexandrum scripsit", Cipolla rightly refers to the *Speculum*, but does not consider this work to be the source of Guglielmo: indeed, he put forward the curious identification of that source with a hypothetical ms. also quoted in the Flores –an earlier Veronese compilation– wherein is mentioned "Aristoteles in Ycono", a term that does not refer to a treatise on images, but to the "yconomicorum libri".

5. In *Mélanges A. Pelzer*, Louvain 1947, pp. 217-74.
6. F. Carmody, *Arabic Astronomical and Astrological Science cit.*, *passim*, refers to the *Speculum* in almost every article, in order to identify translations and works. B. Nardi, *Saggi sull'aristotelismo padovano*, Firenze 1958, p. 29 ff. employed the *Speculum* as a theorical guide to reconstruct the outlines of Pietro d'Abano's astrology.
7. L. Thorndike, 'A Bibliography composed around 1300 A. D. of Works in Latin on Alchemy, Geometry, Perspective, Astronomy and Necromancy', *Zentralblatt für Bibliothekswesen*, LV, 1938, pp. 225-260.
8. L. Thorndike, 'Notes upon some Mediaeval Latin Astronomical and Mathematical Manuscripts at the Vatican Library', *Isis*, XLIX, 1958, p. 36; the ms. Ottob. lat. 1826 that at f. 80v has a marginal note explicitly dated "1333", at f. 85r carries a quotation from the *Speculum* (644b 23ss) on the "calculatio certissima" of the instant of Christ's birth, clearly by the same hand. *Cf.* Thorndike, 'Some little known astronomical and mathematical manuscripts', *Osiris*, VIII, 1948, pp. 62-63, mentioning CLM 2841, a fifteenth-century astrological miscellany that at f. 15r ff. has the *Liber de iudiciis* (inc.: Nota quod omnia quae dicimus in nativitate alicuius ita eadem dicuntur in quaestione, sed non ita proprie...), a work the ms. claims had been attributed to Aristotle in the *Speculum*. Thorndike felt that the edition by Borgnet left "some mystery". The critical text we have established (VI/12-14) solves the mystery: for this extract should be identified with the "secundo tractatu in quo agitur de interrogationibus" of a pseudo-Aristotelian work (*cf.* below the critical commentary which restored this identification) bearing the incipit: "Signorum alia sunt masculini generis", attributed to Aristotle and printed with the title *Liber ad Alconem regem*, Venice, 1509.
9. L. Delisle, *Le Cabinet cit.*, II, p. 90, published a catalog of 1297 where we already find the attribution to Albert: "Tractatus Alberti de continentia librorum astronomicorum et differentia eorum, qui sunt noxii et qui non. [Inc]: *Quoniam quidam libri apud nos*". It is easy to correct the wording of the *incipit*: "occasione quorundam librorum apud quos".
10. R. A. Pack, 'Pseudo-Aristoteles: Chiromantia', *Archives d'histoire doctrinale et littéraire du Moyen Age*, XXXIX, 1972, p. 309: "a naturalium membrorum signis declarari possunt naturales hominum inclinationes affectuum... Est autem haec scientia [physiognomia] necessitatem non imponens moribus hominum, sed inclinationes ex sanguine et spiritibus physicis ostendens, quae retineri possunt freno rationis". Similar anti-deterministic allusions are to be found in the other 'Pseudo-Aristotelian Chyromancy', edited by R.A. Pack

himself *ibid.*, XXXVI, 1969, p. 233: "debes ergo scire quod haec non nunciant aliquem effectum futurum vel venturum, et ideo non debes iudicare quod sic necessario evenirent, sed solum quod hoc solitum est secundum inclinacionem nature, et quod disposicio est ad talia, quorum signa videbis in manu". On the topic of physiognomy the author of the *Speculum* expressed himself explicitly in chap. XVII/17-21; he suspended judgment, though he left it to be understood that he was not opposed to the practice: "forte pars est phisiognomiae quae collecta videtur ex significationibus magisterii astrorum super corpus et super animam, dum mores animi conicit ex exteriore figura corporis; non quia sit una causa alterius, sed quia ambo inveniuntur ab eodem causata."

11. T. Käppeli, *Scriptores O.P. cit.*, II, pp.199-200 and the secondary literature there cit. do not mention this fragment preserved in two Oxford mss. (Corpus Christi College 243 and 283), attributed to Olivier Lebreton or de Tréguier (Trecorensis or Armoricensis) O.P. only by P. Glorieux, *La Faculté des Arts et ses maîtres au XIIIe siècle*, Paris 1917, s.v. and *Id.*, *Rèpertoire cit.*, I, s. v. 46. This Oliverius Brito was listed by Bernardus Guidonis and Laurent Pignon in their catalogs of learned Dominicans: a lecturer at the dominican convent in Angers, he was connected with Giles of Rome; around 1288 he read the *Sentences* at Saint Jacques' in Paris, where he became *magister theologiae* and *regens* in 1291-92; Provincial for France in 1293 and 1294, author of lost commentaries and unedited *Quodlibeta*, he died at Angers in 1296, according to Quetif-Echard, *Scriptores cit,*, p. 448; *Histoire Littéraire de la France*, Paris 1842, pp. 303-304, and P. Glorieux, *La littérature quodlibétique*, Paris 1935, p. 211. The *Philosophia*, still unedited, was however attributed to an older unknown author by R. A. Gauthier, 'Arnoul de Provence et la doctrine de la *phronesis*', *Revue du Moyen Age latin*, XIX, 1963, pp. 139, 143: "Olivier Lebreton devait- ...être un collègue et contemporain d'Arnoul de Provence et de Nicolas de Paris, un maître de la Faculté des Arts de Paris vers 1250"; *cf.* C. Lafleur, *Quatre introduction à la philosophie au XIII siècle. Textes critiques et études historiques,* Montréal-Paris, Vrin, 1988, pp. 53, 391-392, who annonces his edition and underlines the correspondences with *Speculum astronomiae*, *Proem*/2-3, 5-7; XVII/*passim*. I look forward to reading this forthcoming edition to check data and the context of these interesting quotations, which have been very kindly submittted to my attention by Mr. Lafleur.

12. *Cf.* D. Planzer, 'Albertus-Magnus-Handschriften in mittelalterlichen Bibliothekskatalogen des deutschen Sprachgebietes', *Divus Thomas* (Freiberg), X, 1932, pp. 378-408; Planzer studied a catalog of Albertinian works compiled at the end of the fifteenth century in the Carthusian Monastery of Salvatorberg near Erfurt. On the basis of studies he quoted, the scholar pointed out (p. 248) that the ancient catalog of the first Dominican convent in Cologne was unfortunately lost in a fire, together with the greater part of the books there quoted: we are here referring to the Dominican convent in Cologne, where Albert had deposited "libros meos universos librariae communi". Among the ancient catalogues, we should also remember the one edited by A. Werminghoff, 'Die Bibliothek eines Konstanzer Officials [Johann von Kreuzlingen, J. U. D.] aus dem Jahre 1506', *Zentralblatt für Bibliothekswesen*, XIV, 1897, pp. 290-298, listing several Albertinian and Thomist writings, and, together with the the "articulos parisienses" –referring probably to the condemnation of 1277– "Albertum Magnum de defensione astrologiae ac suppositione eiusdem; eundem de signis; de substantia et substantivo secundum Thomam; de judiciis astrorum [Thomae Aquinatis]; eundem Albertum de sensu et sensatu; de ente et essentia [Thomae Aquinatis]; Eugidii [Romani] theureumata cum resolutionibus".

13. P. van Loe, 'De vita et scriptis B. Alberti Magni', in *Analecta Bollandiana*, XIX, Bruxelles 1900, pp. 276-277, § 13: "In Monasterio Praedicatorum Coloniae habetur opus eius [Alberti] solemne *Super Mattheum* propriis manibus suis scriptum. Aliud etiam volumen *De naturis animalium* de manu sua et *Speculum mathematicae* similiter de manu sua". The fate of the Dominican library, much admired by sixteenth-century visitors, is reconstructed by K. Löffler, *Kölnische Bibliotheksgeschichte im Umriss*, Köln 1923, pp. 11 and 13; "sie brannte am 2. März 1659 ab, wodurch wichtige Manuskripte von Albert dem Grossen und Thomas von Aquin zu grunde gegangen sein sollen..."; still identifiable are, however, two Albertinian autographs that, together with the *Speculum astronomiae*, were worshipped in the fifteenth-century by Peter of Prussia, and were proudly shown to J. J. Björnstahl during his visit in 1774: "Bei den Dominikanern besah er 'zwei, wie man sagt, von Albertus Magnus geschriebene Manuskripte auf sehr feinem seidnem Zeuge oder dem feinsten Kalbpergamene; das eine ist im Quart mit dem Titel *De animalibus*, das andere in Folio". From this description, it is possible to recognize the ms. of the *De animalibus*, that has been authenticated and published by Stadler in the critical ed. Münster 1915-16 *cit.* (Köln, Historisches Archiv der Stadt, W 8° 258), and the Commentary to Matthew preserved in the same archive (ms. W 4° 259); *cf.* Alberti Magni *De vegetabilibus*, ed. H. Stadler-C. Jessen, Berlin 1867, p. 672; H. Ostlender, 'Das Kölner Autograph des Matthaeus Kommentars Alberts des Grossen', *Jahrbuch des kölnische Geschichtesvereins*, XVII, Köln 1935, pp. 129-42; Id., 'Die Autographe Alberts des Grossen', in *Studia albertina. Festschrift für B. Geyer* (= Beiträge. Supplementband IV), Münster 1952, pp. 3-21. Ostlender mentioned the ms. of the work on the eucharist once owned by the Dominicans in Cologne and then lost, as well as the exemplars copied from lost autographs, such as the *Priora, Perihermeneias, Metaphysica*. Lastly, see B. Geyer, *Prolegomena* to the critical edition from the autograph of the *De natura et origine animae*, in *Opera omnia cit.*, XII, Münster 1955, p. VII, n. 1, as well as the bibliography there quoted.
I wish wholeheartedly to thank Prof. von den Brincken, Stadt. Oberarchivar at Cologne, who kindly sent me the photocopies of Löffler and of the present ms. catalog of the few Dominican and Albertinian manuscripts still preserved in the Historisches Archiv, where, according to him, there is no trace of the *Speculum astronomiae*.
14. Petrus de Prussia, *Alberti Magni Vita* [1486], quoted from the edition in the appendix to Albertus Magnus, *De adhaerendo Deo*, Antwerp 1621. *Cf.* the studies on this and the previous biographies (*cit.* above ch. 2 n. 25), and in particular the coat of arms established by Scheeben, *Les écrits d'Albert cit.* app. 264 ff., 272, 285-87, who attributed the *Legenda*, ed. by van Loe, considered authoritative by Petrus de Prussia, to an anonymous author living in Cologne slightly before Petrus. According to Scheeben, the archetype of this *Legenda* could help in reconstructing the alleged *Legenda I*, that could perhaps go back to Gottfried von Duisburg, Albert's last secretary in Cologne.
15. *Cf.* above ch. 2, n. 25.
16. B. Nardi, 'Le dottrine filosofiche di Pietro d'Abano', *Saggi sull'aristotelismo padovano*, Firenze 1958, pp. 29-37, where he sees in the *Speculum astronomiae* the model of Peter of Abano's astrological theories and a "documento prezioso dell'atteggiamento e del pensiero di un teologo di fronte alla libertà di ricerca [...] e al soverchio zelo teologico".
17. G. Federici Vescovini, '*Albumasar in Sadan* e Pietro d'Abano', in *La diffusione delle scienze islamiche nel Medio Evo europeo cit.*, p. 46.

18. Very recently – after these pages had already been written on the basis of the transcription of the Ms. Paris. lat. 2598 provided in a thesis by R. Pasquinucci under the supervision of prof. E. Garin, Florence, Faculty of Letters, 1964 – a critical edition has been published: Pietro d'Abano, *Il 'Lucidator astronomiae' e altre opere*. ed., introd. e note di G. Federici Vescovini. Padova, Programma e 1 + 1 Editori, 1988. On ff. 100v – 101r of that ms. of the *Lucidator* the subdivisions of astronomy correspond to those of the *Speculum*: "extat una de revolutionibus, alia de nativitatibus, tertia de interrogationibus, reliqua de electionibus; prima siquidem in tres partes [coniunctiones, revolutiones annorum et nativitatum]; mantica, geomantia, ydromantia, aerimantia, piromantia, horospitium, augurium etc." and moreover, the bibliography that follows seems to have been extracted summarily, but exactly, from it.
19. G. Federici Vescovini, 'Peter of Abano and Astrology', in *Astrology, Science and Society. Historical essays*, ed. by P. Curry, Woodbridge/Suffolk 1987, pp. 20-22; *cf. Speculum*, Proemium/ 5. See also Ead.,'Pietro d'Abano e l'astrologia-astronomia', *Centro Intern. di storia dello spazio e del tempo*. Bollettino, n.5 (no date), p.11, n.11, and p.15: Peter of Abano agrees with "l'autore dello *Speculum astronomiae* che Pietro sembra conoscere"; Ead., 'Pietro d'Abano e le fonti astronomiche greco-arabo-latine a proposito del *Lucidator*), *Medioevo*, XI, 1985, pp. 65-96, especially p. 66: "if Peter strongly using Alfargani's *Elementa astronomiae* and Albattani's *De scientia motum astrorum* "risale alle fonti" of the *Sphaera* and of Campanus' *Theorica*, this note can apply as well to the *Speculum astronomiae*, these treatises being its main sources too"; Ead. 'Un trattato di misura dei moti celesti, il 'De motu octavae sphaerae' di Pietro d'Abano', in *Mensura. Mass, Zahl, Zahlensymbolik im Mittelalter* (= Miscellanea mediaevalia, 16/2) Berlin-New York 1984, pp. 280-281; Ead., 'La teoria delle immagini di Pietro d'Abano e gli affreschi astrologici del Palazzo della Ragione a Padova', in *Die Kunst und das Studium der Natur*, hg. v. W. Prinz -A.Beyer, Weinheim, VHC 1987 (= Acta humaniora, 1987), pp. 27 ff.
20. G. Federici Vescovini, 'Peter of Abano' *cit.*, p. 24 n. 16, where a long passage is cited from the Lucidator, ms. Paris. lat. 2598, f. 99a, against "plurimi qui doctrinis phylosophicis indocti sermones illustrissimi et praecipui Ptolemei prave intellectos suscepere"; *cf. ibid.*, p. 29.
21. *Ibid.*, p. 26 n. 23: "Propter primum sciendum quod quidam assignarunt differentiam inter astronomiam et astrologiam, dicentes astronomiam fore illam quae partem motus pertractat, astrologia autem quae iudicia instruit. Sed illud neque ratio construit aut multorum usus persuadet, cum astronomia dicatur ab astro et nomos, lex; astrologia vero a logos quod ratio, et sermo et logia, locutio. Hac autem indifferentia, similiter alterutrumque invenio in alterutro eius partem utramque proferri (*Lucidator*, diff. 1, f. 100ra; and *cf. Conciliator*, diff. 10, propter primum)."
22. *Ibid.*, p. 27.
23. *Ibid.*, p. 27 quoting from *Lucidator*, diff. 1.
24. TH, III, 12.
25. TH, III, 16.
26. S. Caroti, *L'astrologia in Italia*, Roma 1983, p. 196.
27. Id., *La critica contro l'astrologia di Nicole Oresme*, Roma 1979 (= Accademia Nazionale dei Lincei. Memorie. Classe di Scienze morali storiche e filologiche, S. VIII, xxiii, 6), pp. 555-556.
28. *Ibid.*, p. 562.

29. *Cf. Vigintiloquium de concordantia astronomicae veritatis cum theologia*, in Pierre d'Ailly, *De Ymagine mundi*, s. l. ed. [*ca.* 1480], ad verbum III.
30. *Elucidarius astronomicae concordiae cum theologia et hystorica narratione*, caput 2, *ibid.*, where Pierre addresses to the *Speculum* only one internal criticism, complaining that it had placed the ascendent of the horoscope of Christ in one zodiacal sign rather than in another without refuting such a horoscope on principle. *Cf.* TH, IV, p. 105; *cf.* L. Salembier, *Petrus ab Alliaco*, Lille 1886, pp. 177-194; O. Pluta, 'Albert von Köln und Peter von Ailly', *Freiburger Zeitschrift für Philosophie und Theologie*, 32, 1985, pp. 269-270: while visiting Köln in 1414 Pierre studied Albert's writings, "einen der wichtigsten namentlich genannten Autoritäten".
31. *Apologia defensiva astronomiae ad magistrum Johannem cancellarium parisiensem*, cited by TH, IV, p. 112 and n. from ms. Paris BN, lat. 2692, f. 147v: "Concordemus denique cum Alberto Magno, doctore sancti Thomae, in illo praecipuo tractatu suo qui *Speculum* dicitur, ubi hanc materiam plene utiliter pertractat." *Cf.* Pierre d'Ailly, letter to Gerson (November 1419) in Gerson, *Oeuvres complètes*, ed. P. Glorieux, II, Paris 1960, p. 221.
32. Pierre d'Ailly, *Concordantia astronomiae cum theologia... cum historica narratione. Elucidarius*, Wien, E. Ratdolt, 1490, f. a2v. *Cf.* his letter to J. Gerson, November 1419, in Gerson, *Oeuvres complètes cit.*, II, p. 219: "vera astronomia [...] tamquam naturalis quaedam theologia illi supernaturali theologiae et tamquam ancilla dominae subserviens".
33. *Elucidarius cit.*, f. a2v.
34. *Ibid.*, f. a3r verbum 2um.
35. *Ibid.*, f. a3r verbum 2um.
36. *Ibid.*, f. a3r, verbum 3um. *Cf.* also the conclusion of the *Elucidarius*, where the treatment of Christ's horoscope and the zodiacal image of the Virgin are very similar to those in *Speculum*, XII/60-100.
37. P. Tschackert, *Peter von Ailli*, Gotha 1877, pp. 328-331.
38. P. Mandonnet, 'Roger Bacon' *cit.*, p. 320 n. 3.
39. F. Pangerl, 'Studien über Albert', *cit.*, pp. 325-326.
40. J. Gerson, *Opera omnia*, ed. L.-E. Dupin, The Hague 1728, 2nd. ed., I, col. 201: *Tricelogium* (propositio III): "Composuit super hac re magnus Albertus opusculum quod appellatur *Speculum Alberti*, narrans quomodo temporibus suis voluerunt aliqui destruere libros Albumasar et quosdam libros alios. Videtur autem, salvo tanti Doctoris honore, quod sicut in exponendis libris physicis, praesertim Peripateticorum, nimiam curam apposuit, maiorem quam Christianum doctorem expediebat, nihil adiiciendo de pietate fidei; ita et in approbatione quorundam librorum astrologiae, praesertim de imaginibus, de nativitatibus, de sculpturis lapidum, de characteribus, de interrogationibus, nimis ad partem superstitionum ratione carentium determinavit".
41. This work written by Gerson on 7 April 1420 has been entitled *Tricelogium*, because it contains thirty propositions with which to answer to Pierre's *Vigintilogium*. For this reason it should not be cited as *Trilogium*, as it is usually done; *cf.* M. Liebermann, 'Chronologia gersoniana', *Romania*, 74, 1953, pp. 321-322, 337. According to Liebermann (who published another series of his 'Chronologia gersoniana', *ibid.*, 70, 1948, pp. 51-67; 73, 1952, pp. 480-498; 76, 1955, pp. 289-333) Pierre d'Ailly and Jean Gerson "ne rejetaient pas complètement cette science" and they associated in this period to "combattre les penchants du Régent [future King Charles VII] pour la fausse astrologie". On the title of Gerson's *Tricelogium* and on his astrology in general, see also P. Glorieux, 'Introduction'

to J. Gerson, *Oeuvres complètes*, Paris 1960, I, pp. 35, 134; F. Bonney, 'Autour de Jean Gerson. Opinions de théologiens sur les superstitions et la sorcellerie au début du XVe siècle', *Le Moyen Age*, LXXVII, 1971, pp. 85-98.

42. J. Gerson, *Opera omnia cit.*, col. 201.

43. *Ibid.*, col. 289 [Proemium]: "quin etiam theologia scientias omnes alias sibi subditas habet velut ancillas, in quibus si quid pulchrum est, illud approbat et decorat, si quid noxium et turpe, illud abiicit et mundat; porro si quid superfluum est, resecat supplens quicquid fuerit diminutum". This occurs precisely with regard to astrology, which the theologian does not deny to have been a "scientiam nobilem et admirabilem primo patriarcae Adam et sequacibus revelatam". Later on astrology was corrupted "tot vanis observationibus, tot impiis erroribus, tot supertitionibus sacrilegis" by those who have not been able "in ea sobrie sapere ac modeste uti"; so that she is now, according to Gerson: "infamis, ... religioni Christianorum ... pestilens et nociva".

44. *Ibid.*, col. 190 (Propositio II).

45. *Ibid.*, col. 191 (Propositio V).

46. *Ibid.*, col. 191 (Propositio IV).

46. *Ibid.*, col. 191 "Propositio VI: Caelum generale influens esse et remotum et actiones suas in patiente disposito recipi. Deum nedum universaliter et remote, sed singularissime et propinquissime operari. Commentum: Erraverunt hic Astrologi quidam ut Alkindus de radiis stellicis [...] ponendo res inferiores nihil agere, sed tantummodo deferre radiosas influentias coeli; et inde fieri effectus similes numero prius istum quam illum, propter determinationum coeli, ut in productione gradum caliditatis".

47. *Ibid.*, col. 192: "Propositio VII. Caelum effectus nedum varios sed contrarios vel oppositos in inferioribus facere pro diversitate materiae".

48. *Ibid.*, col. 192: "Propositio VII: Caelum cum sideribus et planetis in omnibus suis combinationibus motuum, directionum, retrogradationum, oppositionibus cum reliquis circumstantiis, multo plus ab hominibus ignorari quam sciri"; col. 193: "Propositio IX: Caelum habere commensurabile vel incommensurabile motus signorum, insuper et certos planetas huic vel illi genti dominari proprius incertum est". There Oresme "et post eum Petrus cardinalis Cameracensis" are cited. *Cf.* S. Caroti, *La critica cit.*, pp. 636-644, where is cited a passage from the *Apologetica defensio* of 1414, f. 89, "Albertus magnus utique philosophus, astronomus et theologus ... astronomicam potestatem non sic deprimit, quod eam a Christi nativitate nitatur excludere". From that and the following quotation of Albumasar this text has to be traced to *Speculum*, XII/60-100.

49. *Ibid.*, col 191: "Propositio IV: Coelum virtutes a Deo diversas pro varietate suarum partium, stellarum, planetarum et motuum recepisse; sed eas ab omnibus comprehendi non posse. Commentum: Errant et experientiam negant sentientes oppositum, cum coelum sit sicut horologium pulcherrimum compositum ab artifice summo, cum sit etiam liber sententiosissimus exemplatus ab exemplari libro vitae infinito et aeterno, qui nominatur mundus archetypus".

50. *Ibid.*, col. 191: "Propositio V: Coelum obedire ad nutum Deo glorioso atque ipsum operibus humanae recreationis seu reparationis inferius et subditum esse. Commentum: Erraverunt hic multi astrologi et philosophi qui posuerunt Deum agere de necessitate naturae et qui negaverunt mysterium nostrae redemptionis a seculis absconditum, propter quid nedum coelum corporeum, sed etiam angeli et intelligentiae sunt (sicut dixit Apostolus) 'in ministerium missi', *Hebr.*1, 14".

52. *Adversus doctrinam cuiusdam medici delati in Montepessulano sculpentis in numismate figuram Leonis cum certis characteribus*, *Ibid*., cols. 206-207: "Imaginum quae astrologicae nominantur [*cf.* the third type, which "virtutem nanciscitur solummodo a figura caelesti", see *Speculum*, XI/32 ff.] fabricatio et usus, suspectus est plurimum de superstitione et idolatria seu magica observatione [...] Characteres huiusmodi si habeant vel habere credantur efficaciam, oportet quod hoc sit a causa spirituali, non a pure naturali et corporali, qualis causa est coelum cum suis influentiis in corpora [...] ibi sunt characteres, literae, figurae et dictiones, quae nullum effectum habent naturalem pure corporalem ad curationem morbi renum et similium [...] iuxta quod notetur in speciali S. Thomas qui tribuit Astrologiae quantum rationabiliter dari potest, ad exemplum Alberti Magni magistri sui, consone tamen ad fidem catholicam".
53. *Cf.* above n. 40 and also Mandonnet, 'R. Bacon' *cit.*, p. 320 n. 3.

CONCLUSION

1. *Cf.* above ch. 1 n. 2.
2. O. Pedersen, 'The Origins of the *Theorica planetarum*', *Journal of the History of Astronomy*, XII, 1981, pp. 113-123.
3. G. J. Toomer, *s. v.* 'Campanus', in *Dictionnary of Scientific Biograpy*, III, New York 1971, p. 27: "He had a gift for clear and plain exposition. But although he had a good understanding of his material and made few errors, he can hardly be called an original or creative scientist. His philosophical position was an unreflective Aristotelianism; his mathematics and cosmology were equally conventional for his time. His talent was for presenting the work of others in a generally intelligible form. As such, Campanus was a writer of considerable influence"; "the popularization of the idea of the planetary equatorium [...] is also Campanus' strongest claim to originality".
4. See the opening paper of the 8th International Congress of the SIEPM (Helsinki,1987) Gregory T., 'Forme di conoscenza e ideali di sapere nella cultura medievale', *Giornale critico della filosofia italiana*, LXVII (LXIX), 1988, pp. 37 ff.: "Non a caso i problemi cruciali dell'astrologia coincidono con quelli della teologia, a cominciare dalla conciliazione fra necessità e libero arbitrio, fra l'inflessibile moto dei cieli e la realtà contingente... Con molta chiarezza l'autore dello *Speculum* richiama l'antico problema teologico: 'Et fortassis attingentius intuenti, eadem aut saltem similis genere est ista dubitatio ei dubitationi, quae est de divina providentia; nam in his quae operatur dominus per caelum, nihil aliud est caeli significatio quam divina providentia...Unde in libro universitatis...potuit figurare, si voluit, quod sciebat; quod si fecit tunc eadem est determinatio de compossibilitate liberi arbitrii cum divina providentia et cum interrogationis significatione. Si ergo divinam providentiam stare cum libero arbitrio annullari non possit, neque annullabitur quin stet magisterium interrogationum cum eo.' [XIV/82-95] Del resto tutti i teologi devono fare i conti con l'astrologia, posto che essa rappresenta per tutti – dopo l'acquisizione del sistema aristotelico – la coerente applicazione di una legge fisica universalmente accettata, la causalità dei cieli sul mondo sublunare ('certum est per Aristotelem – ricordava Bacone – quod caelum non solum est causa universalis, sed particularis omnium rerum inferiorum'): di qui le discussioni sui condizionamenti fisiologici del libero arbitrio, la funzione degli angeli, motori dei cieli, nel corso della storia, la difficile distin-

zione fra previsione astrologica e profezia. I quesiti del generale dell'ordine domenicano Giovanni da Vercelli a Roberto Kilwardby e a Tommaso d'Aquino [nonché ad Alberto] sono un tipico esempio dei problemi posti al teologo dalla fisica peripatetica".

5. K. A. Nowotny, 'Einleitung', in H. C. Agrippa, *De occulta philosophia*, hg. u. erläutert v. K. A. Nowotny, Graz 1967, p. 422, where he mantainst that Agrippa has written "kein erschöpfendes Handbuch oder eine Sammlung von incipits [...] wie Albertus Magnus [in dem *Speculum astronomiae* und in dem *Libellus de alchimia*], sondern ein Essay über den Sinn der Sache".
6. P. Hossfeld, 'Die Arbeitsweise des Albertus Magnus in seinen naturphilosophischen Schriften', in *Albertus Magnus Doctor universalis cit.*, p. 201; *cf.* Id., 'Albertus Magnus über die Natur des geographischen Orts', *Zeitschrift für Religions- und Geistesgeschichte*, XXX, 1978, p. 107 on the sources of the *De causis proprietatibus elementorum* and of the *De natura loci.*
7. *Id.*, 'Die Arbeitsweise' *cit.*, p. 194.
8. G. Hissette, review of *Speculum astronomiae*, *Bulletin de théologie ancienne et médiévale*, XII, 1979, p. 484. See also the most recent synthesis by A. de Libera, *Albert le Grand et la philosophie*, Paris, Vrin, 1990, pp. 22 ff. He lists "œuvres et éditions" (p. 18 ff.) "en détaillant seulement le principal et l'autenthique", but contradictory to this program he feels himself surprisingly obliged to included in this short list the *Speculum astronomiae* as a "Pseudo-Albert", p. 12 ff. is mainly anecdotical and never considers the authenthical texts existing both on magic and – more often – on astrology. "L'homme [Albert] connaissait bien les savoir arabes, notammont l'astrologie et [sic] l'alchimie, encore devait-il beaucoup aux livres, et il faut prendre garde que ce lecteur infatigable n'a sans doute pas, autant que l'on imagine, manié lui-même les fioles, dc jous de mouron, d'euphorbe ou de joubarbe, l'urine de garçon vierge, l'eau de fleurs de fève et les écailles d'ablette": which is too much according the results of my research.
9. B. Barker Price, 'The Physical Astronomy and Astrology of Albertus Magnus' *cit.*, p. 179.
10. Our edition of the *Speculum*, here reproduced with an English translation, considerably improves the Jammy-Borgnet text by the use of six mss. and the examination of all others in a series of examples chosen at crucial points in the text. Concerning the method of this edition, I will not argue now with some negative reviewers, as were the late F. Weisheipl, Prof.Hissette and Prof. Pedersen. *Cf.* the precise answer given to Weisheipl's remarks by S. Caroti- S. Zamponi, 'Note', *Annali dell'Istituto e Museo di Storia della Scienza di Firenze*, V/2, 1980, pp.111-117.

Albertus Magnus: *Speculum Astronomiae*

(Latin Text established by S. Caroti, M. Pereira, S. Zamponi and P. Zambelli. English translation by C.S.F. Burnett, K. Lippincott, D. Pingree and P. Zambelli)

This text here, and the selection, p. 275 ff., of the main astrological sources, previously appeared as the edition and historical commentary of Alberto Magno, *Speculum astronomiae*, S. Caroti, M. Pereira, S. Zamponi eds., under the supervision of P. Zambelli, Pisa, Domus Galilaeana, 1977, to which the reader is referred for variants, and for a detailed list of the fifty-three Mss. with their respective contents (pp. 95–175), titles and attributions (pp. 177–181), as well as ancient editions (pp. 183–188) and compendiums(pp. 189–193). There is also a glossary (pp. 197–206) and a and a list of authors and incipits of astrological works quoted in the *Speculum* (pp. 209–210). I need only mention here that the edition of the text was based on the two oldest Mss., indicated in the brief table below as L and P; in other words L, which is paleographically datable to the years 1260–1280 and therefore to the work's composition, and P which is approximately a generation later (end of the 13th Century or early 14th century). Since the two Mss. are entirely independent, their agreement establishes the text. After an analysis of all the Mss. (see below a short list of the complete series) based on fifteen sample passages, we had supplied a collectio variorum of four more recent 15th. Century Mss., two of which Cumont had already used for his partial edition (G and M), and two other which are in the same tradition of the older independent ones (B = P, A = L) although they partake of the widespread process of contamination common to 15th Century Mss.

MANUSCRIPTS OF THE *SPECULUM ASTRONOMIAE*

Arras, Bibliothèque Municipale, ms. 47 (844), fols. 53v-57r.
Bergamo, Biblioteca Civica A. Maj, ms. 1177 (Sigma.II.2; MA 388), fols. 50r-58v.
Berlin, Staatsbibliothek Preussischer Kulturbesitz, ms. lat. folio 192, fols. 142v-147r. **B**
Berlin, Staatsbibliothek Preussischer Kulturbesitz, ms. lat. folio 246, fols. 75v-79r.
Bern, Bürgerbibliothek, ms. 483, fols. 132r-138v.
Bernkastel-Kues, Hospitalbibliothek, ms. 209, fols. 106r-113v.
Bologna, Biblioteca Universitaria, ms. 3649, int.11, fols. 3r-56v.
Boston, F.A. Countway Library of Medicine-Boston Medical Library, ms. 22, fols. 1r-7v.
Bruxelles, Bibliothèque Royale, ms. 926-40, int. 436, fols. 206r-215.
Bruxelles, Bibliothèque Royale, ms. 1022-47, int.1030, fols. 83r-89r.
Cambridge, Trinity College, ms. 1185 (O.3.13), fols. 1r-7v.
Catania, Biblioteca Universitaria, ms. Un. 87 (già 85), fols. 175v-184r..
Città del Vaticano, Biblioteca Apostolica Vaticana, ms. Borghes. 134, fols. 224v-230v.
Città del Vaticano, Biblioteca Apostolica Vaticana, ms. Pal. lat. 1445, fols. 176r-187v.
Città del Vaticano, Biblioteca Apostolica Vaticana, ms. Vat. lat. 4275, fols. 18v-28v.

Darmstadt, Hessische Landes- und Hochschulbibliothek, ms. 1443, fols. 230r-235r.
Douai, Bibliothèque Municipale, ms. 427, int. 1, fols. 1r-5v.
Erfurt, Wissenschaftliche Allgemeinebibliothek, ms. Amplon. Q. 189, fol. 70r-v.
Erfurt, Wissenschaftliche Allgemeinebibliothek, ms. Amplon. Q. 223, fols. 105v-116v.
Erfurt, Wissenschaftliche Allgemeinebibliothek, ms. Amplon. Q. 348, fols. 114vb-125va.
Erfurt, Wissenschaftliche Allgemeinebibliothek der Stadt, ms. Amplon. Q. 349, fols. 98r-108r.
Firenze, Biblioteca Medicea Laurenziana, ms. Ashburnham 210, fols. 178r-183r. **A**
Firenze, Biblioteca Medicea Laurenziana, ms. Pl. XXX.29, fols. 80r-85r. **L**
Firenze, Biblioteca Nazionale Centrale, ms. Magliab. XI 121, fols. 222r-226r.
Gdańsk, Biblioteka Gdańska Polskiej Akademii Nauk, ms. 2224, fols. 136r-140r.
Gent, Bibliotheek der Rijksuniversiteit, ms. 416 (151), fols. 48r-56r. **G**
Groningen, Universiteitsbibliotheek, ms. 103, fols. 3r-13v.
Klosterneuburg, Stiftsbibliothek, ms. CCL 683, fols. 190r-205r.
Kraków, Biblioteka Jagiellonska, ms. BJ 1970, fols. 48 r-57r.
Kraków, Biblioteka Jagiellonska, ms. BJ 2496, fols. 85r-95v.
Leipzig, Universtätsbibliothek, ms. 1467, fols. 104r-110r.
Ljubljana, Narodna in univerzitetna knijžnica, ms. 23, fols. 16v-23r.
London, Institution of Electrical Engineers, ms. Thomson Coll. 5, fols. 1r-43r.
London, British Library, ms. Harley 2378, fols. 183r-184v.
Milano, Biblioteca Ambrosiana, ms. I 65 Inf., fols.82r-95v.
München, Bayerische Staatsbibliothek, CLM 27, fols. 55r-v.
München, Bayerische Staatsbibliothek, CLM 221, fols. 223r-227v.
München, Bayerische Staatsbibliothek, CLM 267, fols. 91r-94v.
München, Bayerische Staatsbibliothek, CLM 8001, fol. 145r.
München, Bayerische Staatsbibliothek, CLM 18175, fols. 125r-133v. **M**
Oxford, Bodleian Library, ms. Ashmole 345, fols. 14v-21r.
Oxford, Bodleian Library, ms. Canonici Misc. 517, fols. 52v-59v.
Oxford, Bodleian Library, ms. Digby 81, fols. 102r-117v.
Oxford, Bodleian Library, ms. Digby 228, fols. 76r-79r.
Paris, Bibliothèque de l'Arsenal, ms. 387 (missing fols.).
Paris, Bibliothèque Nationale, ms. lat. 7408, fols. 120r-136v.
Paris, Bibliothèque Nationale, ms. lat. 7335, fols. 108r-114v.
Paris, Bibliothèque Nationale, ms. lat. 7440, fols. 1r-16v. **P**
Salzburg, Stiftsbibliothek St. Peter, ms. b III 15, fols. 18v-28v.
St. Gallen, Kantonsbibliothek (Vadianische Bibliothek), ms. 412., fols. 2r-15r.
Venezia, Biblioteca Nazionale Marciana, ms. lat. 2337 (1582; XI 71), fols. 1r-19v.*
Venezia, Museo Civico Correr, Fondo Cicogna, ms. 1097 (ex 2289), fols. 1r-22v.
Wien, Österreichische Nationalbibliothek, ms.lat. 5508 (Univ. 3367), fols. 161v-180v.

Other Manuscripts

Albumasar, *Introductorium Maius*, transl; Ioannes Hispalensis; Firenze, Biblioteca Medicea Laurenziana, ms. Pluteo XXIX, 12.

Haly, *De electionibus horarum*, Firenze, Biblioteca Nazionale Centrale, ms. Conv. Sopp. J.X.20 (S. Marco 163).

C. Ptolemaeus, *Liber centum verborum cum expositione Haly*, Firenze, Biblioteca Nazionale Centrale, ms. Conv. Sopp. J.X.20 (S. Marco 163).

Abbreviations: Printed works

	Al-Bitrujî (Alpetragius), *De motibus celorum*, critical edition of latin translation of Michael Scot. Ed. by F. J. Carmody, Berkeley – Los Angeles 1952.
	Albohali, *De iudiciis nativitatum*, Nürnberg 1544.
	Albumasar, *Flores*, Venetiis 1489.
Alfraganus, *Numerus mensium*	Alfraganus, *Compilatio astronomica*, Ferrara 1493 (ed. F. J. Carmody, Berkeley – Los Angeles 1943).
AN	Alkabitius, *Enarratio elementorum astrologiae*, ed. V. Nabod, Colonia 1560.
B '98	C. Brockelmann, *Geschichte der Arabischen Litteratur*, Weimar 1898.
B '37	C. Brockelmann, *Geschichte der Arabischen Litteratur. Erste Supplementband*, Leiden 1937.
Birkenmajer	A. Birkenmajer, *Études d'histoire des sciences et de la philosophie au Moyen Age*, (Studia Copernicana, I), Wrocław-Warszawa-Kraków-Gdánsk 1970.
BL	A. Bouché-Leclercq *L'astrologie grecque*, Paris 1899.
BT	Campano da Novara, *Theorica planetarum; Canon pro minutionibus et purgationibus*, edd. F. S. Benjamin e G. J. Toomer, *Campanus of Novara and Medieval Planetary Theory*, Madison, Milwaukee and London 1971.
C	F. Cumont, *De astronomis antiquis testimonium novum. Ex Alberti Magni (1205-1280 p.C.) « Speculo astronomico » excerpta de libris licitis et prohibitis*, in *Catalogus codicum astrologorum graecorum*, V, Bruxelles 1904.
Ca	F. J. Carmody, *Arabic Astronomical and Astrological Sciences in Latin Translation*, Berkeley-Los Angeles 1956.
GW	*Gesamtkatalog der Wiegendrucke*, Leipzig 1925-1938.
G	R. Grossatesta, *De sphaera*, ed. L. Baur, *Die philosophischen Werke des Robert Grosseteste, Bischofs von Lincoln*, Münster 1912 (= « Beiträge zur Geschichte der Philosophie des Mittelalters », IX).
H	C. H. Haskins, *Studies in the History of Mediaeval Science*, Cambridge, Mass., 1947.
Haly	Haly Albohazen, *De iudiciis astrorum*, Venetiis 1485.

Hain	L. HAIN, *Repertorium bibliographicum*, Stuttgart-Paris 1826-1838.
	MESSEHALLA, *De receptione planetarum*, Venetiis 1493.
Millás-Vallicrosa, *Las traducciones*	J. M. MILLÁS-VALLICROSA, *Las traducciones orientales en los manuscritos de la Biblioteca Catedral de Toledo*, Madrid 1942.
M	J. M. MILLÁS-VALLICROSA, *Estudios sobre Azarquiel*, Madrid-Granada 1943-1950.
N	C. A. NALLINO, *Al-Battânî sive Albategni Opus astronomicum, Pars I^a*, Milano 1903.
P	GIOVANNI PICO DELLA MIRANDOLA, *Disputationes adversus astrologiam divinatricem*, ed. E. Garin, Firenze 1943-1952.
	PS. PTOLEMAEUS, *Liber centum verborum cum expositione Haly; Liber Quadripartiti*, Venetiis 1484.
	PTOLEMAEUS, *Almagestum*, Venetiis 1515.
Sa	G. SARTON, *Introduction to the History of Science*, Baltimore 1927-1948.
Saxl	F. SAXL, H. MEIER, P. MC GURK, *Verzeichnis astrologischen und mythologischen Handschriften*, voll. I-III, 4, Heidelberg 1915-London 1966.
St	M. STEINSCHNEIDER, Zum Speculum astronomicum des Albertus Magnus, über die darin angeführten Schriftsteller und Schriften, *Zeitschrift für Mathematik und Physik*, XVI (1871).
S	H. SUTER, *Die Mathematiken und Astronomen der Araber und Ihre Werke*, «Abhandlungen zur Geschichte der Mathematischen Wissenschaften», X (1900)
Thebit, *Aequator diei*	THEBIT B. QURRA, *De hiis que indigent expositione antequam legatur Almagesti*, in F. J. Carmody, *The Astronomical Works of Thabit B. Qurra*, Berkeley-Los Angeles 1960.
	THEBIT B. QURRA, *De imaginibus*, in F. J. Carmody, *The Astronomical Works* cit.
Thebit, *Ptolemaeus et alii sapientes*	THEBIT B. QURRA, *De quantitatibus stellarum et planetarum et proportio terre*, in F. J. Carmody, *The Astronomical Works cit.*
Th	L. THORNDIKE, *A History of Magic and Experimental Science*, voll. I-III, New York 1923-1934.
TK	L. THORNDIKE-P. KIBRE, *A Catalogue of Incipits of Mediaeval Scientific Writings in Latin*, Cambridge, Mass., 1963[2].
TP	L. THORNDIKE, *Traditional Medieval Tracts concerning engraved astrological Images*, in *Mélanges Pelzer*, Louvain 1947.
TS	L. THORNDIKE, *The «Sphere» of Sacrobosco and its Commentators*, Chicago 1949.
	ZAHEL, *De interrogationibus; De electionibus; Introductorium; Quinquaginta praecepta*, Venetiis 1493.
Zinner	E. ZINNER, *Verzeichnis der astronomischen Handschriften des deutschen Kulturgebietes*, München 1925.

Prooemium

Occasione quorundam librorum, apud quos non est radix scientiae, qui cum sint verae sapientiae inimici, hoc est Domini nostri Iesu Christi, qui est imago Patris et sapientia, per quem fecit et saecula, catholicae fidei amatoribus merito sunt suspecti, placuit aliquibus magnis viris, ut libros quosdam alios, et fortassis innoxios accusarent. Quoniam enim plures ante dictorum librorum necromantiam palliant, professionem astronomiae mentientes, libros nobiles de eadem fetere fecerunt apud bonos, et graves et abominabiles reddiderunt. Quare quidam vir zelator fidei et philosophiae, utriusque scilicet in ordine suo, applicuit animum ut faceret commemorationem utrorumque librorum, exponens numerum, titulos, initia et continentias singulorum in generali, et qui fuerunt eorundem auctores, ut scilicet liciti ab illicitis separentur, et aggressus est ut diceret nutu Dei.

Caput Primum

Duae sunt magnae sapientiae et utraque nomine astronomiae censetur. Quarum prima est in scientia figurae caeli primi et qualitate motus eius super polos aequatoris diei et caelorum sub eo positorum, qui sunt compositi super polos alios extra primos, et ipsi sunt caeli stellarum fixarum atque errantium, quorum figura est velut figura sphaerarum sese invicem continentium; in scientia quoque descriptionum circulorum in eis, quorundam scilicet aequidistantium aequatori et quorundam concentricorum eidem, sed declinantium ab ipso, et aliorum egressae cuspidis et quorundam brevium compositorum super peripherias egressorum et aliorum similiter compositorum super cuspidem aequatoris ad quantitatem egressionis cuspidum egressarum ab ea; et in quantitate uniuscuiusque eorum

---

4) Cf. Hebr. 1, 2.

2-5) Albumasar, *Introductorium Maius*, I, 2 (6, 162-9). 3-5) Alfraganus, *Numerus mensium*, Diff. II. 4-5) Alpetragius, *De motibus celorum*, 4.5. 4-7) Alfraganus, *op. cit.*, Diff. V. 4) Thebit, *Aequator diei*, 1. 9) Thebit, *op. cit.*, 3. 10) Alfraganus, *op. cit.*, Diff. XII. 11) *ibidem*. 13) *ibidem*.

PROEM

On account of certain books, which lack the essentials of science [and] which, since they are hostile to the true wisdom (that is, Our Lord Jesus Christ who is the image of the Father and [His] wisdom, by whom He [the Father] made the secular world), are rightly suspect by the lovers of the Catholic Faith, it has pleased some great men to accuse some other books which are perhaps innocent. For, since many of the previously mentioned books by pretending to be concerned with astrology disguise necromancy, they cause noble books written on the same [subject (astrology)] to be contaminated in the eyes of good men, and render them offensive and abominable. Therefore, a certain man zealous for faith and philosophy, [putting] each in its proper place, of course, has applied his mind towards making a list of both types of books, showing their number, titles, incipits and the contents of each in general, and who their authors were, so that the permitted ones might be separated from the illicit ones; and he undertook to speak according to the will of God.

CHAPTER ONE

There are two great wisdoms and each is defined by the name of astronomy. The first of these deals with [1] the science of the configuration of the first heaven; and with the nature of its motion about the poles of the equator of day [and night]; and with the heavens placed beneath it, which are placed on other poles away from the first. These are the heavens of the fixed and wandering stars, whose configuration is like the configuration of spheres enclosing one another. It also deals with [2] the science of drawing circles on them [the heavens], some of which are equidistant from the equator [i.e.: the tropics] and some concentric with it, but inclined from it [i.e.: the ecliptic]; and others have an eccentric center [i.e.: the eccentric deferents], and some are small circles placed on the circumferences of the eccentrics [i.e.: the epicycles], and others are similarly placed above the center of the [concentric] equator by the [same] amount [i.e.: distance] as the eccentricity of the centers of the eccentrics [is] from it [i.e.: the equants]. [3] And [the first wisdom deals] with the size of each of them

et elongatione a terra, et qualiter moventur planetae motu orbium deferentium et motu corporum in orbibus, et quid accidat eis ex variatione situs, ut sunt proiectiones radiorum invisibiles et eclipses solis et lunae caeterorumque planetarum ad invicem; et in esse eorum in circulo suae augis, ut sunt elevatio, depressio, motus latitudinis, inflexionis et reflexionis, et in circulo brevi, ut sunt directio, statio et retrogradatio; et in esse eorum a sole, ut sunt combustio, esse sub radiis, ortus, occasus atque dustoriah, quae est dexteratio in orientalitate a sole et occidentalitate a luna; in mensura etiam magnitudinis sphaerae terrae tam habitabilis quam inhabitabilis cum universis partibus suis tam terrestribus quam marinis, atque longitudinis diametri eiusdem; et in mensura magnitudinis corporum planetarum et stellarum, quarum probatio fuit possibilis respectu magnitudinis sphaerae terrae, quae est communis eorum mensura; et in elongatione eorum a terra secundum mensuram diametri ipsius; amplius in descriptione accidentium quae accidunt universae terrae ex volubilitate circuli de diversitate diei et noctis, et in ascensionibus signorum in circulis directis, qui sunt hemisphaerii lineae aequinoctialis, et in circulis declivibus, qui sunt hemisphaerii climatum, et in divisione ipsorum climatum per crementum longioris diei secundum quantitatem dimidiae horae aequalis, et in quantitate temporum diei et noctis in singulis climatibus; praeterea de diversitate aestatis quae fit bis in anno ex transitu solis super zenith capitum regionum quae sunt ab aequatore diei usque versus finem secundi climatis; et in descriptione locorum quae sunt post climata, quorum plura teguntur a mari et habent unam diem longiorem una revolutione caeli aut pluribus et unam noctem similiter, eo quod in eis in multo tempore non occidat sol, neque in multo

---

14) Alfraganus, *op. cit.*, Diff. XIII. 15) Alfraganus, *op. cit.*, Diff. XVII. 16-17) Alfraganus, *op. cit.*, Diff. XXVIII, XIX, XXX. 18) Alfraganus, *op. cit.*, Diff. XII; Diff. XVIII. 19) Alfraganus, *op. cit.*, Diff. XXVII. 19-20) Alfraganus, *op. cit.*, Diff. XV. 20-21) Alfraganus, *op. cit.*, Diff. XXIV. 21) Alkabitus, *Enarratio elem. astr.*, III. 22-25) Alfraganus, *op. cit.*, Diff. VIII. 23) Alfraganus, *op. cit.*, Diff. III. 25) Alfraganus, *op. cit.*, Diff. XIX, XXII. 25-28) Thebit, *Ptolemaeus et alii sapientes*, 1. 28) Alfraganus, *op. cit.*, Diff. XXI. 29) Alfraganus, *op. cit.*, Diff. VI. 30-1) Alfraganus, *op. cit.*, Diff. X. 31) Alfraganus, *op. cit.*, Diff. VI. 33) Alfraganus, *op. cit.*, Diff. VIII. 35) Alfraganus, *op. cit.*, Diff. XI. 35-38) Alfraganus, *op. cit.*, Diff. VII.

[i.e.: the circles] and its distance from the earth; and how the planets are moved by the motion of [their] deferent circles and the motion of [their] bodies on the [epicyclic] circles; and what happens to them because of the variation of [their] position[s], so that there are invisible projections of rays and mutual eclipses of the sun and the moon and the other planets. [4] And [it deals] with their [i.e.: the planets] situation on the [deferent] circle of their apogee[s] so that greater distance (elevation), less distance (depression), latitudinal motion in one direction (inflection) and in the opposite direction (reflection) happen there; and with [their (the planets) being] on [their] small circle[s] [i.e.: epicycles], so that direct motion, station[s] and retrogression[s] happen there. [5] [It deals] with their situation with respect to the Sun so that combustion, being under the rays, rising, setting and also dustoriah (that is, being to the right of the Sun in the east and of the Moon in the west) occurs there. [6] Also [it deals] with measuring the size of the sphere of the earth, both how much is habitable and inhabitable, together with all [its] parts both of land and of sea, and with the length of its [i.e.: the earth's] diameter; [7] also [it deals] with measuring the size of the bodies of the planets and the stars, which was made possible by using the size of the earth as their common means of measure. [8] And [it deals] with their distance from the earth according to the size of its [i.e.: the earth's] diameter. [9] Moreover, [it is concerned] with describing the accidents which happen to the entire earth due to the alternation of day and night caused by the spinning of the [equatorial] circle; and with the ascensions of the signs in direct circles (which are hemispheres [related to] the equinoctial line [i.e.: right ascensions]); and in oblique circles (which are hemispheres [made relative to] the climes [i.e.: oblique ascensions]); and with the division of those climes according to the increase of a day [which is] longer by the amount of half an equal hour; and with the length of the times of day and night in each clime. [10] Moreover, [it deals] with the different summer created twice during the year by the transit of the Sun passing over the zenith in the regions lying between the equator [of the day and night] and the end of the second clime; [11] and with the description of the places beyond the climes, many of which are covered by the sea and have a single day and, similarly, a single night [lasting] longer than one or more revolutions of the heavens because the sun does not set in them for a long period of time, nor does it rise for

alio oriatur usque dum perveniatur sub polos super quos movetur caelum primum, ubi totus annus efficitur unus dies cum nocte sua. Haec est una magna sapientia, quam dixi nomine astronomiae censeri, et huic non contradicit nisi qui fuerit contrarius veritati.

Caput Secundum

Ex libris ergo, qui post libros geometricos et arithmeticos inveniuntur apud nos scripti super his, primus tempore compositionis est liber quem edidit Nemroth gigas ad Iohanton discipulum suum, qui sic incipit: *Sphaera caeli etc.*, in quo est parum proficui et falsitates nonnullae, sed nihil est ibi contra fidem, quod sciam. Sed quod de hac scientia utilius invenitur, est liber Ptolemaei Pheludensis, qui dicitur graece *Megasti*, arabice *Almagesti*, latine *Maior perfectus*, qui sic incipit: *Bonum fuit scire etc.*, et in commento Geber super *Almagesti* de eodem agitur satis late, et compendiosius in libro Messehalla *De scientia motus orbis*, qui sic incipit: *Incipiam et dicam quod orbis etc.* Quod autem in *Almagesti* diligentiae causa prolixe dictum est, commode restringitur ab Azerbeel hispano, qui dictus est Albategni, in libro suo, qui sic incipit: *Inter universa etc.*; ibique corriguntur quaedam quae ipse dicit non ex errore Ptolemaei, sed ex suppositione radicum Abracaz accidisse, quae tamen fidem pungere non videntur. Ex his quoque duobus libris collegit quidam vir librum secundum stilum Euclidis, cuius commentarium continet sententiam utriusque, Ptolemaei scilicet atque Albategni, qui sic inci-

---

5) NEMROTH, *Astronomia*, See St. 380; C. 86; H. 338; Th. III 14; *Osiris*, I, p. 684. 9) CLAUDIUS PTOLEMAEUS (*Πτολεμαιδς*); Alexandria, d. 161 ca.; *Almagestum*, transl. Gerard of Cremona. Ed. J. L. HEIBERG, *Opera Omnia*, Leipzig 1898-1907. See St. 381; C. 87; Sa. 1, 272-278;TH. I 106 ss.; Ca. 15-16; H. 105. 10) GEBER AVEN AFFLAH (Jâbir ibn Aflah Abû Muhammad al-Ishabîlî); d. 1145 ca.; *Elementa astronomica*, transl. Gerardo of Cremona; ed. Nürnberg 1534; see Ca. 163. 11) MESSEHALLA (Mashâ allâh ibn Atharî al-Basrî); d. 815 ca.; *De motibus*, transl. perhaps by Gerard of Cremona. See St. 376; B ('37) 391-392; S. 5; C. 87; Sa. 1, 531; Ca. 32 14) ALBATEGNI (Muhammad b. Jâbir b. Sinân al Battânî al-Harrânî Abû Abd Allâh); Ar-Raqqa b. 858 ca., d. 929; *De scientia astrorum*, transl. Plato of Tivoli. Ed. C. NALLINO, *Al-Battani Opus Astronomicum*, Milano 1903. See St . 359; C. 87; sa. 1 602-603; ca. 129-130; H. 11 b. 31.

another long [period of] time, until one arrives at [that point] beneath the poles about which the first heaven is moved, where the whole year becomes one day together with its night. This is one great wisdom, which, as I said, is defined by the name of astronomy, and it cannot be contradicted, save by someone who opposes the truth.

CHAPTER TWO

Therefore, amongst the books found by us written on these [matters], after the geometrical and arithmetical books, the first in time of composition is the book written by Nemroth, the giant, for his disciple Iohanton, which begins thus: "*Sphaera caeli etc.*" ("The sphere of heaven etc."), in which there is not much that is useful and quite a few falsehoods, but nothing that is against the faith, as far as I know. But what is found [to be] more useful concerning this science is the book by Ptolemaeus Pheludensis called *Megasti* in Greek, *Almagesti* in Arabic and *Maior perfectus* (*The greater perfect*) in Latin, which begins in this manner: "*Bonum fuit scire etc.*" ("It was good to know etc."); and the same [subject] is discussed sufficiently extensively in Geber's Commentary on the Almagest, and more succinctly in Messahalla's book, *De scientia motus orbis* (*On the science of the movement of the sphere*), which begins in this manner: "*Incipiam et dicam quod orbis etc.*"("I will begin and say that the sphere etc."). That which due to diligence was said in an extended manner in the Almagest, however, is conveniently summarized by Azerbeel the Spaniard, known as Albategni, in his book which begins thus: "*Inter universa etc.*" ("Among all things etc."); and some things have been corrected there, which he himself says are not caused by Ptolemy's error, but have occurred as a result of using the radicals [i.e.: epoch positions] of Abracaz [i.e.: Hipparchus]. These, however, do not seem to offend the faith. Also from these two books someone has compiled a book in the style of Euclid, whose commentary contains the opinions of both Ptolemy and Albategni, and it begins like

pit: *Omnium recte pholosophantium etc.* Corrigitur etiam apud Thebit motus sphaerae stellarum fixarum in libro, qui sic incipit: *Imaginabor sphaeram etc.*, et apud Ioannem vel Gebum Hispalensem motus Veneris et Mercurii in libro quem nominavit *Flores suos*. Et apud alium quendam ampliatur id quod est super figura kata coniuncta atque disiuncta in libello, qui sic incipit: *Intellexi etc.* Voluit quoque Alpetragius corrigere principia et suppositiones Ptolemaei, consentiens quidem suis conclusionibus, sed affirmans caelos inclinatos non apparere moveri in contrarium motus caeli primi nisi propter incurtationem et posteriorationem, cum non possint assequi vehementiam motus primi; et incipit liber suus: *Detegam tibi secretum etc.*, quem siquidem multi recipiunt amplectentes eum ob reverentiam sententiae Aristotelis ex libro *Caeli et Mundi*, quam assumit, quidam vero indignantur, quod malo suo intellectu ausus fuerit reprehendere Ptolemaeum. Subtilius autem quod de hac scientia invenitur, est liber eiusdem Ptolemaei, qui dictus est arabice *Walzagora*, latine *Planisphaerium*, qui sic incipit: *Cum sit possibile Iesuri etc.*, in quo demonstrat in plano quae contingit in sphaera corporea demonstrari. Sine demonstratione vero habentur per viam narrationis apud Alfraganum Tiberiadem eaedem conclusiones, quae in *Almagesti* demonstratae sunt, in libro suo, qui dic incipit: *Numerus mensium Arabum etc.* Et plura ex eis sub compendio in libro Thebit *De Definitionibus*, qui sic incipit:

---

20) CAMPANUS DE NOVARA, *Almagestum Parvum*; transl. Gerard of Cremona; *Contra* St. 359; S. 119; C. 87; Ca , 164; M. 203-205; *Isis* (50) 39; A. BIRKENMAJER, *Études*, Wrocław-Warzawa-Krakóv, 1970, 142-47. 21) THEBIT BENCHORAT (Thâbit b. Qurra b. Mirwân al-Harrânî Abu'l-Hassan); Bagdag b. 834, d. 901; *De motu octave spere*; ed. F. Carmody, Berkely (Calif.) 1941. See St. 387; B. ('98) 217-218; S. 34-38; C. 88; Sa. 1, 599-600; Th. I 661-667; Ca. 116. 23) GEBER (II, 9); *Flores ex Almagesto*; transl. Gerard of Cremona; TK 1403; *Isis* 50 (1959) 40-42; Millás-Vallicrosa, *Las traducciones ...*, 151; Ca. 163-164. 25) THEBIT BENCHORAT (see II, 21); *De figura sectore*; transl. Gerardo di Cremona; See St. 390; C. 88; Ca. 121-123 . 30) ALPETRAGIUS (Nûr ad-Dîn al-Bitrûîj al-Ishbîlî Abû Ishâq); Seville, d. 1185; *De motibus celorum*; transl. Michel Scotus 1217; ed. F. Carmody, Berkeley (Calif.) 1952. See St. 362; B. ('37) 866; S. 131; C. 88. Sa. 2, 399-400; Ca. 165-166; *Osiris*, XII. 36) CLAUDIUS PTOLEMAEUS (see II, 9); *Planisphaerium*; transl. Hermannus de Carinthia 1143; See St. 382; C. 89; Ca. 18. 40) ALFRAGANUS (Abu'l 'Abbâs Ahmad b. Muhammad b. Kathîr al-Fârganî); d. 863 ca.; *De scientia astrorum*; transl. Ioannes Hispalensis 1137; ed. F. Carmody, Berkeley (Calif.) 1943. See St. 365; B. ('37) 292-3; S. 18; C. 89; Sa. 1, 567; Ca. 113-114.

this:"*Omnium recte philosophantium etc.*" ("Of all of those who philosophize correctly etc.") Also, the motion of the sphere of the fixed stars is rectified by Thebit in the book which begins thus: "*Imaginabor sphaeram etc.*" ("I shall imagine a sphere etc."); and the motion of Venus and Mercury [is rectified] by Ioannes or Geber from Seville in the book which he entitled his *Flores* (*Flowers*). And that [part] concerning the figure of the intersection and the separation of the sector is dilated upon [i.e.: amplified] by someone else in a booklet beginning like this: "*Intellexi etc.*" ("I understood etc."). Also, Alpetragius had wanted to correct the principles and suppositions of Ptolemy. Indeed, he [Alpetragius] agrees with his [Ptolemy's] conclusions, but affirms that inclined heavens seem to be moved in a [direction] contrary to the motion of the first heaven only because of advance and regression, since they [the inclined heavens] are unable to attain the velocity of motion of the first [sphere]. And his book begins: "*Detegam tibi secretum etc.*" ("I will disclose a secret to you etc."). Many accept it [this view], embracing it out of a reverence for the opinion[s] of Aristotle, which he [Alpetregius] supports from the *Liber de caelo et mundo* (Book on heaven and earth); but some [others] are displeased, because he [Alpetregius], with his faulty understanding, has dared to find fault with Ptolemy. One can find something more sophisticated on this science, however, in a book by the same Ptolemy called *Walzagora* in Arabic and *Planisphaerium* (The planisphere) in Latin, which begins thus: "*Cum sit possibile Iesuri etc.*" ("Since it is possible, Iesurus etc."), in which he shows that what happens on [the surface of] a solid sphere can be demonstrated on a plane. The same conclusions as those which are demonstrated in the Almagest, moreover, are given, without graphic demonstration, in the form of a narration by Alfraganus Tiberiades, in his book which begins like this: "*Numerus mensium Arabum etc.*" ("The number of the Arabic months etc."). And [there is] more concerning this [topic] in a summary form in Thebit's book, *De definitionibus* (On definitions), which begins in this

Aequator diei etc. Exercitium autem ad inveniendum loca planetarum, et capitis et quaedam alia, est in libro qui dicitur *Liber canonum Ptolemaei*, quem non puto fuisse Pheludensem, sed alium ei aequivocum, qui fuit unus ex regibus Aegypti, et sic incipit: *Intellectus climatum etc.* Et constitutus est super annos Aegyptiorum, qui dicuntur *Childomiz*, ad mediam diem civitatis Alexandriae, cuius longitudo est ab occidente quinquaginta unius graduum et tertiae unius, latitudo vero triginta unius graduum. Post quem composuit *Canones* Machometus Alchoarithmi super annos Persarum, qui dicuntur *Gezdagerd*, ad mediam diem civitatis Arim cuius longitudo est ab oriente et occidente aequalis et latitudo eius est nulla; et post illum scripsit librum *Auxigeg*, hoc est cursuum, Humenid magister filiae regis Ptolemaei, quem vocavit *Almanach*; et hic quidem pro diuturnitate temporis his diebus satis ab exquisitae calculationis veritate declinat. Sed qui perfectius hoc tractavit, fuit Azarchel Hispanus in libro suo, qui sic incipit: *Scito quod annus lunaris etc.*, cuius radices constitutae sunt super annos Arabum, qui dicuntur Machometi, ad mediam diem civitatis Toleti, cuius longitudo est ab occidente viginti octo graduum et medietatis unius, latitudo vero ab aequatore quadraginta graduum fere. Et multi multos libros canonum ad civitates suas super annos domini conscripserunt, ut est ille qui est ad mediam noctem civitatis Massiliae, alius ad mediam diem Londoniarum, et alius ad mediam

---

42) THEBIT BENCHORAT (see II, 21); *De hiis qua indigent expositione antequam legatur Almagesti*; ed. F. Carmody, Berkeley (Calif.) 1941; See St. 387; C. 89; Ca. 118. 45) CLAUDIUS PTOLEMAEUS (see II, 9); *Praeceptum canonis*; See St. 382; C. 89; Th. III 15; Cad. 20; A. A. BJÖRNBO-S. VOGL, *Alkindi, Tideus und Ps. Euklid*, «Abhandl. z. Gesch. d. Math. Wiss.», Leipzig 1912, 22; A. VAN DE VYVER, *Les prèmieres traductions latines (Xe et XIe siècles) de traités arabes sur l'Astrolabe*, Extrait du Ièr Congres International de Geographie Historique, T. II, Memoires, Bruxelles 1931 = *Osiris*, I, 687-689. 49) ALCHORARITHIMI (Abû 'Abd Allâh Muhammad b. Mûsâ al-Kwqârizmî), d. 846; *Canones*; transl. Adelard of Bath 1126 (?). Ed. A. BJÖRNBO-R. BESTHORN, Copenhagen 1918. See St. 375; B. ('98) 215-216, ('37) 381-382; S. 10; C. 89; Sa. 1 563-564; Ca. 46-47; H. 22; M. 249. 54) AMMONIUS, *Canones super tabulas Humeniz philosopohi summi egipciorum*; Sciendum est quod Humeniz phiosophus summus egipciroum, magister filiae regis Ptolemaei ...; cfr. MILLÁS-VALLICROSA, *Estudios sobre Azarquiel*, 235-237, 379-393; cfr. ZINNER, 635-640; St. 365-367. 57) ARZARCHEL (Ibrâhîm b. Yahyâ an-Naqqâsh az-Zarquâli al-Qurtubî, Abû Ishâq); Cordoba d. 1100; *Lectiones tabularum secundum Arzarchelis*; ed. Venezia 1547. See St. 367; B. ('98) 472-473. ('37) 862; S. 109; C. 90; Sa. 1, 758-759; Th. III 15; Ca. 157; M. 59. 61-64) cfr. BIRKENMAJER, pp. 228 ss.

manner:"*Aequator diei etc.*" ("The equator of the day [and night] etc.") Also practice in finding the positions of the planets and of the ascending node and certain other things is [described] in the book called *Liber canonum Ptolemaei (The Book of the canons of Ptolemy)*, whom I don't think was [Ptolemaeus] Pheludensis, but was someone else with the same name who was one of the kings of Egypt, and it begins thus: "*Intellectus climatum etc.*" ("An understanding of the climes etc."). And its [calculations] are based on the years of the Egyptians, which are called "of Childomiz", using noon in the city of Alexandria, whose longitude is 51 1/3° from the west, and latitude is 31° [north]. Following [him], Machometus Alchoarithmi composed his *Canons* [based] on the years of the Persians, which are called "of Gezdagerd", [and are calculated] for noon in the city of Arim, whose longitude is equidistant from the east and west [limits] and whose latitude is zero. And after him, Humenid, the teacher of the daughter of King Ptolemy, wrote a book *Auxigeg*, that is, "*Of the motions [of the planets]*", which he called *Almanach*; and this, due to the length of time [which has elapsed], is quite far from the truth of calculation carried out for these [present] days. But the one who more perfectly handled this [matter] was Azarchel of Spain in his book, which begins thus: "*Scito quod annus lunaris etc.*" ("Know that the lunar year etc."), whose roots [i.e.: epoch positions] are based on the years of the Arabs, which are called "of Machometus", and [are calculated] for noon in the city of Toledo, whose longitude is 28 1/2° from the west, and latitude is approximately 50° [north] from the equator. And many have written many books of canons for their own cities based on the years of Our Lord. For example, there is one for midnight at the city of Marseilles, another for noon at London, and another for noon at Toulouse (which is placed on the same meridian

diem Tolosae, quae est sub eodem meridiano cum civitate Parisius, cuius longitudo est ab occidente quadraginta graduum et quadraginta septem minutorum fere, latitudo vero quadraginta novem graduum et decima unius. Sunt praeterea libri necessarii de hac parte scientiae, qui per viam narrationis demonstrationem planisphaerii imitantur, ut est ille quem transtulit Ioannes Hispalensis, qui sic incipit: *Astrologicae speculationis etc.*; et alius Hermanni, incipiens: *Hermannus Christi pauperum etc.*; et alius Messehalla, qui sic incipit: *Opus Astrolabii etc.*; et iterum alius secundum Ioannem Hispalensem de utilitatibus et opere astrolabii, qui sic incipit: *Primum capitulum in inventionibus etc.* Isti sunt utiliores ex libris astronomiae de motu, qui in latina lingua inveniuntur. *Perspectiva* enim Aristotelis ad supra dicta non descendit. Et isti sunt libri, qui si aspectibus virorum desideriorum subtracti fuerint, magna pars et valde nobilis philosophiae erit sepulta saltem ad tempus, donec scilicet consilio saniori resurgat, quia sicut dicit Thebit filius Chorae «non est lumen geometriae, cum evacuata fuerit astronomia». Et iam sciunt inspectores praedictorum librorum quod in eis non invenitur etiam unicum verbum, quod sit, vel esse appareat, contra fidei catholicae honestatem; neque fortasse iustum est quod hi qui numquam eos attigerunt, ipsos iudicare praesumant.

Caput Tertium

Secunda magna sapientia, quae similiter astronomia dicitur, est scientia iudiciorum astrorum, quae est ligamentum naturalis philo-

---

70) Ioannes Hispalensis, *Astrolabium*; ed. Millás-vallicrosa, *Las traducciones*, 261; St. 377; C. 90; Ca. 169-170. 71) Hermannus Contractus, *De mensura sive de compositione astrolabii*; ed. *Isis*, XVI (1931) 203-212. See St. 371; C. 90; H. 51-52. 72) Messehalla (see II, 11), *De compositione astrolabii*; See St. 376; C. 90; ca. 23. 73) Ioannes Hispalensis; Toledo, between 1130-1151; *Astrolabium*; See St. 374; C. 90; *Speculum*, XXXIV (1959), 20-38. 75) Alhazen, *Perspectiva* (?). cfr. St. 367; V. Rose, *Aristoteles Pseudepigraphus* p. 376. 79-80) Thebit Benchorat, *De imaginibus, Prooemium*.

2-3) Albumasar, *Introductorius*, transl. Ioannes Hispalensis, I, 1. 12-13) Claudius Ptolemaeus, *Liber centum verborum cum expositione Haly*, prooemium.

as the city of Paris, whose longitude is approximately 40° 47' from the west, and latitude 59 1/10° [north]. There are also necessary books on this part of the science which provide a [graphic] demonstration of the planisphere through narration, such as that one translated by John of Seville, which begins like this: "*Astrologicae speculationis etc.*" ("Of astrological speculation etc."); and another by Hermannus, beginning: "*Hermannus Christi pauperum etc.*" ("Hermannus, of the paupers of Christ etc."); and another by Messahalla, which begins like this: "*Opus astrolabii etc.*" ("*The workings of the astrolabe etc.*"); and again another on the uses and workings of the astrolabe, which, according to John of Seville, begins like this: "*Primum capitulum in inventionibus etc.*" ("The first chapter on inventions etc."). These are the more useful books of astronomy concerning motion found in the Latin language; for the *Perspectiva* of Aristotle does not descend to the [subjects] mentioned above. And these are the books, which if they are removed from the sight of men wanting [to study them], a great and truly noble part of philosophy will be buried at least for a certain time, that is, until it would rise again due to a sounder attitude; for, as Thebit, the son of Chora, says: "there is no light in geometry when astronomy has been removed". And the readers of the aforementioned books already know that not even a single word is found in them that might be or might seem to be against the honour of the catholic faith; nor, perhaps, is it fair that those who have never touched these [books], should presume to judge them.

CHAPTER THREE

The second great wisdom, also called astronomy, is the science of the judgements of the stars, which provides a link between natural philosophy

sophiae et metaphysicae. Si enim sic ordinavit Deus altissimus sua summa sapientia mundum istum, ut ipse qui est Deus vivus, Deus caeli non vivi, velit operari in rebus creatis, quae inveniuntur in his quatuor elementis inferioribus, per stellas surdas et mutas sicut per instrumenta (et nos habemus unam scientiam metaphysicam, quae docet nos in rerum causis causatorem causarum considerare, et aliam naturalem, quae docet nos in rebus creatis creatorem creaturarum experiri), quid desideratius concionatori quam habere mediam scientiam, quae doceat nos qualiter mundanorum ad hoc et ad illud mutatio caelestium fiat corporum mutatione? Numquid et haec una est ex praecipuis probationibus, quod non sit nisi unus Deus gloriosus et sublimis in caelo et in terra, si videlicet motus inferior motui superiori oboedit? Si enim essent diversa principia, aut haberet participem in caelo aut in terra, ut essent regnum caeli et regnum terrae diversa, non est verisimile quod esset haec oboedientia fixa permanens absque nutu. Nunc autem ex ista scientia convincitur evidenter, quod dicta oboedientia stet atque immutabiliter perseveret: quare tanto provocat hominem ad Deum ardentius diligendum, quanto per ipsam attingentius omnium princeps atque principium declaratur. Non enim diligetur incognitus, neque cum sit primus, cognoscetur per prius, neque per seipsum, cum sit incomprehensibilis. Restat ergo quod per posterius, per suos scilicet gloriosos effectus. Hi autem sunt homo et ordinatio universi ad ipsum, videlicet supercaelestium, ut praebeant ductum rationalibus, et elementorum, in quibus sumptus rationalium mensurentur, quam universi ordinationem nulla scientia humana perfecte attingit, sicut scientia iudiciorum astrorum.

Quod ut liquidius appareat, descendam ad partes eius, commemorans quasi omnes libros laudabiles, quos de ea pauper latinitas ab aliarum linguarum divitiis per interpretes mendicavit.

and metaphysics. For if God, most high in his supreme wisdom, has ordered this world in such a manner, that He who is the living God [and] the Lord of a heaven which [itself] is not living, should wish to operate through the created things found in these four inferior elements, using the mute and deaf stars as if they were instruments; and if we have one metaphysical science which teaches us how to consider the causer of causes amongst the causes of things; and another, natural science which teaches us to experience the creator of creatures amongst the created things, [then] what could be more desirable to a thinking man than to have a middle science which might teach us how this and that change in the mundane world is effected by the changes in the celestial bodies? And if the inferior motion obeys the the superior motion, is this not one of the primary proofs that there is only God, glorious and sublime in heaven and on earth? For, indeed, if there were different principles, or if He had an associate [either] in heaven or on earth so that the Kingdom of Heaven and the Kingdom of Earth were different, it is improbable that this obedience [of the inferior to the superior] would be fixed, persisting without deviation. Now, however, it is clearly proven by means of this science [astrology] that the obedience referred to exists and perseveres without change; whereby, the more fittingly He is shown to be the Prince of all things and their beginning by this means [astrology], so it more intensely provokes men to love God. For if He is unknown, He is not loved; nor, since He is the first, can He be known by what is prior; nor can He be known through His own essence, since He is incomprehensible. It remains, therefore, that He is known by what is posterior, namely, by His glorious effects. These are man and the ordering of the universe up to Himself, namely: the [ordering of the] supercelestial beings so that they provide guidance to rational beings; and [the ordering] of the elements, in which the material aspects of the rational beings are measured. No human science attains this ordering of the universe [as] perfectly as the science of the judgements of the stars does. So that this should be more obvious, I shall descend to its parts, mentioning, almost all of the praiseworthy books which latin culture, impoverished in this [subject], has begged from the riches of other languages by means of translators.

Caput Quartum

Dividitur itaque ista scientia in duas partes. Prima est introductoria et versatur circa principia iudiciorum. Secunda vero expletur in exercitio iudicandi, et haec iterum divisa est in quatuor partes. Prima est de revolutionibus. Secunda de nativitatibus. Tertia de interrogationibus. Quarta de electionibus horarum laudabilium, cui parti supponitur pars illa quae est de imaginibus, de qua dictum est: «sublimitas astronomiae est imaginum scientia». Sed isti parti associantur illi libri maledicti necromantici, de imaginibus, praestigiis et characteribus, annulis et sigillis eo quod simulationis gratia sibi mutuant quasdam observationes astronomicas, ut sic se reddant aliquatenus fide dignos; quorum venenum in sequentibus patefaciam nutu Dei, sed nunc revertar ad partem introductoriam et caeteras per ordinem, secundum quod promisi.

Caput Quintum

Principia iudiciorum, in quorum scientia consistit introductoria, sunt naturae signorum essentiales, secundum quas dicuntur calida, frigida, humida, sicca, mobilia, fixa, communia, masculina, foeminina, diurna, nocturna, imperantia, oboedientia, se diligentia, et odio habentia, concordantia in ascensionibus aut in fortitudine aut in itinere; et quid sit in divisione eorum ex regionibus, civitatibus atque locis, ex arboribus atque seminibus, ex animalibus quadru-

---

8) Thebit Benchorat, *De imaginibus*, 180.

2) Alkabitius, *Enarratio Elementorum Astrologiae*, Praefatio, 1. 3) Alkabitius, *op. cit.*, Diff. I, 8, 1. 2-3) Albumasar, *Introductorius*, II, 4; VI, 1. 3-4) Alkabitius, *op. cit.*, Diff. I, 9-10, 7. 5) Alkabitius, *op. cit.*, Diff. I, 48, I; Ptolemaeus, *Liber Quadripartiti*, cap. 13; Albumasar, *op. cit.*, II, 6. 4-5) Albumasar, *op. cit.*, II, 8-9; Alkabitius, *op. cit.*, Diff. I, 9, 5. 5-6) Albumasar, *op. cit.*, VI, 4-5. 6) Alkabitius, *op. cit.*, Diff. I, 9, 5. 7) Alkabitius, *op. cit.*, Diff. I, 78, 1; Albumasar, *op. cit.*, VI, 9. 8) Albumasar, *op. cit.*, VI, 23. – Albumasar, *op. cit.*, VI, 22.

CHAPTER FOUR

This wisdom, then, is divided into two parts. The first [part] is introductory and is concerned with the principles of [astrological] judgements. But the second part is fulfilled in the exercise of making judgements; and this [second part] is further divided into four sections. The first is concerned with revolutions [of the years]; the second with nativities; the third with interrogations; [and] the fourth with choosing favourable hours – to which that section which deals with images is subjoined, of which it is said: "the most sublime part of astronomy is the science of images". But those cursed necromantic books on images, illusions and characters, rings and sigils are associated with this part [of the science] because they [i.e.: the necromancers] borrow certain astronomical observations for themselves for the purpose of simulation in order to render themselves [as] slightly credible. By the will of God, I will disclose their poison in what follows, but for now let me return to the introductory part and the other [sections] in [their] order, as promised.

CHAPTER FIVE

The principles of judgements, which make up the introductory [part] are: [1] the essential natures of the signs, according to which they are said to be hot, cold, humid, dry, mobile, fixed, common, masculine, feminine, diurnal, nocturnal, commanding, obedient, loving and hating each other, agreeing in their rising-times or in their strenghth or in their paths [i.e.: straight or crooked]; [2] and what regions, cities, and places, trees and sown plants, quadrupedal animals, birds and reptiles they [i.e.: the signs]

pedibus, avibus atque repentibus, ex membris quoque corporis humani, ex infirmitatibus, et ex quibusdam quae secundum aptitudinem vel ineptitudinem pertinent ad mores animi; et naturae signorum accidentales, secundum quas dicuntur anguli, succedentia angulorum atque cadentia, ex divisione duodecim domorum circuli et quartarum, secundum quas dicuntur corporea et incorporea, quibusdam colorata coloribus, ascendentia et descendentia, longa et brevia. Caeterum naturae planetarum in semetipsis, secundum quas dicuntur calidi, frigidi, humidi, sicci, fortunae et infortunae, hoc est operantes iussu Dei effectum et destructionem, masculini, foeminini, diurni, nocturni; et in esse eorum a sole, secundum quas dicuntur zamin, aut combusti, aut sub radiis, aut orientales, sive occidentales, atque almugea et quaedam alia; et in esse eorum ex circulo, secundum quas dicuntur directi, stationarii et retrogradi; et in esse eorum ad invicem, secundum quas dicuntur coniuncti, aspicientes se, separati, frustrati, refrenati, prohibiti, abscissi a lumine, vacui cursus, recipientes aut recepti, transferentes, colligentes, pulsantes, reddentes, retribuentes, obsessi, ferales, fortes aut debiles. Praeterea participatio planetarum cum signis per dignitates essentiales, ut sunt domus, exaltatio, terminus, triplicitas, facies, neuhahar, haiz et augmentum fortunae, peregrinatio, deiectio, puteus; et per dignitates accidentales, ut sunt gaudia planetarum, dominium horae diei et noctis; amplius partium proiectio quae ex tribus significatoribus colliguntur, duobus scilicet naturalibus et tertio locali.

9-10) ALBUMASAR, *op. cit.*, VI, 12. 10) ALBUMASAR, *op. cit.*, VI, 14 . 11-12) ALKABITIUS, *op. cit.*, Diff. I, 160, 4. 11-16) ALBUMASAR, *op. cit.*, VI. 25-30. 13) ALKABITIUS, *op. cit.*, Diff. I, 112 s.n. 14) ALKABITIUS, *op. cit.*, Diff. I, 184. 15) ALKABITIUS, *op. cit.*, Diff. I, 159, 2. 16-19) ALBUMASAR, *op. cit.*, VII, 1 (311, 22-27). 16-26) ALBUMASAR, *op. cit.*, VII, 1-6 . 18) PTOLEMAEUS, *op. cit.*, cap. 5 19) PTOLEMAEUS, *op. cit.*, cap. 6, 7. 20-21) ALKABITIUS, *op. cit.*, Diff. III, 302, 3; 302, 4. 21) ALKABITIUS, *op. cit.*, Diff. III, 301, 1; PTOLEMAEUS, *op. cit.*, cap. 23 23-24) ALKABITIUS, *op. cit.*, Diff. III, 315-316, 1. 24) ALKABITIUS, *op. cit.*, Diff. III, 318-319, 14; 318, 12; 317, 8. 25) ALKABITIUS, *op. cit.*, Diff. III, 319, 15-16; 316, 2; 317, 9. 26) ALKABITIUS, *op. cit.*, Diff. III, 316, 3. 27) ALKABITIUS, *op. cit.*, Diff. I, 29-30, 1; ALBUMASAR, *op. cit.*, I, 1. 27-29) ALBUMASAR, *op. cit.*, See 1-22 . 28) ALKABITIUS, *op. cit.*, Diff. I, 30, 1; 37-38, 1; 57-58 s.n.; 39, 40, s.n.; 64, 1; 193. 29) ALKABITIUS, *op. cit.*, Diff. I, 97, s.n.; 95, s.n. 30) ALKABITIUS, *op. cit.*, Diff. I, 184-185, s.n.

have in their division; and also [what] parts of the human body, illnesses, and certain things which pertain to the character of the soul with respect to its aptitude or inability [they have]; [3] and the accidental natures of the signs according to which they are called cardines, succedents [of the cardines] and cadents which result from the division of the twelve houses [i.e.: places] of the circle and of the quadrants, according to which they are called corporeal and incorporeal, coloured with certain colours, rising and setting, long and short. [There are] also the natures of the planets in themselves, according to which they are said to be hot, cold, humid, dry, benefic and malefic (that is, bringing about [good] effects and destruction by the command of God), masculine, feminine, diurnal, nocturnal; [4] and their [i.e.: the planets'] situation with respect to the Sun, according to which they are called zamin, or combust, or under the rays, or oriental, or occidental, and also almugea and certain other things; [5] and their [i.e.: the planets'] situation on the [epicyclic] circle, according to which they are said to be direct, stationary and retrograde; [6] their [i.e.: the planets'] relation to each other, according to which they are called conjunct, aspecting each other, separated, frustrated, restrained, prohibited, cut off from light, empty in motion, receiving or received, transferring, collecting, repelling, rendering, distributing, beseiged, feral, strong or weak. Moreover, [7] the participation of the planets with the signs [is to be considered] through [their] essential dignities, such as the house, exaltation, term, triplicity, face, neubarhar, haiz, and augment of fortune, peregrination, dejection and pit; and through [their] accidental dignities such as the joys of the planets, the lordship of the hour of the day and of the night; [and] furthermore, the projection of lots which are calculated from three signifiers, two [of which] are natural and the third local.

Caput Sextum

Circa ista principia consistit pars introductoria, et unus liber qui invenitur super hoc est liber Ptolemaei, qui dicitur graece *Tetrastin*, arabice *Acharbe*, latine *Quadripartitus*, et incipit: *Iuxta providam philosophorum assertionem etc.*, excepto quod tertia pars est de his quae ad nativitates pertinent. Alius liber super hoc est liber Geazar, qui dictus est Albumasar, quem vocat *Maiorem introductorium*, et est in eo confirmatio per rationem, qui sic incipit: *Laus Deo etc.* Et alius liber Abdilaziz, quem vocat Alkabitium, et est absque confirmatione per rationem, qui sic incipit: *Postulata a Deo etc.* Est et *Introductorius* Zahelis, qui sic incipit: *Scito quod signa sunt duodecim etc.* Et alius Aristotelis, qui sic incipit: *Signorum alia sunt masculini generis etc.*, excepto quod in secundo tractatu agitur de interrogationibus. Et alius qui simili modo incipit, et est Ptolemaei ad Aristoxenum. Et alius Ioannis Hispalensis, qui sic incipit: *Cinctura firmamenti etc.*; et in principio libri *Novem iudicum* etiam de eodem tractatur, qui sic incipit: *Caelestis circuli etc.*, et in initiis similiter multorum aliorum librorum.

---

4) Claudius Ptolemaeus (see II, 9). *Quadripartitum*; See St. 382; C. 91; Ca 18-19; H. 111; M. 197. 8) Albumasar (Abû Ma'shar Ja'far b. Muhammad al-Balkî, or Ja'far b. Muhammad b. 'Umar); b. ca. 786, d. 885-6; *Introductorius*; transl. Ioannes Hispalensis. See St. 360; B. ('98) 221-222, ('37) 394-396; S. 29; C. 91; Sa. 1, 568-569; TH. 1, 649-652; ca. 88-101. 10) Alkabitius abdylaziz ('Abd al-'Azîz b. Uthmân b. Ali abu's Saqr al-Qabîsi' al-Misrî) Mosul. d. 967 ca.; *Liber introductorius*; transl. Ioannes Hispalensis. Ed. Venezia 1485-1486; ed. Nabod, Coloniae 1560. See St. 361; B. ('37) 399; S. 60; C. 92; Sa. 1, 669; Th. II, 77; Ca. 144-145. 11) Zahel Bembriz (Sahl b. Bishr b. Habîb b. Hânî - or Hâ yâ-al-Isrâ' îlî, Abû Uthmaân); Khurasan d. 840 ca.; *Introductorium*; Ed. Venezia 1484. See St. 388; B. ('37) 396; S. 15; C. 92; Sa. I, 569; TH. II, 390; Ca. 40-41; M. 162; N. 1, lxxvii. 12) Claudius Ptolemaeus (see II, 9), *Doctrina data filio suo* See St. 383; C. 92; TK 1504; ca. 17 (*De iudiciis ad Aristonem*); ed. Venezia 1509; attribuited to Aristotle with the title: *Liber ad Alconem regem*. 15) Cfr. above, VI, 12. 16) Ioannes Hispalensis (see II, 70) *Epitome totius astrologiae*; ed. Nürnberg 1548. See St. 374; C. 92. 17) *Liber novem iudicum*. Ed. Venezia 1509. See St. 394; C. 92; Ca. 103-105.

CHAPTER SIX

The introductory part is comprised of these principles, and one book found on this is the book by Ptolemy, called *Tetrabiblos* in Greek, *Alarbe* in Arabic and *Quadripartitus* (*The four parts*) in Latin, and it begins: "*Iuxta providam philosophorum assertionem etc.*" ("According to the foreseeing assertion of the philosophers etc.") – except the third part [of the book] is about things which pertain to nativities. Another book on this [subject] is the book by Geazar, known as Albumasar, which he calls *Maior introductorius* (*The greater introduction*); and there is in it a rational demonstration [of astrology], which begins like this: "*Laus deo etc.*" ("Praise be to God etc."). And there is another book by Abdilaziz, who is called Alcabitius, which does not have a rational demonstration, which begins in this way: "*Postulata a Deo etc.*" ("[A long life] having been demanded of God etc."). And there is the *Introductorius* (*Introduction*) of Zahel, which begins in this manner: "*Scito quod signa sunt duodecim etc.*" ("Know that the signs are twelve [in number] etc."). And there is another by Aristotle beginning thus: "*Signorum alia sunt masculini generis etc.*" ("Some of the signs are of the masculine gender etc.") – except that the second treatise [in the book] deals with interrogations. And [another] book which begins in the same way is by Ptolemy [and addressed] to Aristoxenus. And [there is] another [one] by John of Seville, which begins like this: "*Cinctura firmamenti etc.*" ("*The belt of the firmament etc.*"); and the same [subject] is dealt with at the beginning of the *Liber novem iudicum* (*The Book of the nine judges*), which begins: "*Caelestis circuli etc.*" ("Of the celestial circle etc."), and similarly at the beginnings of many other books.

Caput Septimum

Pars autem de revolutionibus divisa est in tres partes. Una est de centum viginti coniunctionibus planetarum et eorum eclipsibus. Secunda de revolutione annorum mundi et tertia de temporum mutatione. Prima ergo consistit in coniunctionibus duorum planetarum in uno signo, et sunt viginti unum coniunctiones. Et trium planetarum, et sunt triginta quinque coniunctiones. Et quatuor planetarum, quae sunt similiter triginta quinque coniunctiones. Et quinque planetarum, quae sunt iterum viginti unum coniunctiones. Et sex planetarum, quae sunt septem coniunctiones. Et omnium, quae est una. Haec sunt in universo centum viginti, quarum praecipue considerat eas quae trium sunt altiorum; consistit etiam in eclipsibus omnium planetarum ad invicem, et praecipue luminarium, de quibus in *Libro coniunctionum* Albumasaris agitur, qui sic incipit: *Scientia significationum etc.* Et apud Messehalla in quodam libello, qui duodecim capitula continet, vocaturque *Epistola Messehalla* et sic incipit: *Quia dominus altissimus etc.*

Secunda vero pars, quae est de revolutione annorum mundi, consistit in scientia significatoris hora introitus solis in primum minutum signi Arietis, qui dicitur dominus anni, hoc est dispositor iussu Dei, ex cuius scientia et aspectu planetarum ad eum ex impedimento quoque et fortuna singulorum cum scientia partium latitudinis eorundem in signis duodecim et eorum ortu atque occasu, directione quoque et retrogradatione, indicatur quid operetur Deus gloriosus et sublimis in eodem anno per stellas sicut per instrumenta super divites quorundam climatum et in universitatem vulgi eorum ex gravitate vel levitate annonae, ex guerra vel pace, ex terraemotu et diluviis, ex scintillis et prodigiis terribilibus, et caeteris esse quae accidunt in hoc mundo; nec non et quid eveniat de operibus stellarum fixarum in revolutione anni mundi, quidque signi-

---

8-9) Ptolemaeus, *Liber centum verborum cum expositione Haly*, v. 50. 13) Albumasar (see VI, 8); *Liber de magnis coniunctionibus*; transl. Ioannes Hispalensis; ed. Augsburg 1489; see St. 360; C. 93; ca. 91-92. 17) Messehalla (see II, 11); *De rebus eclipsium*; transl. Ioannes Hispalensis; ed. Venezia 1493. See St. 379; C. 93; Ca. 30; *Osiris*, (XII), 1956, 62-66. 18-31) Albumasar, *Flores*, I, 1-7.

CHAPTER SEVEN

The section on revolutions, however, is divided into three parts. The first [part] deals with the 120 conjunctions of the planets and with their eclipses. The second deals with the revolution of the years of the world (*anni mundi*); and the third with the mutation of time. The first [part], therefore, deals with the conjunction of planets in one sign: and there are 21 conjunctions between two planets; and there are 35 conjunctions of three planets; and there are likewise 35 conjunctions of four planets; and there are 21 conjunctions of five planets; and there are 7 conjunctions of six planets; and only one conjunction of all the planets. These are 120 altogether, of which he [the astrologer] particularly considers those [conjunctions] of the three superior [planets]. [This part] contains the mutual eclipses of all the planets, and especially [those] of the luminaries. Albumasar writes [about this subject] in his *Liber conjunctionum* (Book of conjunctions), which begins thus: "Scientia significationum etc." ("The science of the significations etc."). And [it is also discussed] by Messahalla in a certain booklet containing twelve chapters, which is called *Epistola Messahalla* (*The Letter of Messahalla*), and [it] begins in this way: "Quia dominus altissimus etc." ("Because the supreme Lord etc.").

The second part, however, concerning the revolution of the years of the world, consists [1] in the knowledge of the signifier at the hour of the entrance of the Sun into the first minute of the sign of Aries; [this signifier] is called the Lord of the Year; (that is, the disposer by the command of God). And [2] from the knowledge of the [Lord of the Year] and the aspect of the planets to it, [and] also from the impediment and the [good] fortune of each one [of the planets], together with the knowledge of the lots and their latitude amongst the twelve signs and their rising and setting, [and] also their direct and retrograde motion. [All this] indicates what God, glorious and sublime, will produce in a given year, using the stars as if they were instruments, on the rich men of some climes and on the whole of their common populace with respect to the high or low price of grain, war or peace, earthquake and floods, falling stars and terrible prodigies, and other events which happen in this world; [3] as well as what may come to pass due to the effects of the fixed stars in the revolution of the year of the world;

ficet caput et cauda, et stellae quae dicuntur cometae, de quibus agitur in *Libro Florum* Albumasar, qui sic incipit: *Oportet te primum scire etc.* Et in *Libro Experimentorum* eiusdem, qui sic incipit: *Scito horam introitus etc.* Et in *Libro Revolutionum* Messehalla qui sic incipit: *Custodiat te Deus etc.* Et in libro Ioannis Hispalensis, qui dicitur *Prima pars artis* pro maiori parte, et incipit: *Quoniam huic arti etc.* Et in quibusdam aliis libris minus utilibus.

Pars autem tertia, quae est de temporis mutatione, consistit in accidentibus planetarum et causis eorum super impressiones altas in aere superiori et inferiori, et in anni differentiis, et quartis eius humidis atque siccis, et in scienta roris et pluviae et horarum eorum in locis terrae per viginti octo mansiones lunae, et per directiones et retrogradationes planetarum et latitudines in signis dextrorsum atque sinistrorsum, in portis lunae duodecim et praecipue in apertione ipsarum. Amplius in scienta flatus ventorum et partium eorum, de quibus agitur in libro Alkindi, qui sic incipit: *Rogatus fui etc.* Et in libro Gaphar, quem puto fuisse Geazar Babylonensem, qui sic incipit: *Universa astronomiae iudicia etc.*; et in *Libro Temporum Indorum*, qui sic incipit: *Sapientes Indi etc.* Et in libro *Quadripartito* Ptolemaei per loca, et in parte libri Ioannis Hispalensis, quem dixi

---

32) ALBUMASAR (see VI, 8) *Flores*; transl. Ioannes Hispalensis; ed. Augsburg 1488. See St. 361; C. 94; Ca. 92-94. 33) ALBUMASAR (see VI, 8) *De revolutionibus annorum mundi sive liber experimentorum*; transl. Ioannes Hispalensis; ed. Augsburg 1489. See St. 361; C. 94; Ca. 94. 35) MESSEHALLA (see II, 11); *In revolutionibus annorum mundi*; ed. 1484 s.l.; See St. 379; C. 94; ca. 25-26; *Osiris*, (XII), 1956, 66-67. 36) IOANNES HISPALENSIS (see II, 70); *Epitome totius astrologiae (Quadripartitum)*; 1142; ed. Nürnberg 1548. See St. 347; C. 94; Ca. 169. 46) ALKINDUS (Ya'qûb ibn Ishâq ibn as-sabbâh al-Kindi, abû Yûsuf); *De pluviis*; transl. Agozont; partial ed. Venezia 1507. See St. 362; B. ('98) 209-210, ('37) 372-374; S. 23; C. 94; Sa. 1, 559-560; TH. I; 642-649; Ca. 79-81; H. 77; G. HELLMANN, *Die Wettervorhersage im ausgehenden Mittelalter*, «Beitr. zur Gesch. d. Meteorologie», Berlin 1917, 16. 48) GAFAR (Ja'far Indus). *De imbribus*; transl. Ugo Santiliensis; ed. Venezia 1507. See St. 369; C. 94; Ca. 85-88; TH. I 652; H. 77; L. THORNDIKE, *The Sphere of Sacrobosco and its Commentators*, Chicago 1949, 58. 49) GAFAR (see above); *De pluviis et ventis*. See St. 369; C. 94; Ca. 87-88; TH. I 652; HELLMANN, *cit.*, 201. – Cfr. VI, 4.

[4] and what the the ascending and descending [lunar] nodes and the stars which are called comets may signify. These things are treated in a book called *Liber Florum* (*The Book of Flowers* by Albumasar, which begins thus: "*Oportet te primum scire etc.*" ("First you ought to know etc."); and in the *Liber experimentorum* (*The Book of Experiments*), which begins in this way: "*Scito horam introitus etc.*" ("Know the hour of the entrance etc."); and in the *Liber revolutionum* (*The Book of Revolutions*), by Messahalla, which begins like this: "*Custodiat te Deus etc.*" ("May God keep you etc."). And for its larger part in the book by John of Seville, called *Prima pars artis* (*The first part of the art*), which begins: "*Quoniam huic arti etc.*" ("Since to this art etc."), and also in some other less useful books.

The third part, on the other hand, dealing with the change of the seasons, consists of [1] the accidents of the planets and their effects on the impressions from on high on the superior and inferior air; [2] the difference in the year and its humid and dry quarters [i.e.: seasons]; [3] and the knowledge of dew and rain and their hours in the places, of the earth according to the 28 mansions of the moon; and by means of the direct and retrograde motions of the planets and their latitudes to the right or to the left in the signs; [4] and with the twelve gates of the moon and especially with opening them up. In addition, [it deals] with [5] the knowledge of the blowing of the winds and their directions. On this subject, there is the book written by Alkindi beginning thus: "*Rogatus fui etc.*" ("I have been asked etc."); and [there is] the book by Gaphar (who, I think, was Geazar the Babylonian) which begins like this: "*Universa astronomiae iudicia etc.*" ("All the judgements of astronomy etc."); and the *Liber temporum Indorum* (*The Book on the Times of the Indians*), which begins in this manner: "*Sapientes Indi etc.*" ("*The wise men of India* etc."); and [there are] some passages of Ptolemy's *Liber Quadripartitus* (*The four parts*); and a section of John of

superius *Primam partem artis* vocari. In his ergo tribus particulis pars de revolutionibus consummatur.

Caput Octavum

Nativitatum vero pars docet in nativitatibus, quorum significatores nutritionis liberi fuerint, eligere locum hylech ex luminaribus et parte fortunae, ex gradu quoque ascendentis et gradu coniunctionis seu praeventionis, quae fuerit ante nativitatem; eligere quoque alchochoden ex dominis quatuor dignitatum ipsius loci hylech, quae sunt domus, exaltatio, terminus atque triplicitas, aspicientis scilicet et eius praecipue qui fuerit aspectus propior; et per directionem gradus hylech ad loca concisionis, donationem quoque annorum alchochoden cum augmento et diminutione ex aspectu planetarum ad eum, iudicare quantitatem vitae nati, non quantum scilicet ipsum oporteat vivere de necessitate, sed ultra quod vita eius non protenditur ex natura; et cum hoc dirigere gradum ascendentis et gradum lunae ad eventus corporis ex infirmitate et sanitate, gradum vero medii caeli et gradum solis ad esse eius regno, et gradum partis fortunae ad acquisitionem divitiarum, patri quoque aspicere ex sole et domino quarti, matri autem ex luna et domino medii caeli, partem etiam hylech dirigere sicut dirigitur locus hylech, nisi quia dirigitur retrorsum. Amplius scire modos directionis ex divisore qui est algebutam, et ex domino radiorum et ex recipientibus dispositiones ipsorum. Docet etiam revolvere annos nati ex signo profectionis ad maiora esse et ex ascendente revolutionis ad minora, et demum iudicare dignitates nati et eius accidentia per commixtionem almutam super ascendens, cum almutam autem super quaedam loca ex circulo ex quatuordecim modis, qui significant effectum et destructionem iussu

51) Cfr. VI, 16; VII, 34.

2-3) Albohali, *De iudiciis nativitatum*, I. 3-5) Albohali, *op. cit.*, II. 4) Alkabitius, *Enarratio elementorum astrologiae*, p. 376. 5-6) Alkabitius, *op. cit.*, p. 379. 5-8) Albohali, *op. cit.*, III. 8-11) Albohali, *op. cit.*, IV. 13-14) *ibidem*. 14-15) Albohali, *op. cit.*, VI. 15-16) Albohali, *op. cit.*, VII. 16-19) Albohali, *op. cit.*, cap. XVI, XVII, XVIII; XIX.

Seville's book, mentioned above, called *Prima pars artis* (*The first part of the art*). The section [of astrology] dealing with revolutions, therefore, is made up of these three sections.

CHAPTER EIGHT

The part on nativities teaches us about nativities of those for whom there are signifiers of the growth of a child. [It teaches us how] [1] to select the place of the hylech from amongst the luminaries and the lots of fortunes, and also from the degree of the ascendent and the degree of the conjunction or opposition [of the Sun and Moon], which preceded the birth; [2] also to choose [the place of] the alchochoden, [which is] from amongst the lords of the four dignities of the place of the hylech (which are [its] house, exaltation, term and triplicity, that is, the one which aspects [the hylech], and especially that one whose aspect is more appropriate; [3] and to judge the length of life of the native by means of the prorogation of the degree of the hylech to the place of the cutting off [and] also by means of the gift of years of the alchochoden together with the increase and decrease [resulting] from the planets' aspect to it, not certainly how long he must live by necessity, but [the time] beyond which his life is not extended by nature, and [4] together with this [a] to prorogate the degree of the ascendent and the degree of the Moon for the occurences of disease and of health in the body, but [b] the degree of Mid-Heaven and the degree of the Sun for his being in rulership, and [c] the degree of the Lot of Fortune for his acquisition of riches, [and] also [d] for [his] father to look from the Sun and the lord of the fourth place, but for his mother from the Moon and the lord of the Mid-Heaven, [and] also [5] to prorogate the lot of the hylech just as the place of the hylech is prorogated, except that it is prorogated retrogressively, [and] [6] in addition, to know the modes of the prorogation from the "divider", that is, the algebutam, and from the lord of the rays and from the recipients of their dispositions. It also teaches [one how] to revolve [i.e.: to calculate] the years of the native from the sign of the starter [for determining] the more important events [in the life of the native] and from the ascendent of the revolution for the less important events; and, finally, to judge the dignities of the native and his accidents by means of mixing together the almutam and the ascendent; and with the almutam, however, in certain places on the circle of the fourteen [different] ways which indicate [good] effect or destruction by the command of God. These

Dei, quorum fit mentio in tertia parte libri Ptolemaei, qui *Quadripartitus* inscribitur, de quibus plenius agitur in libro Aomar Tiberiadis, qui sic incipit: *Scito quod definitiones nativitatum etc.* Et in libro Albohali, qui sic incipit: *Iste est liber in quo exposui etc.* Et in libro Ioannis Hispalensis, qui dicitur *Secunda pars artis*, et sic incipit: *Primum est considerandum etc.*

CAPUT NONUM

Pars iterum interrogationum docet iudicare de re de qua facta fuerit interrogatio cum intentione radicali, utrum scilicet perficiatur, an non. Et si sic, quid sit causa illius et quando erit hoc. Et si non, quid prohibet, ut non fiat, et quando apparebit, quod fieri non debeat; hoc videlicet iudicare per complexionem significatoris domini interrogationis cum domino rei quaesitae, aut cum planeta fortuna, aut cum recipiente, vel recepto, complexionem dico ex coniunctione vel aspectu, translatione quoque et collectione aut omnino praeter haec, eo quod circulus sit secundum intentionem interrogantis in eadem hora; quod si significatores interrogationis aequales fuerint in fortuna et malo, auxiliari cum ascendente coniunctionis seu praeventionis quae fuerit ante interrogationem, et cum almutam super ipsum gradum coniunctionis seu praeventionis ipsius, qui est animodar in nativitatibus, praecipue si aliquam in ascendente interrogationis habeat dignitatem; quod si et tunc aequales fuerint significatores,

---

26) Cfr. VI, 4. 28) AOMAR ALFRAGANUS TIBERIADIS ('Umar Muhammad ibn al-Farrukhân at Tabari Abû Bakr), d. 815 ca.; *De nativitatibus*; transl. Ioannes Hispalensis 1127; ed. Venezia 1503. See St. 373; B. ('37) 392; S. 7. C. 95; Sa. 1, 567; TH. II, 74; Ca. 38-39; M. 200. 29) ALBOHALI ALCHAIT (Yahyâ b. Ghâlib Abû 'Ali al-Khaiyât), m. 835; *De iudiciis nativitatum*; transl. Ioannes Hispalensis 1135; ed. Venezia 1509. See St. 538; B. ('37) 394; S. 9; C. 95; Sa. 1, 569; Ca. 49-51; *Speculum*, (34) 1959, 32 31) IOANNES HISPALENSIS (see II, 70); *Epitome totius astrologiae (Quadripartitum), Pars secunda*; See St. 374; C. 95; Ca. 169. 3-5) MESSALLACH, *De receptione plantarum*, Prologo. 9-11) ZAHEL., *De interrrogationibus*. 13) ALKABITIUS, *Enarratio elementorum astrologiae*, 360. 14) ***Enarratio***, 381.

subjects are mentioned in the third section of Ptolemy's book entitled the *Quadripartitus* (*The four parts*); a fuller discussion appears in the book of Omar Tiberiades which begins thus: "*Scito quod definitiones nativitatum etc.*" ("Know that the definitions of the nativities etc."); and in Albohali's book, which begins in this manner: "*Iste est liber in quo exposui etc.*" ("This is the book in which I have explained etc."); and in the book by John of Seville called *Secunda pars artis* (*The second part of the art*), and [which] begins thus: "*Primum est considerandum etc.*" ("First is to be considered etc.").

CHAPTER NINE

Again, the section devoted to interrogations teaches [one how] to make judgements concerning that thing about which the interrogation has been made with a radical intention – [to know], namely, whether it will come to pass or not. And if [the answer is] positive, what might be its cause and when will it occur? And if [the answer is] negative, what prevents it from happening, and when will [something] happen that ought not to happen – that is, to judge this by means of the involvement of the signifier, which is the lord of the interrogation, with the lord of the thing asked, or with a benefic planet, or with a recipient or one received (I mean involvement through conjunction or aspect, also tranference and collocations and – [what is] altogether beyond these – because the [zodiacal] circle at that hour is in accordance with the intention of the interrogator. But, if the signifiers of the interrogation are equal in [good] fortune and bad [fortune], [it teaches one how] to help [the situation] with [a] the ascendent of the conjunction or opposition which occurred before the interrogation, [b] and [with] the {almutam} of the degree of that conjunction or opposition (which is the {animodar} in nativities), especially if it has some dignity in the ascendent of the interrogation. But if the signifiers are even, [it teaches one] to postpone the interrogation to another moment, or perhaps rather to

differe in aliud tempus, vel fortassis potius supersedere eo, quod Dominus voluit celare a nobis.

Super quibus invenitur liber Messehalla, qui *De receptionibus* inscribitur, et sic incipit: *Invenit quidam etc.* Et liber *De interrogationibus* Zahel Israelitae, quem vocant *Iudicia Arabum*, qui sic incipit: *Cum interrogatus fueris etc.* Liber quoque Gergis *De significatione planetarum in domibus*, qui sic incipit: *Sol cum fuerit in ascendente etc.* Et liber Messehalla *De inventione occultorum*, qui sic incipit: *Scito quod aspiciens etc.* Et alius eiusdem *De interpretatione cogitationis*, qui sic incipit: *Praecipit Messehalla etc.* Et alius Zahel *De significatore temporis*, qui sic incipit: *Et scito quod tempora excitant motus etc.* Praeterea *Liber novem iudicum* et itidem *Liber trium iudicum* et secundus tractatus ex libro Aristotelis, et similiter Ptolemaei ad Aristoxenum, qui superius nominati sunt, et liber Ioannis Hispalensis, quem vocat *Tertiam partem artis*, qui sic incipit: *Est sciendum etc.*

---

20) MESSEHALLA (see II, 11), *De receptione planetarum sive de interrogationibus*; transl. Ioannes Hispalensis; ed. Venezia 1484. See St. 379; C. 96; Ca. 26-27; *Osiris*, (XII) 1956, 50-53. 22) ZAHEL (see VI, 11), *De iudiciis (De interrogationibus)*. See St. 389; C. 96; Ca. 41. 23) MESSEHALLA (see II, 11) *De septem planetis*; ed. 1509; See St. 370; Ca. 29-30; TH. II, 718-719; III, 16; *Bibliotheca Mathematica*, (1905) 237. 25) MESSEHALLA (see II, 11) *De occultis*. See St. 379; Ca. 33-35; *Osiris*, (XII) 1956, 54-56. 26) MESSEHALLA (see II, 11), *De cogitionibus ab intentione*; ed. Venezia 1493. See St. 379; C. 96. Ca. 28-29; *Osiris*, (XII) 1956, 53-54. 27) ZAHEL (see VI, 11) *De significatione temporis ad iudicia*; ed. Venezia 1493. See St. 389; C. 96; Ca. 44. 28) *Liber novem iudicum* (see VI, 17). – *Liber trium iudicum*. See Ca. 105-106. 29-30) Cfr. VI, 12; VI, 15. 32) IOANNES HISPALENSIS (see II, 70) *Epitome totius astrologiae (Quadripartitum), Pars tertia*. See St. 374; C. 97.

abandon that interrogation because the Lord had wished to conceal [the answer] from us.

Concerning these things [there] is found the book by Messehalla, which is entitled *De receptionibus* (*On receptions*), which begins like this: "*Invenit quidam etc.*" ("Someone discovered etc."); and the book *De interrogationibus* (*On interrogations*) by Zahel the Israelite, which they call *Iudicia Arabum* (*The Judgments of the Arabs*), which begins like this: "*Cum interrogatus fueris etc.*" ("When you are asked etc."); also [there is] the book of Gergis, *De significatione planetarum in domibus* (*On the significance of the planets in the houses*), which begins in this manner: "*Sol cum fuerit in ascendente etc.*" ("When the Sun is in the ascendent etc."); and the book by Messehalla, *De inventione occultorum* (*On finding hidden things*), which beings in this way: "*Scito quod aspiciens etc.*" ("Know that looking etc."); and another by the same [author], *De interpretatione cogitationis* (*On the interpretation of thought*), which begins thus: "*Praecipit Messehalla etc.*" ("Messehalla teaches etc."); and another by Zahel, *De significatore temporis* (*On the signifier of the time*), which begins like this: "*Et scito quod tempora excitant motus*" ("And know that times produce motions etc."). Moreover, [there are] the *Liber novem iudicum* (*The book of the nine judges*) and also, the *Liber trium iudicum* (*The book of the three judges*) and the second treatise from Aristotle's book, and similarly, [from] Ptolomy's [book addressed] to Aristoxenus, [both] mentioned above, and the book by John of Seville, which he calls *Tertia pars artis* (*The third part of the art*), which begins like this: "*Est sciendum etc.*" ("It must be known etc.").

Caput Decimum

Rursum pars electionum docet eligere horam laudabilem incipiendi aliquod opus ei cuius nativitas nota fuerit per convenientiam domini rei cum significatore nativitatis eiusdem; quod si fuerit eius ignota nativitas, accipere ei interrogationem certissimam, eo quod homo quando interrogat, iam pervenit ex nativitate sua ad bonum seu ad malum quod significavit eius nativitas; et loco nativitatis ipsam interrogationem accipere pro radice, eo quod cum nativitates sint res naturales, interrogationes sunt res similes naturalibus.

Quorum siquidem mentio agitur in *Libro electionum* Zahel, qui sic incipit: *Omnes concordati sunt etc.*, et in *Libro electionum* Haly, qui sic incipit: *Rogasti me carissime etc.* Caeterum sunt quidam libri, qui de universis praedictis partibus sparsim tractant, ut est *Liber centum verborum* Ptolemaei, qui sic incipit: *Mundanorum etc.* Et *Liber quinquaginta praeceptorum* Zahel, qui sic incipit: *Scito quod significatrix Luna etc.* Et *Liber capitulorum ad Mansorem*, qui sic incipit: *Signorum dispositio est ut dicam etc.*

4-7) Zahel, *De electionibus*, pars I. 8-9) Haly, *De electionibus horarum*, I, 4. 11) Zahel (see VI, 11) *De electionibus*, See St. 389; C. 97; Ca. 41. 12) Haly Embrani ('Ali ibn Ahmad al-'Imrânî), Mosul d. 955 ca.; *De electionibus horarum*: transl. Abraham bar Hiyya and Plato de Tivoli; part. ed. J. M. Millás-Vallicrosa, *Las traducciones* ..., 328-339. See St. 370; S. 56; C. 97; Sa. 1, 632; Ca. 137-139; H. 11, b. 30; 14) Pseudo Ptolemaeus (see II, 9). *Liber centum verborum cum expositione haly*; transl. Ioannes Hispalensis; ed. Venezia 1484. See St. 383; C. 97; Ca. 16; A. Pelzer in *Arch. Franc. Hist.*, (XII) 1919, 60; A. Björnbo, *Alkindi* ..., cit, 103. 16) Zahel (see VI, 11) *Quinquaginta praecepta*, See St. 389; C. 97; Ca. 41. 17) Rhazes (Muhammad b. Zakarîyâ ar-Râzî Abû Bakr), Bagdad b. 865, d. 932; *Iudicia (Capitula) Almansoris*; transl. Plato de Tivoli 1136; ed. Milano 1481. See St. 362; B. ('98) 233-235; ('37) 417-421; C. 98; Sa. 1, 609-610; TH. II, 752; Ca. 132-134.

CHAPTER TEN

Again, the section on elections teaches [one how] to choose the favourable hour for beginning any project for one whose nativity is known, by using the agreement of the lord of the matter with the signifier of his nativity; but if his nativity is unknown, [it teaches one] to take a very certain interrogation [as a basis] for it, on the grounds that when a man makes an interrogation, he has already, because of his nativity, come to that good or bad [fortune] which his nativity has signified; and to accept that interrogation itself as a basis in place of his nativity because while nativities are natural things, interrogations are things similar to natural [things].

These things are mentioned by Zahel in the *Liber electionum* (*Book of elections*), which begins like this: "*Omnes concordati sunt etc.*" ("All have agreed etc."); and in the *Liber electionum* (*Book of elections*) by Haly, which begins thus: "*Rogasti me carissime etc.*" ("Dearest one, you have asked me etc.").

But there are certain books, which discuss all these abovementioned parts [of astrology] in passing, such as Ptolemy's *Liber centum verborum* (*Book of 100 Statements*), which begins like this: "*Mundanorum etc.*" ("Of earthly things etc."); and Zahel's *Liber quinquaginta praeceptorum* (*Book of 50 Precepts*), which begins thus: "*Scito quod significatrix Luna etc.*" ("Know that the Moon, the signifier, etc."); and the *Liber capitulorum ad Mansorem* (*Book of Chapters* [*addressed*] *to Mansor*), which begins in this way: "*Signorum dispositio est ut dicam etc.*" ("The disposition of the signs is as I shall say etc.").

Caput Undecimum

Parti autem electionum dixi supponi imaginum scientiam, non quarumcumque tamen sed astronomicarum, quoniam imagines fiunt tribus modis. Est enim unus modus abominabilis, qui suffumigationem et invocationem exigit, quales sunt *Imagines* Toz Graeci et Germath Babylonensis, quae habent stationes ad cultum Veneris, quales sunt *Imagines* Balenuz et Hermetis, quae exorcizantur per quinquaginta quatuor nomina angelorum, qui subservire dicuntur imaginibus lunae in circulo eius, et forte sunt potius nomina daemonum, et sculpuntur in eis septem nomina recto ordine pro re bona et ordine transverso pro re cuius expectatur repulsio. Suffumigantur etiam pro bona re cum ligno aloes, croco et balsamo, et pro mala re cum galbano, sandalo rubeo et resina, per quae profecto spiritus non conguntur, sed quando Dominus permittit peccatis nostris exigentibus ut decipiant homines, exhibent se coactos. Haec est idololatria pessima, quae, ut reddat se aliquatenus fide dignam, observat viginti octo mansiones lunae et horas diei et noctis cum quibusdam nominibus dierum, horarum et mansionum ipsarum. A nobis longe sit iste modus: absit enim ut exhibeamus creaturae honorem debitum creatori.

Est alius modus aliquantulum minus incommodus, detestabilis tamen, qui fit per inscriptionem characterum per quaedam nomina exorcizandorum, ut sunt quatuor annuli Salomonis et novem candariae et tres figurae spirituum, qui dicuntur principes in quatuor plagis mundi, et Almandal Salomonis, et sigillum ad daemoniacos. Amplius septem nomina ex libro Muhameth et alia quindecim ex eodem et rursum nomina ex *Libro Institutionis*, qui dicitur Razielis, videlicet terrae, maris, aeris atque ignis, ventorum, et mundi cardinum, signorum quoque et planetarum et angelorum eorum, secundum quod singula in diei et noctis triplicitatibus diversa nomina sortiuntur. Hic modus etiam a nobis longe sit; suspectus enim est, ne saltem sub ignotae linguae nominibus aliquod lateat, quod sit contra fidei catholicae honestatem.

Isti sunt duo modi imaginum necromanticarum, quae nobile

CHAPTER ELEVEN

As I have said, the science of images is added to the part on elections; not any of them [i.e.: the images] whatsoever, however, but only the astronomical ones, since images are made in three ways. One way is abominable – [that] which requires suffumigations and invocation, such as the images of Toz the Greek and Germath the Babylonian, which have stations for the worship of Venus, [and] the images of Balenuz and Hermes, which are exorcized by using the 54 names of the angels, who are said to be subservient to the images of the Moon in its orbit, [but] perhaps are instead the names of demons, and seven names are incised on them in the correct order to affect a good thing and in inverse order for a thing one wants to be repelled. They are also suffumigated with the wood of aloe, saffron and balsam for a good purpose; and with galbanum, red sandlewood and resin for an evil purpose. The spirit is certainly not compelled [to act] because of these [names and fumigations], but when God permits it on account of our own sins, they [the spirits] show themselves as [if there were] compelled to act, in order to deceive men. This is the worst [kind of] idolatry, which, in order to render itself credible to some extent, observes the 28 mansions of the Moon and the hours of day and night along with certain names [given] to these days, hours and mansions themselves. May this method be far from us, for far be it that we show that [sort of] honour to the creature which is due [solely] to the Creator.

There is another method [of making images] that is somewhat less unsuitable ([but it is] nevertheless detestable), which is effected by means of inscribing characters which are to be exorcized by certain names, such as, the four rings of Solomon, and the nine candles and three figures of the spirits (who are called the princes of the four regions of the world), and the Almandal of Salomon, and the sigil for those possessed by demons. Further [there are] the seven names from the book of Muhameth, and the other fifteen from the same; and, in addition, [there are] the names from the *Liber institutionis* (*The Book of Instruction*), which is said to be by Raziel, namely of the earth, the sea, the air and the fire, of the winds, and of the cardines of the world, [and] also of the signs and the planets and of their angels, according to which each thing takes a different name in the triplicites of the day and the night. May this method also be far from us; for it is suspected that something lies under the names of the unknown language, that might be against the honour of the catholic faith.

These are the two sorts of necromantic images, which (as I have said)

nomen astronomiae (sicut dixi) sibi usurpare praesumunt; et ex eis, iamdiu est, libros multos inspexi, sed quoniam eos abhorrui, non extat mihi perfecta memoria super eorum numero, titulis, initiis aut continentiis sive auctoribus eorundem; spiritus enim meus numquam requiescebat in illis, bene tamen volebam transeundo vidisse, ut saltem non ignorarem qualiter esset miseris eorum sectatoribus irridendum, et haberem de suo unde repellerem excusationes eorum, et quod potissimum est, ut super consimilibus de caetero non tentarer, cum persuasiones suas invalidas non admittendas censerem. Et libri quidem ex eis quos possum modo ad memoriam revocare, sunt ex libris Hermetis *Liber praestigiorum*, qui sic incipit: *Qui geometriae aut philosophiae peritus expers astronomiae fuerit etc.*; *Liber Lunae*, qui sic incipit: *Probavi omnes libros etc.*, cui adiungitur liber Balenuz *De horarum opere*, qui sic incipit: *Dixit Balenuz qui et Apollo dicitur: Imago prima etc.* Et liber eiusdem *De quatuor imaginibus ab aliis separatis*, qui sic incipit: *Differentia in qua fiunt imagines magnae etc.* Ex libris quoque Hermetis est *Liber imaginum Mercurii*, in quo sunt multi tractatus, unus de imaginibus Mercurii, alius de characteribus eius, alius de annulis, alius de sigillis, quorum inceptionum non recolo, nisi illius de sigillis, qui sic incipit: *Dixit expositor huius libri: Oportet quaerentem hanc scientiam etc.* Post istos est *Liber Veneris* habens similiter plures tractatus, scilicet de imaginibus, de characteribus, de annulis, de sigillis, quorum inceptionum similiter non recolo, nisi illius de annulis, quae est talis: *Mentio decem capitulorum atque annulorum Veneris etc.* Et hos sequitur *Liber Solis*, qui sic incipit: *Lustravi plures imaginum scientias etc.* De isto non vidi nisi singularem tractatum de characteribus,

---

45) Hermes, *De imaginibus qui praestigiorum dicuntur*. See St. 371; C. 100. 47) Hermes, *De lunae mansionibus liber*; See St. 372; C. 100; Ca. 64; TP. 238-241. 48-49) Belenus, *De imaginibus*. See St. 369; TP. 242. 50) Belenus, *De quatuor imaginibus ab aliis separatis*. See St. 369; C. 100; TH. II, 234-235; TP. 242. 55) Hermes, *Liber imaginum Mercurii* (1. IV). See St. 372; TP. 244. 59) Cfr. TP. 244. 60) Hermes, *Liber Solis (De imaginibus et horis)*. See Ca. 61-63; TH. II, 223 e segg.; TP. 244-246.

have presumed to usurp the noble name of astronomy for themselves; and a long time ago I inspected many of these books, but since I shrank with horror from them, I do not have perfect memory regarding their number, titles, incipits or contents or their authors. In fact, my spirit was never tranquil when dealing with these [matters]; all the same, I wanted to observe them well whilst passing over them so that, at least, I might not be ignorant of how to ridicule their wretched believers, and [so that] and I might have [something] taken from their own [work] with which I might repell their excuses, and – what was most important – so that I would not be tempted concerning similar things from another [source] when I had judged that the [necromancer's] invalid arguments should not be accepted. And, in fact, amongst those books which I can remember now, [there are those] from the books of Hermes, the *Liber praestigiorum* (*Book of Illusions*), which begins thus: "*Qui geometriae aut philosophiae peritus expers astronomiae fuerit etc.*" ("Whoever is skilled in geometry and philosophy without knowing astronomy etc."); the *Liber Lunae* (*Book of the Moon*), which begins in this manner: "*Probavi omnes libros etc.*" ("I have tested all the books etc."), to which the book of Balenuz is joined, *De horarum opere* (*On the work of the hours*), which begins like this: "*Dixit Balenuz qui et Apollo dicitur: Imago prima etc.*" ("Balenuz, who is also called Apollo, said: The first image etc."); and in his book, *De quatuor imaginibus ab aliis separatis* (*On the four images separated from the others*), which begins like this: "*Differentia in qua fiunt imagines magnae etc.*" ("The difference in which the great images are made etc."). And from amongst the books of Hermes there is the *Liber imaginum Mercurii* (*The Book of the Images of Mercury*), which contains several treatises: one on the images of Mercury, another on its characters, another on [its] rings, another on [its] sigils – whose incipits I do not remember, save that one on sigils, which begins in this manner: "*Dixit expositor huius libri: Oportet quaerentem hanc scientiam etc.*" ("The expositor of this book said: He who seeks this wisdom ought etc."). After these, there is the *Liber Veneris* (*Book of Venus*), which similarly has many treatises, such as on images, on characters, on rings [and] on sigils, whose incipits I likewise do not remember, except for the one on rings, which is the following: "*Mentio decem capitulorum atque annulorum Veneris etc.*" ("*The mention of the ten chapters and the ten rings of Venus etc.*"). And these are followed by the *Liber solis* (*Book of the Sun*), which begins like this: "*Lustravi plures imaginum scientias etc.*" ("I have examined several sciences of images etc."). Of this I have seen only a single treatise on characters, and possibly, there are others as in the cases of those [mentioned] above,

et fortasse sicut in superioribus sunt alii, sed non translati. Trium etiam superiorum planetarum non vidi nisi singulares tractatus, *Librum* scilicet *Imaginum Martis*, qui sic incipit: *Hic est liber Martis quem tractat etc.*; et *Librum Iovis*, qui sic incipit: *Hic est liber Iovis quem tractat etc.*; et *Librum Saturni*, qui sic incipit: *Hic est liber Saturni quem tractat Hermes Triplex etc.* Hos septem libros sequitur quidam, qui sic incipit: *Tractatus octavus in magisterio imaginum etc.* Et ipsae sunt de his quae referuntur ad Hermetem. Est et unus liber *De septem annulis septem planetarum*, qui sic incipit: *Divisio lunae quando impleta fuerit etc.* Ex libris vero Toz Graeci est liber *De stationibus ad cultum Veneris*, qui sic incipit: *Commemoratio historiarum etc.* Et liber *De quatuor speculis* eiusdem, qui sic incipit: *Observa Venerem cum pervenerit ad Pleiades etc.* Et alius *De imaginibus Veneris*, qui sic incipit: *Observabis Venerem cum intraverit Taurum etc.* Et ex libris Salomonis est liber *De quatuor annulis*, quem intitulat nominibus quatuor discipulorum suorum, qui sic incipit: *De arte eutonica et ydaica etc.* Et liber *De novem candariis*, qui sic incipit: *Locus admonet ut dicamus etc.* Et liber *De tribus figuris spirituum*, qui sic incipit: *Sicut de caelestibus etc.* Et liber *De figura Almandal*, qui sic incipit: *Capitulum in figura Almandal etc.* Et alius parvus *De sigillis ad daemoniacos*, qui sic incipit: *Capitulum sigilli gandal et tanchil etc.* Et ex libris Mahometh est *Liber septem nominum*, qui

---

64) HERMES, *De imaginibus Martis*. See Ca. 60; TP. 246. 65) HERMES, *Liber Iovis*. See Ca. 61; TP. 246. 67) HERMES, *Liber Saturni (De imaginibus Saturni)*. See Ca. 61; TP. 246. 68) Cfr. Ca. 58-63. 71) HERMES, *De septem annulis septem planetarum*. See Ca. 63; TP. 247. 72) GEZ (Toz?) GRAECUS, *Libri imaginum de stationibus ad cultum Veneris*. See St. 388; TH. II, 226^n; TP. 247^n, 248. 74) TOZ GRAECUS, *De quatuor speculis*. See St. 388; TH. II, 226 ^{n}; TP. 248-9. 75) TOZ GRAECUS, *De imaginibus Veneris*. See St. 388; TP. 248. 78) SALOMO, *De quatuor annulis*. See St. 386; C. 101; TP. 250. 79) SALOMO, *De novem candariis*, See St. 386; TP. 251. 80) CAUDAS ASTROLOGUS, *Liber de tribus figuris sprituum*. See St. 386; TP. 251. 81) SALOMO, *Liber de figura Almandal*. See St. 386; TP. 251; *Speculum*, II 1927, 326-331. 82) CAUDAS ASTROLOGUS, *De figura Almandal*. See St. 386; TP. 251.

but they are not translated. I have also seen only single treatises for each of the three superior planets: namely, the *Liber imaginum Martis* (*The Book of the Images of Mars*), which begins thus:"*Hic est liber Martis quem tractat etc.*" ("This is the book of Mars, which [Hermes] writes etc."); and the *Liber Iovis* (*Book of Jupiter*), which begins thus: "Hic est liber Iovis quem tractat etc." ("This is the Book of Jupiter, which [Hermes] writes etc."); and the Liber Saturni (Book of Saturn), which begins thus: "*Hic est liber Saturni quem tractat Hermes Triplex etc.*" ("This is the Book of Saturn, which the threefold Hermes writes etc."). These seven books are followed by a certain one which begins in this manner: "*Tractatus octavus in magisterio imaginum etc.*" ("The eighth treatise on the teaching of images etc."). And they are from amongst those which are attributed to Hermes. And there is also one book, *De septem annulis septem planetarum* (*On the seven rings of the seven planets*), which begins thus: "*Divisio lunae quando impleta fuerit etc.*" ("The division of the Moon when it is full etc."). From amongst the books of Toz the Greek, [there] is the book *De stationibus ad cultum Veneris* (*Concerning the stations for the worship of Venus*), which begins like this: "*Commemoratio historiarum etc.*" ("The recollection of the histories etc."); and his book, *De quatuor speculis* (*On the four mirrors*), which begins in this manner: "*Observa Venerem cum pervenerit ad Pleiades etc.*" ("Observe Venus when she reaches the Pleiades etc."); and another, *De imaginibus Veneris* (*On the images of Venus*), which begins like this: "*Observabis Venerem cum intraverit Taurum etc.*" ("You will observe Venus when she enters Taurus etc."); and from amongst the books of Salomon, there is the book, *De quatuor annulis* (*On the four rings*), which he entitles with the names of his four disciples, which begins like this: "*De arte eutonica et ydaica etc.*" ("On eutonic and ydaic art etc."); and the book *De novem candariis* (*On the nine candles*), which begins in this manner: "Locus admonet ut dicamus etc." ("The place suggests that we should say etc."); and the book *De tribus figuris spirituum* (*On the three figures of the spirits*), which begins thus: "*Sicut de caelestibus etc.*" ("Just as concerning the heavens etc."); and the book *De figura Almandal* (*On the figure of Almandal*), which begins thus: "*Capitulum in figura Almandal etc.*" ("The chapter on the figure of Almandal etc."); and another small one *De sigillis ad daemoniacos* (*On the sigils for those possessed by demons*), which begins like this: "*Capitulum sigilli gandal et tanchil etc.*" ("The chapter on the sigil of gandal and tanchil etc."). And from amongst the books by Mahometh, there is the *Liber septem nominum (Book of the seven names), which begins like this: "Dixit Mahometh filius Alhalzone etc.*" ("Mahomet, the son of

sic incipit: *Dixit Mahometh filius Alhalzone etc.* Et *Liber quindecim nominum*, qui sic incipit: *Haec sunt quindecim nomina etc.* Est et unus liber magnus Razielis, qui dicitur *Liber institutionis*, et sic incipit: *In prima huius prooemii parte de angulis tractemus etc.* Est et alius *De capite Saturni*, qui sic incipit: *Quicumque hoc secretissimum etc.*, cuius auctorem ignoro. Cum his libris inveniuntur scripti duo ex libris Hermetis, quos non putant necromanticos, sed potius naturales, quorum unus est *De quibusdam medicinis in coniunctionibus planetarum*, qui sic incipit: *Quando Saturnus iungitur Iovi etc.* Alius est *De quatuor confectionibus* ad capienda animalia silvatica et lupos et aves. Et ipse est liber Hermetis ad Aristotelem, qui sic incipit: *Dixit Aristoteles: Vidistine o Hermes etc.* Sed qui omnium pessimus invenitur, est liber quem scribit Aristoteles ad Alexandrum, qui sic incipit: *Dixit Aristoteles Alexandro regi: Si vis percipere etc.* Hic est quem quidam vocant *Mortem animae*. Isti sunt libri quos modo ad memoriam revoco, licet plures viderim ex illis, scilicet de imaginibus, quas dixi fieri cum suffumigationibus, invocationibus, exorcizationibus et characterum inscriptionibus, qui sunt duo modi imaginum necromanticarum, ut dixi.

Tertius enim modus est imaginum astronomicarum, qui eliminat istas spurcitias, suffumigationes et invocationes non habet, neque exorcizationes aut characterum inscriptiones admittit, sed virtutem nanciscitur solummodo a figura caelesti; ut si fuerit imago destructionis alicuius speciei ab aliquo loco, de qua scilicet fuerimus requi-

---

84) Cfr. TP. 253. 85) *Liber quindecim nominum*. See TP. 253. 87) Raziel, *Liber magnus de institutionibus*. See St. 384-386; TP. 253-254. 88) *De capite Saturni*. See C. 102; TP. 254-255. 92) Hermes, *De quibusdam medicinis in coniunctionibus planetarum*. See St. 373; C. 102; Ca. 69; TP. 247. 95) Hermes, *De quatuor confectionibus ad capienda animalia silvatica*. See TP. 247-248. 97) *Mors anime*. See TP. 255.

Alhalzone, said etc."); and the *Liber quindecim nominum* (*The Book of the Fifteen Names*), which begins thus: "*Haec sunt quindecim nomina etc.*"("These are the fifteen names etc."). And there is one large book by Raziel called the *Liber institutionis* (*The Book of Instruction*), and it begins like this: "*In prima huius prooemii parte de angulis tractemus etc.*" ("In the first part of this preface we will deal with cardines etc."). And another, *De capite Saturni* (*On the head of Saturn*), which begins like this: "*Quicumque hoc secretissimum etc.*" ("Whoever this most secret etc."), whose author I do not know. There are two books written by Hermes found in addition to these books, which they do not consider to be necromantic, but rather [dealing only with the] natural, one of these is the *De quibusdam medicinis in coniunctionibus planetarum* (*Concerning certain medicines in the conjunctions of the planets*), which begins thus: "*Quando Saturnus iungitur Iovi etc.*" ("When Saturn is conjoined with Jupiter etc."); [and] the other [one] is the *De quatuor confectionibus* (*On the four recipes*), [which are] for catching wild animals and wolves and birds. And this is the same as the book by Hermes [addressed] to Aristotle, which begins in this manner: "*Dixit Aristoteles: Vidistine o Hermes etc.*" ("Aristotle said: Have you seen, Hermes, etc."). But the worst of all these is that written by Aristotle to Alexander, which begins thus: "*Dixit Aristoteles Alexandro regi: Si vis percipere etc.*" ("Aristotle said to Alexander the King: If you want to perceive etc."). This is the one which some call the *Mors animae* (*The death of the soul*). These are the books which I can remember now, although I may have seen many others of them, that is on images which, as I said, were are made with suffumigations, invocations, exorcisms and the inscription of characters, which are the two types of necromantic images, as I said.

The third type is [that] of astronomical images, which eliminates this filth, does not have suffumigations or invocations and does not allow exorcisms or the inscription of characters, but obtains [its] virtue solely from the celestial figure; such as if there should be an image for eliminating some species from some place, concerning which [image], we have received a re-

siti, accepta primum interrogatione cum numero certissimo a quo nihil cadat exiguum vel plurimum, si significatores significaverint abscissionem, fundatur imago sub ascendente simili illius speciei, aut sub ascendente interrogationis eiusdem, infortunato ascendente et domino eius a domino domus mortis, vel a planeta infortuna per oppositionem aut quadratum aspectum absque ulla receptione inter eos; infortunato quoque domino domus domini ascendentis et Luna et domino domus Lunae et parte fortunae et domino eius et domino horae, remotisque fortunis ab ascendente et eius angulis et a triplicitatibus ascendentis, et sit Luna in ascendente facie et signo; postquam ergo perfecta fuerit imago cum quibusdam aliis conditionibus quae observandae sunt, sepelienda est in medio loci a quo fuganda est ipsa species, posito in ventre imaginis de terra ex quatuor quadrantibus eiusdem loci. Si vero fuerit imago cuius opere quaeritur dilectio et profectus, fiat e contrario horum quae dixi, addito quod forma eius sculpenda est sub hora electa et habebit effectum iussu Dei a virtute caelesti, eo quod imagines quae inveniuntur in hoc mundo sensibili ex quatuor elementis, oboediunt caelestibus imaginibus, quarum quaedam sunt prope res inventas nomine et creatione, quaedem vero mirabiles longe sunt a nobis surguntque in aestimatione rationali cum profundatione intellectus.

Super istis imaginibus invenitur unus liber Thebit filii Chorae, qui sic incipit: *Dixit Thebit Benchorat: Dixit Aristoteles qui philosophiam etc.*, in quo sunt imagines super fortuna et impedimento, super substantia et negotiatione, super principatu et proelatione et super coniunctione atque separatione; et ipsae sunt imagines astronomiae, quarum nomine se insigniunt necromanticae ante dictae. Est et alius liber, qui sic incipit: *Opus imaginum Ptolemaei etc.*, qui sicut est inutilis est, cum nihil sit ibi nisi sub quo ascendente

---

108) THEBIT, *De imaginibus*. See 36 p. 186. 110-117) THEBIT, *op. cit.*, II, 10 p. 182; I, 6 p; 181. 117-121) THEBIT, *op. cit.*, I, 5-9, p. 181. 130) THEBIT (see II, 21); *De imaginibus*; transl. Ioannes Hispalensis; ed. F. CARMODY, Berkeley (Calif.) 1941. See St. 387; C. 104; Ca. 124. 135) PSEUDO-PTOLEMAEUS (see II, 9); *De imaginibus super facies signorum*; See St. 384; C. 104; TP. 256-259.

quest, [first], when the interrogation was received [in accordance] with a most certain numerical datum from which nothing [either] small or large should be lacking, if the signifiers show a cutting off, the image is cast under an ascendent similar to that species, or under the ascendent of the interrogation itself; when the ascendent and its lord are injured by the lord of the house of death, or by a malefic planet through opposition or quartile aspect without any reception between them, or when the lord of the house of the lord of the ascendent and the Moon and the lord of the house of the Moon and the lot of fortune and its lord and the lord of the hour are injured, and benefic [planets] are removed from the ascendent and its cardines and from the triplicity of the ascendent, and let the Moon be in the ascendent decan and sign. After, therefore, the image has been completed, along with certain other conditions which must be observed, it should be buried in the middle of [that place] from which the species itself is to be banished, with some earth taken from the four quarters of the same place put in the belly of the image. But if the image is made in order to attain love and profit, let it be made according to the opposite [way] to what I have said, with the addition that its shape is to be engraved under an elected hour; and it will have a [good] effect from the celestial virtue by the command of God, because [the images] found in this sensible world [made] from the four elements obey the celestial images [i.e.: the constellations] of the heavens – some of which are close to things found existing in their naming and their creation [in our world], but certain ones are miraculous and are far from us and rise up in rational estimation, by means of the profundity of [our] intellect.

On these images there is found one book by Thebit, the son of Chora, which begins like this: "*Dixit Thebit Benchorat: Dixit Aristoteles qui philosophiam etc.*" ("Thebit Benchorat said: Aristotle said, who [will read] philosophy etc."), in which there are images for good fortune and for impediment, for wealth and business, for rulership and governorship and for joining together and separating; and these are astronomical images, under which [title] the aforementioned necromantic images [falsely] present themselves, as I said before. And there is another book, which begins in this manner: "*Opus imaginum Ptolemaei etc.*" ("Ptolemy's work on images etc."), which, is virtually useless as it stands since [there is] nothing in it except for the ascendent under which each image must be made. But, if the

sint imagines singulae faciendae, quod si tacite conditiones necromanticae sunt, intolerabilis est, sicut et caeteri maledicti, quos nullus sanae mentis excusare praesumit.

Caput Duodecimum

Quoniam autem occasione eorum, ut dictum est, multi libri praenominati, et fortassis innoxii accusantur, licet accusatores eorum amici nostri sint, veritatem tamen oportet, sicut inquit Philosophus, nihilominus honorare, protestor tamen quod si aliquid dicam quo velim uti in defensione eorum, quoniam determinando non dico, sed potius opponendo vel excipiendo et ad determinationis animadversionem determinatoris ingenium provocando.

Primum itaque volo reverti ad partem illam revolutionum, quae est de temporum mutatione, cuius necessitas ex praedictis apparet, videlicet ex oboedientia motus inferiorum ad motum superiorum, nec habet unde impediatur eius necessitas, cum neque libero arbitrio sit subiecta, sed soli voluntati sui conditoris, qui ab initio providit sic, et ab ipso solo averti potest, ut apud quem solum plenitudo potestatis habetur, cum tamen nolit avertere, non est enim eius consilium mutabile sicut unius ex pueris aut ancillis, sed vult illud durare usque ad terminum ab ipso ei impositum, sicut clamant Ptolemaeus et Albumasar, et ab ipso solo notum, quando scilicet ex praecepto suo stabit motus, sicut et coepit ex ipsius praecepto, in quo solo ille utilis Aristoteles invenitur errasse (nihilominus regratiandus in mille milium aliorum). Et iam scimus, quod non est causa in circulo quae non sit sapienter disposita nutu Dei; cum ergo sapiens non poentiteat quod providit sapientissimus, non est eius avertere seu mutare. Quod si pars ista scientiae iudiciorum astrorum, quae scilicet est de mutatione temporum, debet stare, numquid et stare partem illam quae est de principiis oportebit? Cum praesertim ipsa

3-5) Cfr. Aristoteles, *Ethica nicomachea*, I, 4 (1096 a 11). 17) Ptolemaeus, *Almagestum*, f. aii; 18) Albumasar, *Introductorius*; transl. Ioannes Hispalensis; I, 2; I, 5; cfr. *infra*. 11. 51-52.

conditions are secretly necromantic, it is intolerable, as are also the other cursed [books] which no one of a sane mind presumes to excuse.

CHAPTER TWELVE

However, the occasion having been [provided] by them, as has been said, many of the aforementioned books, [some being] perhaps innocent, stand accused [and] even though their accusors may be our friends, we must, nevertheless, honour the truth, as the Philosopher says. I swear, however, that if I say anything that I wish to use in defense [of these books], I do not speak as in a determination [i.e.: conclusion], but instead [I speak] in opposition, offering exceptions [to present opinion, so as] to provoke the mind of those who are reaching a decision to pay [careful] attention [to the criteria he is using] for his conclusion[s].

Therefore, I would like to return first to that section on revolutions which deals with the change of the seasons, whose necessity is clear from what has been said above; that is, the obedience of the motion of the inferior beings to the motion of the superior ones. The necessity of this has nothing by which it may be impeded, since it is not subjected to free will, but only to the will of its Creator, who preordained [it] in this manner from the beginning; [and] it can be averted by Him alone, since in Him alone is there the plenitude of power. Yet since He does not wish to avert it (for His will is not changeable, as is that of children or servants), but He wants this to last up until the end which has been imposed on it by Him alone (as Ptolemy and Albumasar proclaim) – and [that] which is known by Him alone, that is, when the motion [of the heavens] will be stopped as it began by His command. In this matter alone, we find that the [normally] useful Aristotle has erred (nevertheless, he is to be thanked for a million other [ideas]). And we already know, that there is no cause in the circle [of heaven] which is not wisely disposed by the will of God; therefore, since a wise man should not regret that which the Wisest [One] has provided; it is not in [this man's hands] to avert or change it. But if this section of the science of the judgements of the stars concerning the change of the seasons ought to be saved, shouldn't that part concerning its principles be saved as well? Especially since if that [latter section] were destroyed, all of

destructa, omnes aliae destruantur; sed et in ipsa quid invenitur contrarium veritati sanae? Quod apud Albumasar, qui de ea tractat, plenissime reprehensione dignius invenitur, est illud quod dicit in tractatu primo sui libri, differentia quinta, capitulo de secta tertia, scilicet quod planetae sunt animati anima rationali; sed quod dicit, dicere recitando videtur, cum dicat Aristotelem hoc dixisse, licet non inveniatur in universis libris Aristotelis quos habemus, et forte illud est in duodecimo aut in decimo tertio *Metaphysicae*, qui nondum sunt translati et loquuntur de intelligentiis, sicut ipse promittit. Quod autem sententiae de rationalitate planetarum non consentiat Albumasar, apparet ex hoc quod in eodem capitulo dicit: «Planetis etsi sint animae rationales, non eligunt tamen, nec indigent electione propter longitudinem eorum ab impedimentis». Et apertius ex hoc quod dicit in eodem tractatu, differentia tertia, ubi habetur versus finem: «Et non secundum electionem eiusdem solis fuit introitus eius in hanc partem, neque effectus ipsarum rerum et earum corruptio, sed per adventum eius per motum naturalem in ipsam partem». Et infra probat quod motus circuli est a virtute primae causae immobilis et aeternae. Unde et dicit: «Benedicendum est nomini eius, et ipsum exaltare, non Solem». Quid ergo meruit liber suus? Quem si revolveris, invenies multa bona, mala autem nulla, quod sciam. Illic invenies tractatu primo, differentia secunda, capitulo de aptatione temporum, quoniam stabit motus, ubi dicit: «Planetae non corrumpuntur, neque recipiunt augmentum, neque diminutionem, neque effectum, neque detrimentum usque ad tempus quod Deus voluerit». Et in eodem differentia quinta, capitulo de prima secta, «Efficitur ex motibus planetarum naturalibus atque durabilibus effectus naturalis et durabilis, qui fit usque ad tempus quod Deus voluerit». Quod similiter testatur Ptolemaeus in *Almagesti* capitulo primo dictionis primae, ubi habetur: «Nos autem laboravimus, ut in amore scientiae sempiternorum manentium usque ad terminum, quem eorum conditor eis imposuit, in sequentibus huius nostri libri addamus, etc.».

Invenies quoque apud Albumasar longe elegantius scilicet testi-

31) ALBUMASAR, *op. cit.*, I, 5. 38) ALBUMASAR, *op. cit.*, I, 5, cap. *de tertia secta.* 41) ALBUMASAR, *op. cit.*, I, 3. 45) *ibidem.* 49) ALBUMASAR, *op. cit.*, I, 2; cfr; *supra* I. 18. 53) ALBUMASAR, *op. cit.*, I, 5. 56) PTOLEMAEUS, *op. cit.*, f. aii. cfr. *supra* I. 18.

the other [sections] would be destroyed [as well]. But also what can be found in it that is against the wholesome truth? That which appears [in the writings] of Albumasar, who discusses this, is found to be more worthy of the reprehension to the fullest extent; [namely], that which he claims in Book I, chapter 5 in the section on the third sect, [where he says] that the planets are animated by a rational soul; but what he says [here] he seems to say [as if he were] quoting, since he says that Aristotle has said it (even though it is not found in any of Aristotle's books that we have – but perhaps it is in the twelfth or thirteenth [book] of the *Metaphysics* which have not yet been translated and [which] discuss the intellegences, as the author himself promises). But, [the fact] that Albumasar does not agree with [this] opinion regarding the rationality of the planets is apparent from the fact that he [also] says, in the same chapter, that: "Even if the planets had rational souls, they nevertheless do not choose nor [do they] need choice on account of their distance from impediments." And [it is] more clear from what he says in the same treatise, third chapter, where near the end he writes: "The entrance of the Sun into this degree [of the heavens], was not made according to a choice by the Sun itself, nor is the effecting of things nor their corruption, but simply, through its arrival into this degree [of the heaven] by means of its natural motion." And further on, he proves that the motion of a circle is due to the virtue of the immobile and eternal first cause. And for that reason he also says: "One must bless His name, and exalt Him, not the Sun". Therefore, what is the merit of his book? If you study it carefully, you will discover many good things, [and] none, however, that are bad, as far as I know. You will [also] find in the Book I, chapter 2, in the section on the suitability of the times when the motion will cease, where he says: "The planets are not subject to corruption, nor to increase or decrease in size, effect nor detriment until that time which God wishes". And in the same treatise, in chapter 5, [in] the section on the first sect, [he says]: "By means of the natural and everlasting motions of the planets, a natural and lasting effect is realized, which will continue up until the time which God wishes". Ptolemy gives a similar testimony in chapter 1 of Book I of the *Almagest*, where it is written: "We, however, have laboured so that, in [our] love for the knowledge of the eternal things which will last until the end which their Creator has imposed on them, in the following [chapters] of this book of ours we should add etc.".

You will also find in Albumasar something far more elegant; namely, a

monium fidei et vitae aeternae, quae non acquiritur nisi per fidem, tractatu sexto, differentia vicesimasexta, ubi assignat causam, quare nona domus est domus fidei, et dicit: «Domus quoque nona vocata est domus peregrinationis et motionis fidei quoque atque operum bonorum, propter reversionem eius ad Iovem etc.». Et infra: «Rursum Iuppiter et Venus sunt fortunae. Fortunae autem duae sunt species, quarum una est fortuna huius mundi et altera fortuna futuri saeculi, et fortuna futuri saeculi dignior est fortuna huius mundi, et hoc quaeritur per fidem; et quia Iuppiter est plus fortuna quam Venus, ideo facta est ei significatio super fidem per quam quaeritur fortuna futuri saeculi, quae est dignior, et facta est Veneri significatio super fortunas huius mundi, ex ludis et gaudio et laetitia». Quid ergo in his meruit liber suus? sed et quid meruit, si scriptum est in eo ab initio figuratam esse in caelo nativitatem Iesu Christi de Virgine, etiam cum expressione nominis ab angelo nuntiati? In tractatu namque sexto, differentia prima, in capitulo de ascensionibus imaginum quae ascendunt cum Virgine, invenitur: «Et ascendit in prima facie illius (scilicet Virginis) puella quam vocat Celchuis Darostal; et est virgo pulchra atque honesta et munda prolixi capilli, et pulchra facie, habens in manu sua duas spicas, et ipsa sedet super sedem stratam, et nutrit puerum, dans ei ad comedendum ius in loco qui vocatur Abrie. Et vocat ipsum puerum quaedam gens Iesum, cuius interpretatio est arabice Eice. Et ascendit cum ea vir sedens super ipsam sedem. Et ascendit cum ea stella virginis etc.». Etiam scimus quod sub ascendente eiusdem partis caeli, scilicet Virginis, natus fuit Dominus Iesus Christus, cum hoc quod aequatio motus octavae sphaerae in tempore nativitatis eiusdem fuit octo graduum et triginta septem minutorum et duorum secundorum secundum calculationem certissimam, et quod ipsa tunc erat minuenda de locis planetarum inventis per canones; non quia subiaceret stellarum motui aut earum iudicio natorum desideratissimus, qui creaverat ipsas stellas, sed quia cum extenderet caelum sicut pellem, formans librum universitatis, et dedignaretur opus facere incompletum, noluit litteris eius deesse, ex eis quae secundum providentiam suam in libro aeternitatis sunt

63) ALBUMASAR, *op. cit.*, VI, 26. 78) ALBUMASAR, *op. cit.*, VI, 1. 93) Cfr. Ps. 103 «Deus ... extendens caelum sicut pellem».

testament of faith and of eternal life not acquired save by faith. [This appears] in the sixth treatise, chapter 26, where he explains the reason why the ninth house is known as the house of faith, and he says: "The ninth house also called the house of pilgrimage and of movement [by reason] of faith, of good works, on account of its reversion to Jupiter, etc." And further on [he says]: "Furthermore, Jupiter and Venus are benefic. There are two kinds of fortune, however, one of which is the fortune of this world and the other is the fortune of the hereafter; and the fortune of the hereafter is more worthy than the fortune of this world; and this is attained by faith. And because Jupiter is more benefic than Venus, it is for this reason that signification of faith by means of which the fortune of the hereafter is sought (the more worthy [fortune] is his [i.e.: Jupiter's]); and signification of the fortunes of this world [such as] games, joy and happiness is given to Venus." What, then, did his [i.e.: Albumasar's] book deserve in these [matters]? But, also, what was its value if it was written in it that the birth of Jesus Christ from the Virgin, as well as the utterance of the Name announced by the angel, was figured in the heaven from the beginning [of time]? For, in treatise six, chapter 1, in the section on the rising of the images which ascend with [the sign of] Virgo, [the following] is found: "And in first decan [i.e.: of Virgo], there arises a girl whom he calls Celchuis Darostal; and she is a beautiful, honorable, pure virgin with long hair, and a beautiful face, holding two ears of wheat in her hand; and she sits on a covered bench, and she nurses a male child, giving broth to him in a place which is called Abrie. And a certain people call this child Jesus, which is translated as "Eice" in Arabic. And also rising there with her is a man sitting on the same seat. And there rises with her the star of the virgin ... etc." Also we know that under the ascendent of this part of heaven, that is of Virgo, the Lord Jesus Christ was born, and also that the equation of the motion of the eighth sphere at the time of His birth was 8° 37' and 2" according to the most certain calculation[s], and that it was then subtracted from the positions of the planets [which were] found by using [astronomical] tables; not because the most desired of natives who created the stars himself was subject to the motion of the stars or to their judgement, but because when He spread out the heavens, just like vellum, to form the book of the universe, He refused to make the work incomplete; He did not wish there to be missing from its letters which were written according to His providence in the book of eternity – even what was furthest

scripta, etiam illud elongatissimum a natura quod de Virgine nasceretur, ut profecto per hoc innueretur homo naturalis et verus, qui non naturaliter nascebatur, non quod caeli figura esset causa quare nasceretur, sed potius significatio, immo ad vero verius, ipse erat causa quare modus admirandae suae nativitatis significaretur per caelum. Haec et multa alia notabilia poteris reperire, si diligenter revolveris librum illum. Quod si forte cum his aliquod suspectum, quod non memini me vidisse, invenire contingat, corrigatur, potius quam, multa utilia cum uno relegando, indiscrete damnetur. Quid iterum meruit liber Abdilaziz, quem vocat Alkabitium, qui similiter cum iniquis deputatus est? Si sunt in textu eius nomina ignotae linguae, statim subduntur in littera interpretationes eorum; quod si forte aliquorum interpretationes defuerint, paratus est vir earum copiam exhibere.

Caput Decimum Tertium

De partibus autem revolutionum, quae sunt de revolutione annorum mundi et de coniunctionibus et eclipsibus planetarum, sicut de illa quae est de mutatione temporum, potest dici. Si enim ex figura revolutionis anni, aut eclipsis, aut coniunctionis, quae significat sectam, significatur terraemotus sive diluvium, aut scintillae, aut super divites et universitatem vulgi guerra vel pax, fames sive mortalitas, caeterum apparitio alicuius magni prophetae sive haeretici, aut ortus horrendi schismatis universalis vel particularis, secundum quod providit Deus altissimus, quid ad arbitrium liberum? Numquid est in potestate hominis talia immutare? Apparet quod et istae partes stare meruerint, neque reprehensione indigeant, nisi aliud obstet quod nondum audivi fuisse propositum contra eas.

Ad nativitates me transfero, quae pars caeteris acrius videtur pungere liberum arbitrium, ut etiam appareat quod invecem se

---

4 ss.) cfr. cap. VII.

removed from nature: that [He would be] born from a Virgin – in order that by this means He might be recognized as a natural and true human being, who was not born in the natural manner not because the figure of heaven was the cause of His birth, but rather because [it was] a sign; or rather, as is truer than the truths, He Himself was the cause by which the manner of His miraculous birth was signified by means of the heavens. You could discover this and many other notable [things], if you went through the book carefully. But if, perhaps, something suspect happens to be found in additon to this, which I do not remember seeing, it should be corrected, rather than be indiscriminately condemned because many useful things [happen to be] bound with one [erroneous statement]. Again, what has the book by Abdilaziz, whom he calls Alcabitius, which was similarly included amongst the iniquitous books deserved? If there are names in an unknown language in his text, their meanings are immediately added to the text itself; but if perhaps the meanings of some [of these words] should be missing, [there is a] man prepared to supply them.

CHAPTER THIRTEEN

One can say the same things about the sections dealing with revolutions (which discuss the revolution of the years of the world and the conjunctions and eclipses of the planets) as [has been said] about that [section] which addresses the change of the seasons. For if an earthquake and a flood, or fires – as far as rich men and all the common people are concerned – war or peace, famine and death, as well as the appearance of some great prophet or heretic, or the rising of a horrible universal or particular schism are indicated by the figure of the revolution of the year, or of an eclipse, or of a conjunction (which signifies a [religious] sect) in accordance with that for which God, the most high, has provided, what has that to do with free will? Is it in a man's power to change such things? It seems that these parts ought to be preserved as well, and that they do not demand reprehension, unless due to some other criticism that I have not yet heard proposed against them.

I turn now to nativities, which seems to be the section that offends free will more severely than the other parts, so [much so] that it even seems

destruant, neque ullatenus se posse compati videantur, praecipue quantum ad partem illam quae pertinet ad mores animi. De scienda namque quantitate vitae nati per gradum hylech et planetam alchochoden iam dictum est, quod non iudicatur quantum oporteat vivere de necessitate, sed ultra quod vita eius non protenditur ex natura: abbreviari enim possunt dies hominis, non augeri. De sciendis autem eventibus corporis nati ex infirmitate et sanitate, quaeritur quid prodest homini si malum sibi futurum praesciat, cum illud praepedire non possit; quod si potest, est ergo falsum magisterium astrorum et fallitur aspiciens in eodem. Ego autem dico, quod omnis operatio causae agentis supra rem aliquam est secundum proportionem materiae recipientis ipsam operationem, ut unus idemque ignis operatur in luto arefactionem atque liquefactionem in cera. Unde si homo ex astrorum magisterio praescierit quod in aestate futura ex operatione caeli passurus sit ex superfluitate caloris et siccitatis, in multo tempore ante per exhibitionem diaetae potest mutare complexionem suam, donec declinet ad latus frigiditatis et humiditatis, ut operatio caeli adveniens, quae si ipsum in media consistentia invenisset, ad latus aegritudinis ex calore superfluo et siccitate traxisset, dum ipsum in opposito latere invenit, potius reducit ad medium sanitatis. Hac ergo via potuit removeri in toto aut in parte impedimentum praescitum; nec tamen frustrata fuit caeli operatio, sed perfecta; non enim caeli operatio, sed operationis qualitas est remota, iuxta quam intentionem loquitur Ptolemaeus in verbo quinto, ubi dicit: «Potest astrologus plurimum avertere de operatione stellarum, cum fuerit sciens naturae agentis in eum, et praeparaverit ante suam descensionem recepturum sustinere valentem». De scienda quoque acquisitione substantiae quaestio est ad quid prosit ei congaudere antequam veniat, quia sub illa spe fortassis aliquid temere attentabit. Et ego iterum dico, quoniam bonum futurum potest, si presciatur, augeri et effici melioris profectus iuxta verbum eiusdem Ptolemaei octavum, quod est: «Anima sapiens potest adiuvare caelestem operationem, quemadmodum

---

19 ss.) Cfr. cap. VIII. 29) HALY, *Glosa super verbum quintum* (PTOLEMAEUS, *Liber centum verborum cum expositione Haly*); 40) PTOLEMAEUS, *op. cit.*, v. 5. 46) HALY, *Glosa super verbum octavum* (PTOLEMAEUS, *op. cit.*). 47) PTOLEMAEUS, *op. cit.*, v. 8.

that [the two] might destroy each other mutually, nor do they seem to be compatible in any respect, especially as far as that part which pertains to the character of the soul is concerned. For, with regard to the knowledge of the length of the native's life by means of the degree of the hylech and the planet which is the alchochoden, it has already been said that the judgement cannot be made about how long he [i.e.: the native] ought to live by necessity, but about the time beyond which his life is not be extended naturally: for the days of a man can be shortened, but not increased. Also, with regard to the knowledge of the occurrences of health and sickness in the body of the native, one might ask how man profits if he foresees the evil which will befall him, since he is not able to prevent it; because if he could, then the profession of the stars would be false and whoever looked into it would be deceived. But, I say that every operation of a cause acting on some thing is [determined] in accordance with the proportion of material receiving that operation, as [for example], one and the same fire effects in drying mud and melting wax. So that, if a man knew in advance, from the teaching of the stars, that he would experience in a future summer a superfluity of heat and dryness from the operation of the heavens, he could change his temperament a long time beforehand by organizing his diet until it declines on the side of coldness and wetness, so that the operation of heaven, which, had it found him in a moderate temperament would have drawn him to the side of illness from the excess of heat and cold, now, since it finds him in the opposite disposition, it brings him back to a moderate temperament instead. In this way, therefore, an impediment that is foreseen could be removed, totally or in part; and yet the operation of heaven is not frustrated, but is perfected: for the operation of heaven was not removed, but the quality of the operation, in accordance with the intention of Ptolemy expressed where he says, in the fifth sentence [of the *Centiloquium*]: "An astrologer can avert a great deal from the operation of the stars when he is informed about the nature acting upon him, and has prepared [himself] before its descent [so that] the one who will receive [the influence] will be able to bear it". Also, regarding the knowledge of the acquisition of wealth, the question is: what advantage is there in rejoicing in [wealth] before it comes, since under [the influence of] that hope [of riches] he might try [to do] something rashly. And I say again, that a future good, if it is foreseen, can be increased and its advantages optimized, according to the eighth sentence of the same [Centiloquium] of Ptolemy, which is: "The knowing soul can help the celestial operation, just

seminans virtutem per cultum et purgationem». Sed quid de moribus animi respondendum, nisi quia non iudicatur natus castus, aut incestus, aut iracundus, aut patiens et talia, nisi secundum aptitudinem et ineptitudinem? Unde nihilominus eliget hoc aut illud, sed ex opere caeli est quod, ad eligendum id ad quod aptus est, citius inclinetur. Quod si propter hoc condemnetur ista scientia, eo quod liberum arbitrium destruere videatur hoc modo, certe eadem ratione non stabit magisterium medicinae; numquid enim ex eius magisterio iudicatur quis secundum causas inferiores aptus ad huiusmodi vel ineptus? Quod si magisterium medicinae destruatur, multum erit utilitati reipublicae derogatum; eo vero stante, non videntur habere quid contra partem nativitatum allegent.

Caput Decimum Quartum

Ad interrogationes transeo, et illae quidem quae fiunt de praesentibus, non videntur habere dubitationem, ut quando quaeritur de absente, utrum sit vivus vel mortuus; aut de rumoribus, utrum sint veri vel falsi; et de epistola recepta, a cuiusmodi persona exierit, utrum scilicet a rege aut ab alio; et de muliere quam scimus peperisse, cuiusmodi prolem peperit, masculum scilicet an foeminam; et de homine, qui profitetur alchimiam, utrum sit veritas operis apud eum. Talia enim, quorum veritas determinata est in partem alteram apud rerum naturam, nihil est mirum si significentur per caelum. Sed illae, quae de futuro sunt, merito dubitationem admittunt; neque enim super rebus necessariis aut impossibilibus interrogationibus indigemus; illarum tamen, quae fiunt super futuris possibilibus, maiorem habent dubitationem quaedam quam aliae, ut illae quae sunt de rebus quae penitus libero arbitrio sunt subiectae. Nam quaedam res sunt possibiles et futurae, quas nihilominus non potest cuiusquam arbitrium impedire; ut est interrogatio de gravitate vel levitate annonae, utrum futura sit in eodem anno, licet hoc certius possit ex anni revolutione cognosci; ut est interrogatio utrum quis acquiret multam substantiam ex magisterio suo, aut ex negotiatione,

as the sower [can help] the strength [of plants] through [careful] cultivation and weeding."

But how should one respond to those questions concerning the character of the the soul, except [by saying] that the native is not judged to be chaste or impure, wrathful or patient, and so on save according to his aptitude or lack of aptitude? Hence, nonetheless, he chooses this or that [conduct]. But it belongs to the operation of the heavens whether he is inclined more readily towards choosing that for which he has an aptitude. But if this science is condemned for this reason – because it seems to destroy free will in this way – then, certainly, the profession of medicine cannot be preserved for the same reason, since, surely, it is judged from its profession who, due to inferior causes, is fit or unfit for something? But, if the profession of medicine were destroyed, it would detract greatly from the public good; but as long as this profession is preserved, [the accusors] can alledge nothing against the section on nativities.

CHAPTER FOURTEEN

I pass on to interrogations. And those made concerning present affairs, do not appear subject to doubt –as are those made concerning someone who is absent, [such as] whether he is alive or dead; or about whether rumours are true or false; or from what kind of person a received letter has come – whether from a king or some other person; and about a woman whom we know has already given birth, [asking] what kind of child was born, that is, [whether it is] masculine or feminine; and whether the work of a man who professes alchemy is true or not. For there is nothing surprising if such things as those cases whose truth is determined [already] by the other direction [i.e.: the past] as an aspect of the nature of things are signified by the heavens. But those [questions] concerning the future justifiably admit uncertainty, since we don't need to ask about matters which are necessary or impossible ; nevertheless, some of those [interrogations] concerning future possibilities, have greater uncertainty than others; such as, those which concern things which are completely subjected to free will. For some things are possible and future, which, nonetheless, no one's will can impede: such as a question concerning the high or low price of grain in the coming year (although this can be known more certainly from the revolution of the year); [or a question about] whether someone might acquire wealth from his profession or from business; or whether a

aut utrum quidam vir adipiscetur hoc regnum vel illud et caetera huiusmodi. Talia namque accidunt homini ex significatione suae nativitatis; quia cum ipse interrogaverit de seipso, et fuerit motus per caelum cum intentione radicali, iam pervenit ex nativitate sua ad bonum seu ad malum quod significavit eius nativitas. Sollicitudo enim hominis in hora interrogationis erit secundum habitudinem circuli, et est circulus in eadem hora secundum intentionem ipsius. Quare oportet figuram caeli horae interrogationis esse proportionalem figurae caeli suae nativitatis, alioquin non est intentio radicalis et hoc etiam potest ex figura interrogationis perpendi. Unde etsi esset nota eius nativitas, non esset nobis necessaria interrogatio super similibus, cum nativitas sit radix fortior, iuxta illud Haly: «Nativitates sunt res naturales, et interrogationes sunt res similes naturalibus». In illis ergo quorum nativitates ignotae sunt, necessario ponitur interrogatio pro radice, et eadem ratione, qua salvatur nativitatum scientia, salvantur et interrogationes super huiusmodi nutu Dei. Interrogationum vero de rebus futuris possibilibus, quae arbitrio subiacent libero, duo sunt modi. Sunt enim interrogationes facti ut; quid fiet de aliquo? Et sunt interrogationes consilii ut: quid melius fieri conveniat, hoc an illud? Et illae quae sunt consilii, non destruunt, immo potius rectificant et dirigunt arbitrii libertatem, ut est interrogatio de negotiatione, utrum sit mihi utilis vel non; et de duabus rebus, quam illarum emere sit melius; et de via quam intendo arripere, utrum ire sit mihi melius an morari. Talia destruere plus esset contra liberum arbitrium quam pro eo, quia oportere consiliare et negotiari est unum ex mediis urgentioribus per quae ostenditur non omnia esse ex necessitate, sed quaedam a casu atque ad utrumlibet. Determinare autem de interrogationibus facti, qualiter maneant cum arbitrii libertate, difficillimum est, ut est interrogatio de substantia ab aliquo petenda: utrum scilicet eam dabit vel non dabit? Nam si millesies significatum fuerit quod non dabit, nihilominus poterit eam dare. Similiter si quod dabit, semper poterit et non dare; alioquin non remaneret electio apud eum, nisi quia non inve-

---

24-25) Cfr. cap. X, 11 ed il testo citato. 29) Cfr. cap. IX, 3 ed il testo citato. 32-34) Cfr. cap. X, 8-9 ed il testo citato. 48-74) Cfr. ALBUMASAR, *Introductorium Maius*, transl. Ioannes Hispalensis; I, V cap. *De tertia secta*.

certain man might acquire this or that kingdom, and so on in this manner. Because such events do happen to a man due to the signification of his own nativity, because when he asked [questions] about himself, [he] was moved by heaven according to a radical intention, [namely, that] due to his nativity, he has already come to that to the good or evil which his nativity signified. For the concern of a man at the hour of the interrogation, will be in accordance with the situation of the [zodiacal] circle [of his nativity]; and the circle at that hour is in accordance with his [own] intention. Wherefore, the figure of heaven at the hour of the interrogation ought to be proportional to the figure of heaven for his nativity, otherwise there is no radical intention; and this can also be judged from the figure of the interrogation. Whence, if his nativity were known, an interrogation concerning similar things would not be necessary because the nativity is the stronger root according to this [saying] of Haly: "Nativities are natural things, and interrogations are things similar to natural [ones]". In the case of those people whose nativities are unknown, therefore, the interrogation is necessarily regarded as the root. And for the same reason for which the science of nativities should be preserved, so interrogations made about this kind of subject should be preserved, by the will of God.

There are two kinds of interrogations about contingent things which are subjected to free will. For there are questions of fact (such as, what will happen concerning something?). And there are questions of advice (such as, would it be more convenient if this or that happened?). And those [questions] about advice do not destroy the freedom of the will, but, on the contrary, they rectify and direct it (such as, with a question about whether a [business] negotiation might be useful to me or not; or about which of two things it might be better to buy; and about a route that I intend to take, whether it might be better to proceed or to delay). To destroy such [things] would be [a decision] more against free will than for it, because to have to take advice and to negotiate is one of the most persuasive means by which it is demonstrated that everything does not happen due to necessity, but that some things [happen] by chance and they [could go] either way. However, it is extremely difficult to determine how interrogations of fact might be reconciled with the freedom of the will, such as [for example], in an interrogation about money being sought from someone: that is, whether he will give it or not. Now, even if it were signified in a thousand ways that [this person] will not give [money], he is, nevertheless,always free to give it. Similarly, if [it were indicated] that he will give it, he will always be able not to give it; otherwise, the choice will not be left

nitur astrologus affirmasse quod dabit, sed quod per figuram interrogationis significatum est eum daturum, et de eo quod significatum est, adhuc restat quaestio utrum erit. Nam si non erit, magisterium est falsum; si vero erit, ergo non potest non esse, vel forte istud non sequitur; nam de contingenti quod erit, et de quo verum est dicere quoniam erit, antequam sit semper possibile est de esse et non esse; sed quando est, iam non potest non esse, sicut de eo quod est album nunc, et de quo prius erat verum dicere quoniam erit hoc album; non sequitur ergo antequam esset, non potuit non esse, sed quod non potest non esse quando est. Omne enim contingens sive sit natum in pluribus, sive in paucioribus, sive ad utrumlibet, semper antequam sit, potest et esse et non esse, ut dictum est, licet non aequaliter quaedam eorum; sed quando est iam revertitur ad naturam necessarii, non quod prius fuerit necessarium, sed quod necessario est quando est. Non enim idem est esse necessario quando est, et simpliciter esse ex necessitate. Antequam ergo sit potest non esse, et tamen erit, quia non est necesse illam potentiam ad actum reduci. Similiter de eo de quo significatum est quoniam non erit in tempore determinato, et de quo verum est dicere, quoniam non erit tunc, nihilominus semper ante hoc potest esse, et tandem revertitur ad naturam impossibilis. Et haec est sententia Albumasaris, a qua tamen famosus Aristoteles in aliquo declinare videtur, cum non concedat quod prius sit verum dicere. Me autem nihilominus sic dixisse non piget, sed in his negativis quae absque temporis determinatione significata sunt non similiter, ut est illud de quo verum est dicere, quoniam numquam erit, quia non revertitur ad naturam impossibilis, quin semper possit esse usquequo cesset motus, quia ex hoc iam non poterit non esse.

Et fortassis attingentius intuenti, eadem aut saltem similis genere est ista dubitatio ei dubitationi, quae est de divina providentia; nam in his quae operatur Dominus per caelum, nihil aliud est caeli significatio quam divina providentia. In his vero quorum nos sumus principium, nihil prohibet etiam caelo non causam, sed signi-

open to him not unless because no astrologer is found to have affirmed that he will give it, but because it was indicated by the figure of the interrogation that he will be about to give it – and there still remains the question of whether that which is signified will be realized. For if this does not happen then the profession [of interrogations] is false; but, if it does happen, then it could not not be, or perhaps that does not follow. For, concerning something for which it is contingent that it will be (and about which it is true to say that it will be), before it is, it is always possible for it both to be and not to be. But when it [actually] is, [then] already it can not not be (as, for example, with regard to that which is white now and about that which it was previously true to say that it would be white). It does not follow, therefore, [that] before it is, it can not not be, but when it is, it can not not be. For every contingent being, [regardless of] whether it is produced to a greater or lesser extent by nature, or on either side, before it is, it can always both be and not be (as was said), although some of them [can] not equally [be and not be]; but when it is, it already returns to the nature of what is necessary, not because it was necessary previously, but because it necessarily is when it is. For it is not the same thing to be necessarily as to be simply "by necessity". It can not be, therefore, before it is; and yet, it will be because it is not necessary that potentiality [must] be reduced to actuality. Similarly, regarding that about which it is signified that it will not be at a determined time, and about which it is true to say that it will not be then [at that time], nevertheless, it can always be before that [time], and [up until that time when] it finally reverts back to the nature of the impossible. And this is the opinion of Albumasar, from which the famous Aristotle seems to depart to some extent, since he [i.e.: Aristotle] does not concede that it may be true to say [something will or will not be] beforehand. I do not regret having said this, but the situation is not the same in the case of those negatives which are signified without the determination of time, as it is about [that] which it is true to say that it will never be because it does not revert [back] to the nature of the impossible; instead, it always could be [up] until the motion ceases, because already from this it will not be able not to be.

And perhaps, someone considering [this matter] more closely, will have the same uncertainty or one similar in kind to that uncertainty which concerns divine providence; since in those things which God operates by means of the heavens, the indication of heaven is nothing other than divine providence. In those things, indeed, which we initiate, nothing prevents [the fact] that there is also not a cause in heaven, but a signification. For of the

ficationem inesse: duarum enim partium contradictionis, quarum alterutram potest homo eligere, sciebat Deus ab aeterno quam illarum eligeret. Unde in libro universitatis, quod est caeli pellis, sicut praedictum est, potuit figurare, si voluit, quod sciebat; quod si fecit, tunc eadem est determinatio de compossibilitate liberi arbitrii cum divina providentia et cum interrogationis significatione. Si ergo divinam providentiam stare cum libero arbitrio annullari non possit, neque annullabitur quin stet magisterium interrogationum cum eo. Qualiter autem non annulletur de divina providentia, relinquendum arbitror negotio altiori; verumtamen non volo dicere, quod quaecumque non latent divinam providentiam, sint etiam cognita apud caelum, longe enim est caelum inferius. Unde sicut dictum est, cum fuerint significatores aequales in fortuna et malo, consilium magisterii astrorum est supersedere, quia Dominus voluit celare a nobis.

Caput Decimum Quintum

De electionibus vero est quaestio minus difficilis, non enim libertas arbitrii ex electione horae laudabilis coercetur, quin potius in magnarum rerum inceptionibus electionem horae contemnere est arbitrii praecipitatio, non libertas. Etiam invenitur quidam dixisse de huiusmodi hominibus, quod quotquot ex eis salvari contigerit, eos Deus non ut homines, sed ut iumenta salvabit. Caeterum in hoc concordati sunt omnes philosophi, quod cum sciverimus horam impraegnationis alicuius mulieris, sciamus per eam quid fiat de foetu donec inspiretur et quid usquequo egrediatur ab utero, et quid forte usque ad obitum. Neque enim iudicaverunt astrologi per nativitates, nisi quia hora impraegnationis vix potest certificari. Unde inquit Ptolemaeus horam nativitatis esse secundum iudicium. Quare ergo uxore regis seu principis aut magnatis existente in optimis conditionibus, non eligemus viro eius horam suscipiendi ex ea liberum, si creator universae generationis annuerit, ut scilicet eveniant nato bona, quae ex serie

98-101) Cfr. cap. IX, 16-18.

7-10) Haly, *De electionibus horarum*, I, 1.

two sides of a dilemma from which man can choose one or the other, God knew from eternity which of these he [i.e.: the man] would choose. For which reason, in the book of the universe, which is the vellum of heaven (as was said before), He was able to configure, if He wished, what He knew; [but] if He did this, then the compatibility of free will with divine providence or with the indication of an interrogation is the same. Therefore, if it cannot be denied that divine providence co-exists with free will, it cannot be denied that the profession of interrogations co-exists with it as well. But I think [the question of] how it might not be denied with respect to divine providence should be left to another, more elevated investigation. Nevertheless, I do not wish to say, that whichever [of those] things that are not hidden by divine providence might also be recognized in heaven; for heaven is greatly inferior. For that reason, as was said, when the signifiers are equal in [good] fortune and evil, the counsel of the profession of the stars is to abandon [the interrogation], since God wished to keep it hidden from us.

CHAPTER FIFTEEN

The question of elections is certainly less difficult; for the freedom of the will is not coerced by the choice of a favourable hour, but instead, it is a precipitation of the will, not [its] freedom, to disregard the choice of the hour for the beginnings of important matters. Also someone is known to have said about such men, that as many of them as might happen to be saved, God will save them not as men, but as beasts. But all philosophers are in agreement on this point, [namely,] that when we know the hour of the impregnation of some woman, we may know, by means of that hour, what might happen with regard to the foetus until [the time when] it is quickened or what [will happen] until it is delivered from the uterus, and, perhaps, what [will happen], until its death. For astrologers have not judged [these things] by means of nativities, only because the exact hour of conception can rarely be verified. For which reason Ptolemy says that the hour of the nativity is made according to judgement. Why, therefore, when the wife of a king or prince or magnate exists in the optimum conditions, do we not choose for her husband the hour for getting a child from her, if the Creator of all generation allows, so that good things might come

librorum nativitatum astrologus futura praenuntiat? Quare iterum non eligemus horam pharmacum exhibendi, si sciverimus quod ascendens et significatores in signis ruminantibus, et praecipue in Capricorno, provocant vomitum? et cum hoc si sciverimus, quod oportet eos proiici ab aspectu utriusque infortunae, Saturni videlicet et Martis, et quod Saturnus constrigit medicinam, Mars vero educit usque ad sanguinem, et cum hoc etiam sciverimus quod Luna existente cum Iove operationis purgatorii minuitur angustia. Rursum in magisterio chirurgiae, quare non cavebo facere incisionem in membro, Luna existente in signo habente significationem super illud membrum. Tunc est enim membrum valde rheumaticum, et dolor rheuma provocat. Et audeo dicere me vidisse ex hoc quasi infinita inconvenientia accidisse. Vidi hominem peritum astrorum et medicinae, qui pro periculo squinantiae minuerat sibi de brachio, Luna existente in Geminis qui habent significationem super brachia, et absque ulla manifesta aegritudine, excepta modica brachii inflatione mortuus est in die septimo. Scivi quoque quendam patientem fistulam iuxta caput longaonis fuisse incisum, Luna in Scorpione, qui significat super partes illas, a quodam misero chirurgico, qui erat ignarus utriusque magisterii, medicinae scilicet et astrorum, et absque venae incisione, aut alia causa rationabili, inter manus eum tenentium inventus est mortuus ipsa hora, fuitque caeli operationi adscriptum, cum non videretur ab aliqua causa interficiente subito accidisse, ut sunt oppilationes ventriculorum cerebri, aut laesio spiritualium seu defectus. Quod si inveniatur inter electiones aliquid apparens frivolum, ut est indumenta nova induere Luna in Leone, etc., attendi debet quod Ptolemaeus, cum esset vir tantae auctoritatis, non dixit hoc nisi ut significaret maiora, et per hoc innuit quod signa fixa utilia sunt ad res quarum stabilitatem volumus, sicut et domus quae dicuntur anguli; signa vero mobilia, sicut et domus cadentes ab angulis, ad res cito mutabiles, quarum recessio expectatur et non e converso.

17-41) Cfr. CAMPANUS, *Canon pro minutionibus et purgationibus* (BT) 23-24) Cfr. PTOLEMAEUS, *Liber centum verborum cum expositione Haly*, v. 19. 26-28) Cfr. PTOLEMAEUS, *op. cit.*, v. 20. 42-43) PTOLEMAEUS, *op. cit.*, v. 22.

to him when he is born which the astrologer could predict as [being] about to happen from the series of books of nativities? Again, why do we not choose the hour to employ medicine, when we know that the ascendent and signifiers in the ruminating signs (and especially in Capricorn) provoke vomiting? And if we know in addition to this that they ought to be struck by the aspect of both malefic [planet], that is, Saturn or Mars, because Saturn fixes medicine and Mars draws it to the blood; and also if we know together with this that when the Moon is conjunct with Jupiter the difficulty of the operation of purging is reduced. Again, in the profession of surgery, why shall I not take care not to make an incision in a limb when the Moon is in a sign which has significance over that limb? For at that time, the limb is very rheumatic and pain provokes rheum. And I have the courage to say that I myself have seen as it were an infinite number of inconveniences happen as a result of this. I have seen a man who was an expert in astronomy and medicine, who due to the threat of angina bled himself from [his] arm while the Moon was in Gemini, which has significance for the arms, and without any apparent illness, except for a moderate inflamation of the arm, he died seven days later. I also knew a certain patient who was suffering from an ulcer near the head of his gut [and] was cut open by some miserable surgeon who was completely ignorant of both professions (namely, medicine and of the stars) whilst the Moon was in Scorpio (which has significance over those parts), and without the cutting of a vein or some other reasonable cause, he was found dead in the arms of the men who were holding him within that very hour; and [his death] was attributed to the operations of heaven, since it did not seem to have occurred due to any cause that kills suddenly, such as obstructions in the ventricles of the brain or a lesion or failure of the air passages [i.e.: respiratory system]. But if something apparently frivolous is found amongst the elections (such as that one should put on new garments when the Moon is in Leo, etc.), it must be noted that Ptolemy, since he was a man of great authority, did not say this [sort of thing] except [in order] to indicate more important issues, [and] by this [statement] he meant that fixed signs are useful for matters which we want stabilized; such as, the [astronomical] houses which are callled cardines; but the mobile signs, such as houses cadent from the cardines are [useful] for things that change quickly whose departure is expected and not the other way around.

Caput Decimum Sextum

Partem vero quae est de imaginibus astronomicis propter vicinitatem quam habent ad necromanticas, non defendo aliter quam secundum quod superius in earum capitulo dictum est, eas nancisci virtutem a figura caelesti iuxta verbum Ptolemaei nonum, quod ibi tactum est, scilicet quod imagines quae sunt etc., et nisi quia nihil prohibet eas defendere secundum quod possunt negari vel defendi. Esto itaque exempli gratia, quod cum praedictis conditionibus fundatur imago ad scorpiones fugandos ab aliquo loco, si Deus voluerit, non videtur esse exorcismus aut invocatio, si dicatur in fusione illius «haec est imago destructionis scorpionum a loco illo quamdiu fuerit in eo imago servata». Non videtur iterum inscriptio esse characterum, si in dorso eius sculpatur hoc nomen: «Destructio»; sicut et in imaginibus ad amorem hoc nomen «Amor» scriberetur in ventre et in ante scilicet, neque si in fronte eius scribatur hoc nomen «Scorpius», quod est nomen speciei fugandae, et in pectore nomen ascendentis et nomen eius domini qui est Mars et nomen Lunae. Quis iterum cultus exhibetur ei si in medio loci, a quo ipsam speciem fugare volueris, fuerit imago sepulta capite deorsum et sursum pedibus elevatis? Non commendo eas, sed neque videtur quod absque ratione debeant aliarum iniquitatem portare.

Caput Decimum Septimum

De libris vero necromanticis sine praeiudicio melioris sententiae videtur, magis quod debeant reservari quam destrui: tempus enim forte iam prope est, quo propter quasdam causas quas modo taceo eos saltem occasionaliter proderit inspexisse, nihilominus tamen ab

---

5) Ptolemaeus, *Liber centum verborum cum expositione Haly*, v. 9; 9) Thebit, *De imaginibus*, 1, p. 180-181. 11-12) Thebit, *op. cit.*, II, 17 p. 183. 14) Thebit, *op. cit.*, 69-70 p. 192. 16) Thebit, *op. cit.*, 63 p. 190; 69 p. 192. 17) Thebit, *op. cit.*, 5 p. 181. 19-20) Thebit, *op. cit.*, 7 p. 181.

3-4) cfr. Apoc. 1, 3; 22, 10: «Tempus enim prope est».

CHAPTER SIXTEEN

I do not defend that section concerning astrological images on account of the nearness they have to necromantic [images], beyond what is said above in the chapter devoted to them, [namely] that they take their virtue from a celestial figure according to the ninth sentence of Ptolemy['s *Centiloquium*] which is touched upon there (that is, that "Images which are etc".) And [I would not defend them] unless [it were] that nothing prohibits one from defending them in accordance with that which can be denied or defended. So, let it be that when an image should be cast with the conditions mentioned previously for expelling scorpions from some place, if God should wish it, it does not appear to be an exorcism or an invocation if it is said during its casting: "This is an image for the destruction of scorpions from that place as long as the image is preserved in it." Nor, again, does it seem to be an inscription of characters if the word "Destruction" is engraved on its back [any more than] if the word "Love" is written on the heart and on the back in images for love; nor if on its forehead the word "Scorpion" were inscribed (that is, the name of the species to be banished) and the name of the ascendant or the name of its [planetary] Lord (which is Mars) or the name of the Moon were written on its breast. Again, what cult is shown by this: if in the middle of the place from which you want some species banished, the image were buried with its head downwards and its feet turned upwards? Not that I recommend them, but without any reason, it does not seem that they should carry the iniquity of the other [type of images].

CHAPTER SEVENTEEN

Concerning those books, however, which are [truly] necromantic, without the prejudice of a better opinion, it seems that they ought to be put aside rather than destroyed. For perhaps the time is already at hand, when, for certain reasons about which I am now silent, it will be useful on occasion to have inspected them, but, nevertheless, their inspectors should be

ipsorum usu caveant sibi inspectores eorum. Sunt praeterea quidam libri experimentales, quorum nomina necromantiae sunt conterminalia, ut sunt geomantia, hydromantia, aerimantia, pyromantia et chiromantia, quae ad verum non merentur dici scientiae, sed garamantiae. Sane hydromantia in extis animalium abluendis inspiciendisque fibris, et pyromantia in figura ignis, quo consumitur holocaustum, procul dubio idololatriae speciem non excludunt. In geomantia vero nihil tale invenio, sed confidit in Saturno et domino horae, qui ei pro radice ponuntur, gaudetque numeri ratione fulciri, et multi sunt qui ei testimonium perhibent. Non sic autem de aerimantia, frivola enim est, licet de ratione numeri se iactare praesumat. De chiromantia vero nolo determinationem praecipitem ad praesens facere, quia forte pars est physiognomiae, quae collecta videtur ex significationibus magisterii astrorum super corpus et super animam, dum mores animi conicit ex exteriori figura corporis; non quia sit unum causa alterius, sed quia ambo inveniuntur ab eodem causata.

8-9) Isidorus Hispalensis, *Origines*, 8, 9. 9) Isidorus Hispalensis, *Origines*, 9, 2, 125.

wary of using them. Moreover, there are certain experimental books whose names are coterminous with necromancy, such as [those which treat] geomancy, hydromancy, aerimancy, pyromancy and chiromancy, which really do not deserve to be called sciences, but "garamancies". Of course, hydromancy (dealing with the washing of the interiors of animals and of inspecting [their] fibres) and pyromancy (dealing with the figure of a fire, by which the holocaust is consumed) undoubtedly do not exclude the appearance of idolatry. I find nothing like this, however, in geomancy, since it relies on Saturn and the lord of the hour, which are put down as its root, and it rejoices to be based on the ratio of number; and there are many who bear testimony in its favour. But aerimancy is not like this; as it is frivolous even though it presumes to boast of the ratio of number. I really do not want to make a precipitous determination about chiromancy at the moment, perhaps because it is a part of physiognomy, which seems to be collected from the significations of the profession of the stars over the body and over the soul, while it makes conjectures about the character of the mind from the exterior figure of the body; not because the one might be the cause of the other, but because both are found to be caused by the same thing.

Historical commentary
by S. Caroti, M. Pereira,
S. Zamponi and P. Zambelli:

Sources of the *Speculum astronomiae*

SOURCES: CAPUT I

ALBUMASAR, *Introductorium, transl. Ioannes Hispalensis* I, 2 (Laur. Plut XXIV, 12), diff. II: «Differentia secunda in inventione scientie iudiciorum astrorum. In astris et motibus eorum due sunt sapientie mirabiles in cogitatione et magne in estimatione. Quarum una dicitur scientia totius; que est scientia qualitatis atque quantitatis circulorum altiorum et circulorum planetarum uniuscuiusque singulariter, longitudinis quoque uniuscuiusque circuli ab alio et declinationis eorum ab invicem, et magnitudinis eorum, et quantitatis uniuscuiusque circuli in semetipso ac longitudinis eius a terra».

ALFRAGANUS, *Numerus mensium*, diff. II: «Differentia secunda de hoc quod celum est secundum similitudinem sphaerae et revolutio eius cum omnibus quae sunt in eo est secundum revolutionem sphaerae».

ALPETRAGIUS, *De motibus celorum*, §4: «Et ab illo tempore meo non auferebar a dubitatione illarum positionum, quas etiam abhorret natura, et hoc quia ipse dicit (in collectione tertia tractatus primi) hoc scilicet, et cum eo quod narravimus, tunc pertinet quod sit de summa quam debemus premittere quod motus celorum sunt duo; unus quo movetur totum semper ab oriente ad occidentem secundum unum modum et revolutiones equales, et super circulos equidistantes unum alteri, et revolutiones super duos polos spere rotantis universum equaliter, et nominatur maior istorum circulorum equator diei.»

Ibidem, 5: «Et post hoc parum dixit: Et alius motus est quo moventur plures stellarum que currunt ad diversum motus primi super alios duos polos et non super illos».

ALFRAGANUS , *op. cit.*, diff. V: «Differentia quinta de duobus primis motibus celi quorum unus est motus totius, alter vero stellarum quem videntur habere in orbe signorum. – Quoniam premisi modo narrationem figure celi et terre prosequamur narrando quid nobis videatur de motibus celi. Dicamusque quod inicia motuum qui videntur in celo sunt duo quorum primum est quod movet totum, et fit dies et nox, quia volvit solem et lunam et universa sidera ab oriente in occidentem in unoquoque die ac nocte ... Et motus secundus est qui videtur inesse soli et planetis ab occidente in orientem contra partem primi motus super duos axes alios exeuntes ab axibus primi motus ...».

THEBIT, *Aequator diei*, 1: «Equator diei est circulus maior qui describitur super duos polos orbis super quos movetur ab oriente in occidentem».

THEBIT, *op. cit.*, 3: «Declinatio est arcus circuli meridiei cadens inter orbem signorum et equatorem diei».

ALFRAGANUS, *op. cit.*, diff. XII: «Differentia duodecima de narratione forme orbium stellarum et de compositione eorum et de ordinibus longitudinum eorum a terra. – ... Cuspis autem circuli signorum qui est circulus stellarum fixarum est cuspis terre. Cuspides vero ceterarum stellarum 7 que sunt spere planetarum erraticarum sunt remote a cuspide terre in partibus diversis. Et in unaquaque harum sperarum octo est circulus abscindens speram per duas medietates ab oriente in occidentem et circulus qui abscindit speram stellarum fixarum est cingulus circuli cuius mentio precessit et ad hunc refertur motus diversus equatus qui videtur omnibus planetis ab occidente in orientem. Unusquisque autem egresse circulorum cuspidis vocatur circulus egresse cuspidis».

Ibidem: «... Corpus vero solis est compositum super speram suam cuius cuspis egressa est a cuspide circuli signorum volviturque in eo volutione equali. Et superficies huius circuli egresse cuspidis est in superficie circuli signorum non declinans ab eo. Planetarum autem residuorum corpora non sunt super circulos egresse cuspidis. Sed sunt composita super circulos modicos qui vocantur circuli breves. Cuspides autem horum circulorum brevium sunt composite super circulos egresse cuspidis. Superficies vero utrorumque circulorum, idest egresse cuspidis et brevis declinat a superficie circuli signorum ... Cuspides autem circulorum brevium scilicet compositorum sunt composite super alios circulos egresse cuspidis preter primos quorum mentionem fecimus».

Ibidem: «... Dicamusque quod numerus circulorum circundantium universos motus planetarum atque stellarum sit octo ex quibus septem sunt septem planetis erraticis attributi et octavus qui est superior universis stellis fixis qui est circulus signorum. Et figura horum circulorum est ut figura intra se positorum invicem. Eritque minor omnibus et propior Terre spera Lune et secunda Mercurii, tercia est Veneris, quarta Solis, quinta Martis, sexta Iovis, septima Saturni, octava stellarum fixarum».

ALFRAGANUS *op. cit.*, diff. XIII: «Differentia decimatertia de narratione motuum solis et lune et stellarum fixarum in orbibus suis et in duabus partibus orientis et occidentis qui nominantur motus longitudinis. Soli autem sunt duo motus ab occidente in orientem quorum unus est ei proprius in suo circulo egresse cuspidis quo movetur omni die ac nocte 59 minutis fere et alius est motus tardus qui est spere eius super axes circuli signorum qui est equalis motui spere stellarum fixarum idest in omnibus 100 annis gradu uno. Ex hiis duobus motibus colligitur cursus eius ... ab occidente in orientem».

ALFRAGANUS, *op. cit.*, diff. XVII: «Differentia decimaseptima de orbibus planetarum».

ALFRAGANUS, *op. cit.*, diff. XXVIII, XXIX, XXX: «Differentia vigesimaoctava de eclipsi lune; Differentia vigesimanona in eclipsi solis; Differentia trigesima de quantitate temporis quod est inter eclipses».

ALFRAGANUS, *op. cit.*, diff. XII (after defining «circuli egresse cuspides»): «...Fitque ex hoc quod diximus necessario ut sint in unaquaque sperarum harum loca duo, unus scilicet in longitudine ultima spere a terra, et alius a propiori longitudine. Unus autem istorum locorum qui est longitudo longior, vocatur aux planete. Et alter qui est longitudo minor, vocatur oppositio augis».

ALFRAGANUS, *op. cit.*, diff. XVIII: «Differentia decimaoctava de motu latitudinis stellarum. – Et sequitur quod precessit de narratione nostra, de motibus (i.e.: *add.* planetarum) in longitudine, narrando motus eorum in latitudine, que est declinatio a linea ecliptica circuli signorum in utrisque partibus septentrionis et meridiei».

ALFRAGANUS, *op. cit.*, diff. XXVII: «In hoc quod accidit lune et stellis propinquioribus terre de diversitate aspectus. Et hic narremus quid accidit soli et quidquid est ex planetis sub eo per visionem de diversitate aspectus a locis suis certissimis ex circulo signorum. Dicamus primum quod cum consideramus lineam directam exeuntem a puncto terre qui est punctus circuli signorum usque ad corpus lune vel alium quemlibet planetarum erraticarum transeuntem ad circulum signorum pervenit ad circuli punctum in quo fuerit planeta in longitudine certissime, quo si fuerit planeta in zenith capitis, erit hec linea et linea que egreditur a loco aspectus nostri ad cuspidem planete eadem, ostenditque planetam in hoc loco ex circulo signorum certissime. Si vero non fuerit planeta in zenith capitis fuerintque utreque linee diverse et abscindunt se invicem super cuspidem corporis planete, et erit ipsa que egreditur de loco aspectus nostri que ostendit eum extra locum suum certissimum ex circulo signorum. Nominatur hec diversitas que est inter utraque loca reflexio, sive diversitas aspectus, et erit hec reflexio ex circulo maiori arcus eunte super zenith et super planetam, et est circulus altitudinis».

ALFRAGANUS, *op. cit.*, diff. XV: «Differentia decimaquinta de retrogradatione planetarum in circulo signorum – Dicto de motu planetarum in longitudine narremus quod accidit quinque planetis erraticis de retrogradatione in motu eorum in circulis. Et dicamus primo quod cum planeta fuerit in superiori parti circuli brevis movetur ad orientem ex motu scilicet cuspidis circuli brevis, et ex motu eius in circulo brevi et videtur planeta velocior cursu propter coniunctionem utrorunque motuum in unam partem, cum vero fuerit in inferiori parte eius, erit motus eius versus occidentem contra motum primum. Nunc dicamus quod planeta cum fuerit in utroque latere circuli brevis ab oriente in occidentem, et super locum gradus utrarumque linearum exeuntium a terra ad utrumque latus circuli brevis non videtur motus eius in circulo brevi quantitas apparens in circulo signorum. Eritque id quod videtur in circulo signorum id quo movetur cuspis circuli brevis tantum. Sed cum transierit ex linea contingente versus orientem circulum brevem, fietque tunc inicium motus qui videtur planete in circulo brevi tardior, minuiturque ex hoc circuli brevis motus qui videtur versus orientem, et quanto plus describit planeta in circulo brevi et appropinquaverit longitudini propiori tanto plus videtur motus eius versus orientem tardus quousque equatur quantitas que videtur de motu eius in circulo brevi motui cuspidis circuli brevis. Cumque equalis fuerit uterque motus in duabus partibus diversis non videtur planeta in circulo signorum precedere vel subsequi, vel ire ante vel retro, sed stat immobilis. Deinde augetur motus eius qui videtur in circulo brevi versus occidentem et augetur super alterum motum eius qui est versus orientem. Et tunc videtur planeta retrogardus in circulo signorum iens versus occidentem. Et plus videtur in motu retrogradus cum fuerit planeta in propinquiori longitudine circuli brevis. Cumque transierit longitudinem propinquiorem versus occidentem, fueritque in similitudine longitudinis, a qua inceperit retrogradari versus orientem, equatur similiter ibidem uterque motus, et videtur immobilis in loco suo in circulo signorum donec transeat ipsum locum».

ALFRAGANUS, *op. cit.*, diff. XXIV: «Differentia vigesima quarta in ortu et occasu planetarum et occultationibus eorum de sub radiis solis. – In hoc loco demonstremus ortum planetarum et occasum eorum et occultationem eorum sub radiis solis. Dicamusque quod saturnus, iuppiter et mars sunt tardiores sole. Cumque fuerit unus eorum ante solem, appropinquat ei sol et videtur eius apparitio in occidente vespere, nominaturque occidentalis donec occultetur sub radiis solis. Cumque transierit eum sol per cursum suum et exierit de sub radiis apparebit in oriente mane et nominatur orientalis. Eritque unicuique occasus in vespere et ortus in mane, Venus autem et Mercurius ...»; cfr. «Differentia vigesima sexta in ortu quinque planetarum erraticorum de sub radiis solis».

ALKABITIUS, *Enarratio*, diff. III, p. 301,2: «Ex hoc ductoria [i.e. securitas] planetae, id est, ut sit planeta in suo haim, id est, in parte sibi propria, et aliquo angulorum ascendentis; et aliquod luminarium similiter in loco sibi consimili, in quadrante videlicet in aliquo angulo ita quod sit planeta in die orientalis a sole, in nocte occidentalis a luna. Et omnis planeta dicitur esse in sua ductoria secundum quosdam, cum fuerint inter planetam orientalem et solem LX gradus».

ALFRAGANUS, *op. cit.*, diff. VIII: «Distinctio octava de mensura terre et divisione climatum que habitantur de ea. – ... Cum ergo multiplicaveris portionem unius gradus in rotunditate in summam circuli quod est 360 graduum erit quod collectum fuerit ex hoc rotunditas terre que sunt 20400 millaria, et cum divisa fuerit rotunditas terre per terciam et septimam partem unius tercie erit quod collectum fuerit quantitas dyametri terre, que sunt 6 millia et quingenta millaria fere, videlicet 6491 millaria».

ALFRAGANUS, *op. cit.*, diff. III: «Differentia tercia quod terra cum omnibus suis partibus terrestribus et marinis est ad instar spere. – Convenerunt quoque sapientes quod terra cum universis partibus suis tam terrestribus quam marinis sit similis spere».

ALFRAGANUS, *op. cit.*, diff. XIX, XXII: «Differentia decimanona de numero stellarum fixarum. – ... Et diviserunt quantitates eorum in magnitudine per sex divisiones luminosas ... Differentia vigesimasecunda in mensura quantitatis planetarum ceterarumque stellarum et quantitatis mensure terre erga quantitatem uniuscuisque eorum. – Et patefaciamus post longitudinem stellarum mensura corporum earum ...».

THEBIT, *Ptolomeus et alii sapientes*, 1: «Ptolomeus et alii sapientes posuerunt corpus terre communem mensuram qua metiebantur stellarum corpora, et posuerunt medietatem diametri terre communem mensuram qua stellarum ipsarum a centro terre longitudines mensurabant, sicut fuit possibile mensurare terre diametrum».

ALFRAGANUS, *op. cit.*, diff. XXI: «Differentia vigesimaprima in mensura longitudinis planetarum erraticorum et stellarum fixarum a terra».

ALFRAGANUS, *op. cit.*, diff. VI: «Differentia sexta de esse vel forma quarte habitabilis de terra, et summa eorum que accidunt in ea de revolutione orbis et diversitate noctis et diei. – Et quia ausiliante deo iam premisimus quod debuit premitti de utriusque motibus celi, nunc incipiamus commemorare loca terre habitabilia secundum quod nos novimus, et pervenit ad nos, et universa que accidunt de volubilitate circuli et diversitate noctis atque diei».

ALFRAGANUS, *op. cit.*, diff. X: «Differentia decima de ortibus signorum et diversitate eorum in circulis rectis, qui sunt orizontes circuli equalitatis et circulis declivibus, qui sunt orizontes climatum. – Consequentes precedentium vestigia, narremus ascensiones signorum in circulis rectis et circulis declivibus. Dicamusque prius quod circuli directi sunt qui vadunt super utrosque axes equinoctii diei et ipsi sunt circuli emisperii universarum regionum que sunt sub circulo equinoctii; et ipsi quoque sunt circuli medii diei universorum climatum. Circuli quoque declivi sunt circuli emisperiorum climatum, et nullus ex eis vadit super utrosque axes circuli equinoctii diei ...».

ALFRAGANUS, *op. cit.*, diff. VI: «Dicendoque de circulo emisperii quid sit, quod circulus emisperii sit circulus qui dividit id quod apparet de celo super terram ab eo quod occultatur de eo sub terra. Et eius axis est semper super zenith capitis; et est de circulis maioribus qui dividit celum per medium».

ALFRAGANUS, *op. cit.*, diff. VIII: « ... Versus autem septentrionem longitudo minuitur, quia augmentantur ibidem divisiones spere, eritque quantitas duarum quintarum orbis fere que est 4080 milliarum divisa, que sunt loca huius quarte habitabilis culta per 7 divisiones que sunt 7 climata quorum primi medietas vadit super loca in quibus longitudo maioris diei est horarum 13. Et medietas septimi vadit super loca in quibus longitudo diei maioris est 16 horarum, quia quidquid transierit terminum primi climatis versus meridiem magis ac magis tegitur a mari et eius habitatio rara est».

ALFRAGANUS, *op. cit.*, diff. XI: «Differentia undecima in quantitate temporum noctis et diei et diversitate horarum equalium. – Nunc etiam narremus quantitatem ipsam (idest: quantitatem temporum) diei ac noctis ac diversitatem noctis, diversitatem etiam horarum. Sed primum patefaciemus quantitatem longitudinis uniuscuiusque diei cum nocte sua».

ALFRAGANUS, *op. cit.*, diff. VII: «Differentia septima de proprietatibus divisionum quarte terre habitabilis et commemoratione locorum super que elevatur sol mensibus et non occidit et occidit mensibus et non oritur... Dicamusque quod in locis habitabilibus existentibus inter circulum equinoctialem et locum in quo elevatur axis minus declinatione circuli signorum vadit sol super zenith capitum bis in anno».

CAPUT II

HERMANNUS DE CARINTHIA, *De essentiis*, ed. C.S.F. Burnett, Leiden 1982, p. 348/20–23: «Ex quibus et duo Ionica lingua collegit volumina, in primam Sintasim, in secunda Tetrastim – Arabice dicta Almagesti et Alarba, quorum Almagesti quidem Albateni commodissime restringit, Tetrastim verum Albumasar non minus commode exampliat». *Ibid.*, p. 306, a spelling «genezia» – similar to «gencatici» underlined here above p. 183 – is noticed in Abû Ma'shar and other translations and original works by Herman, as well as in Hugo of Santalla.

THEBIT, *De imaginibus*, p. 180, Prooemium: «Dixit Thebit Bencorat: Dixit Aristoteles: Qui philosophiam et geometriam omnemque scientiam legerit, et ab astronomia vacuus fuerit, erit occupatus et vacuus, quia dignior geometria et altior philosophia est imaginum scientia. Et iam dixit philosophus in secundo tractatu sui libri quia sicut non est motus corpori anima carenti nec vita animato corpori nisi per cibum quo diriguntur et aptantur eius nature, ita non est lumen sapientie cum astronomia evacuata fuerit. Et quemadmodum spiritus non poterit subsistere nisi per cibum quo aptantur nature corporis, ita non est radix scientie apud eum qui philosophia caruerit, nec est lumen geometrie cum vacua fuerit astronomia; sublimitas autem et altitudo astronomie est imaginum scientia».

CAPUT III

ALBUMASAR, *Introductorium*, transl. Ioannes Hispalensis, I, diff. I (Laur. Plut. 29.12, fol 2^{v}): «Secunda vero species est scientia iudiciorum astrorum, hoc est scientia fieri uniuscuiusque planetae et uniuscuiusque circuli et proprietatis significationis eorum et super omne quod generatur, id est oritur et evenit ex fortitudine diversorum motuum et ex naturis eorum in hoc mundo, quod est infra circulum lunae, ex diversitate temporum et corruptione naturarum id est elementorum ... Per primam igitur speciem sapientiae astrorum, quae est scientia totius, significatur secunda species quae est scientia iudiciorum astrorum. Maxima autem pars scientiae iudiciorum astrorum patet et apparet et invenitur».

PTOLEMAEUS, *Liber centum verborum cum epositione Haly*, *Prooemium* (*ms*): «Mundanorum ad hoc et ad illud mutacio celestium corporum mutacione contingit».

CAPUT IV

THEBIT, *De imaginibus*, p. 180, Prooemium: « ... Dignior geometria et altior philosophia est imaginum scientia ... Sublimitas autem et altitudo astronomie est imaginum scientia». Cfr. cap. II.

CAPUT V

ALKABITIUS, *Enarratio elementorum astrologiae*, Praefatio, p. 1, 2: «Cum vidissem conventum quorundam antiquorum ex autoribus magisterii de iudiciis astrorum edidisse libros, quos vocaverunt huius magisterii, id est iudiciorum de astris introductorios» p. 1. 3: «Et nominavi eum [librum] introductorium, et non introduxi ratiocinationes disputationi sive defensioni eorum quae protulimus necessarias, cum sint in libro Ptolemaei, qui appellatur quatuor tractatuum; et in libro meo, quem edidi pro confirmatione magisterii iudiciorum astrorum ...» p. 1–2, 4: «Et divisi eum in quinque differentias. Prima differentia est de esse circuli signorum essentiali et accidentali. Secunda differentia de naturis planetarum septem, et quid illis proprium et quid significent. Tertia differentia est de his, quae accidunt planetis septem in semetipsis, et quid accidat eis ad invicem. Quarta differentia versatur in expositione nominum astrologicorum. Quinta differentia in universitate partium et expositione esse earum in gradibus».

ALKABITIUS, *op. cit.*, diff. I, p. 8, 1: «Nitach, i.e. Circulus signorum dividitur in duodecim partes aequales, secundum divisionem circuli signorum, et hae duodecim partes dicuntur signa».

ALBUMASAR, *op. cit.*, II, 4; VI, 1: «*Quarta*, in ordinatione naturarum signorum».
«*Prima differentia* in naturis signorum et esse eorum, et quid ascendat in faciebus eorum de imaginibus».

ALKABITIUS, *op. cit.*, diff. I, p. 9–10, 7: «Et vocatur illa medietas, quae est ab initio Arietis usque in finem Virginis, medietas calida. Et alia quae est ab initio Librae usque in finem Piscium, vocatur medietas frigida. Et vocatur illa quarta pars circuli, quae est ab initio Arietis usque in finem Geminorum, quarta calida, humida, vernalis, puerilis, sanguinea. Et illa quae est ab initio Cancri usque in finem Virginis, dicitur quarta calida, sicca, aestivalis, iuvenilis, cholerica ...».

ALKABITIUS, *op. cit.*, diff. I, p. 48, 1: «De signis mobilibus, fixis et communibus – Quatuor quoque ex his signis dicuntur esse mobilia, scilicet Aries, Cancer, Libra et Capricornus. Et quatuor fixa scilicet, Taurus, Leo, Scorpius et Aquarius. Reliqua vero quatuor, scilicet Gemini, Virgo, Sagittarius et Pisces, sunt communia. Dicuntur autem mobilia, fixa, vel communia, quia quando Sol ingreditur aliquod istorum, movetur, id est, mutatur tempus; vel figitur, id est, in eodem statu perseverat; aut fit commune, id est medietas illius unius temporis erit ut medietas alterius, Verbi gratia. Quando Sol Ingreditur signum Arietis, tempus mutatur, id est vertitur hyems in ver. Et quando intrat Taurum, figitur idem tempus vernale. Quando vero Sol ingreditur Geminos, fit tempus commune, id est, dimidium veris et dimidium aestatis, et sic de caeteris».

Ibidem, p. 92–93: «De gradibus signorum masculinis et foemininis – Sunt quoque in unoquoque signo gradus, qui proprie dicuntur masculini atque foeminini. Nam ab initio Arietis usque in VIII gradum, dicuntur esse masculini, et ab VIII in IX foeminini, et a IX in XV masculini, et a XV usque in XXII foeminini, et a XXII usque in finem Arietis masculini». Cfr. PTOLEMAEUS, *Liber Quadripartiti*, cap. 13 «In masculinis et foemininis signis – Masculinaeque autem diurnae naturae sex signa iudicaverunt. sex vero residua ad naturam foemininam atque nocturnam rettulere: et unum post aliud disposuere. Ob hoc quod nocti dies adheret et iuxta eam semper existit, et quia sexus masculini sunt prope foemininos: et eis frequenter adiunguntur ... Ob hoc igitur masculino fecere initium Arietis ... Ordo vero signorum haec duo subsequentium est velut praediximus: foemininum scilicet post masculinum et post foemininum masculinum». Cfr. ALBUMASAR, *op. cit.*, II, 6 (78, 12–13): «*Sexta* in scientia significationis signorum mobilium et fixorum atque communium».

ALBUMASAR, *op. cit.*, II, 8–9: «*Octava* in scientia signorum masculinorum ac femininorum. *Nona* in scientia signorum diurnorum ac nocturnorum».

ALKABITIUS, *op. cit.*, diff. I, p. 9, 5: «Et tortuose ascendentia oboediunt directe ascendentibus, hoc est, duo signa quae fuerint unius longitudinis a capite Cancri, oboediunt sibi, ut Gemini Cancro, Taurus Leoni, Aries Virgini, et Pisces Librae, Aquarius Scorpioni, et Capricornus Sagictario».

ALBUMASAR, *op. cit.*, VI, 4–5: «*Quarta differentia*, in signis se diligentibus atque odientibus, et prolongantibus, atque inimicantibus, et directe ascensionis atque tortuose, obedientibus quoque sibi invicem et inobedientibus. – *Quinta differentia*, in signis concordantibus in circulo et ascensionibus, et concordantibus in fortitudine et in itinere».

ALKABITIUS, *op. cit.*, diff. I, p. 9, 5: «Et duo signa que fuerint unius longitudinis a capite Arietis, dicuntur concordantia in itinere, ut Aries et Pisces, Taurus et Aquarius, Gemini et Capricornus, Cancer et Sagittarius, Leo et Scorpio, Virgo et Libra».

ALKABITIUS, *op. cit.*, diff. I, p. 78, 1: «De significationibus signorum – Et unumquodque signum habet propriam significationem in his quae significant ex creatione membrorum et moribus hominis et regionum et seminum et arborum et caetera. Aries habet ex corpore hominis caput et faciem, et ex regionibus ...».

ALBUMASAR, *op. cit.*, VI, 9; VI, 23: «*Nona differentia* in significatione signorum super universas regiones ac provincias terrarum». « *Vicesima tertia differentia*, in signis significantibus arbores ac semina».

ALBUMASAR, *op. cit.*, VI, 22; VI, 12; VI, 14: « *Vicesima secunda differentia* in signis significantibus ⟨species⟩ avium, et omnium quadrupedium, et luporum ac repentium terre, et animalium aque». «*Duodecima differentia*, in divisione eorum que sunt unicuique signo ex membris corporis humani». «*Quartadecima differentia*, in signis significantibus luxuriam et infirmitates».

ALBUMASAR, *op. cit.*, VI, 25–30: «*Vicesima quinta differentia*, in partibus signorum. – *Vicesima sexta differentia*, in angulis circuli et eius quartis, et in xii domibus et in universa significatione earum et causa huius rei. – *Vicesima septima differentia*, in quartis circuli que referuntur ad corporalitatem et incorporalitatem [i.e. que dicuntur esse corporea et incorporea], et cetera. – *Vicesima octava differentia*, in commixtione naturarum angulorum ascendentis sive circuli. – *Vicesima nona differentia,* in coloribus quartarum circuli, et xii domorum. – *Tricesima differentia* in quartis circuli ascendentibus et descendentibus, longis ac brevibus».

ALKABITIUS, *op. cit.*, diff. I, p. 160, 4: «Et ascendens et quarta et septima et decima alamed, id est, quas nos angulos vocamus, ut pulchrius sonet. Et secunda domus, octava et quinta et undecima, succedentes angulis vocantur. Tertia autem et sexta, nona et duodecima cadentes ab angulis dicuntur».

ALKABITIUS, *op. cit.*, diff. I, p. 112 s.n.: «De esse circuli accidentali – Sed quia auxiliante Deo iam protulimus esse circuli signorum essentiale, nunc proferamus accidentale. Nam circulus figuratur in omni hora tali figura, quae dividitur in quatuor partes, quas dividit circulus hemisphaerii et circulus meridiei, id est, circulus medii caeli qui facit medium diem. Et unaquaeque pars istarum partium dividitur in tres partes inequales, secundum ascensiones signi ascendentis, atque hoc modo dividitur circulus in duodecim partes, quae vocantur domus ...».

ALKABITIUS, *op. cit.*, diff. I, p. 184: «De coloribus duodecim domorum – Et significant etiam duodecim domus colores; et hi sunt: nam domus ascendens vel prima et septima sunt albae; II et XII virides; III et XI croceae; IIII et X rubeae; V et IX mellitae, id est habent mellis colorem; VI et VIII nigrae sunt».

ALKABITIUS, *op. cit.*, diff. I, p. 159, 2: «Et illae duae partes, quae sunt a medio caeli usque ad ascendens, et ab ascendente usque ad quartam domum, faciunt medietatem, quae vocatur medietas ascendens. Reliquae partes, quae sunt a quarta domo usque ad septimam, et inde ad medium caeli, constituunt medietatem descendentem vocatam».

ALBUMASAR, *op. cit.*, VII, 1: «*Differentia prima*, in esse planetarum in ⟨suis essentiis⟩. – Iam narravimus in precedentibus naturas planetarum calidas et frigidas, humidas quoque ac siccas, et proprietatem eorum in fortuna et in infortuna, masculinitatem quoque eorum ac femininitatem, ⟨diuturnitatem⟩ etiam ac nocturnitatem eorum, et res alias preter istas.»

ALBUMASAR, *op. cit.*, VII, 1–6: «*Differentia prima*, in esse planetarum in suis essentiis. – *Differentia tertia*, in esse planetarum ex quartis circuli et domibus eorum, et quantitate eorum corporum. – *Differentia iiii*, in coniunctione planetarum ad invicem, et in commixtione qualitatum eorum, et quis illorum sit fortior aut debilior. – *Differentia v*, in aspectu planetarum ad invicem, et eorum coniunctione atque separatione, et reliquum esse eorum quod hoc sequitur de his que congruunt sibi. – *Differentia vi*, in fortuna planetarum et eorum fortitudine ac debilitate et eorum impedimento, ac corruptione Lune».

PTOLEMAEUS, *op. cit.*, Cap. 5: «In fortunis et in infortunis. Duas stellarum erraticarum, Jovem scilicet ac Venerem, Lunam etiam secundum prisce viros auctoritatis fortunas esse dixere. Eo quod earum est complexio temperata et quod eis multum caloris ac humoris inest. Opera quidem Saturni atque Martis operibus praedictarum stellarum naturaliter contraria fore testati sunt; eo quod horum alter per frigiditatem intensam et alter per siccitatem intensam operatur ...».

PTOLEMAEUS *op. cit.*, Cap. 6: «In masculinitate et foemininitate. – Cum partium item naturae duo sint prima genera masculinum scilicet et foemininum. Cumque ex viribus praedictis propriae foeminina vis sit ex humida substantia; eo quod haec qualitas in foeminis generaliter invenitur, et residua qualitas in maribus proprie semper reperiatur, ab antiquis concinne dictum est Lunam et Venerem quibus multum inest humiditatis foemininas esse. Solem autem atque Saturnum, Iovem etiam ac Martem masculinos. Mercurium vero quoniam siccitatem et humiditatem equaliter operatur, in utroque genere societatem habere dixerunt». PTOLEMAEUS, *op. cit.*, cap. 7: «In diurnis et nocturnis. – Quoniam item duo sunt spatia ex quibus tempus efficitur. Quorum alterum dies est qui proprie masculinitati propter calorem eiusque vim efficacem atque moventem, alterum vero nox est qui feminiçat propter illius humiditatem atque quietem. Lunam ac Venerem nocturnas, Solem autem et Iovem diurnos esse dixere. Mercurium vero velut praediximus utrisque socium fecere et eum in oriente diurnum, in occidente vero nocturnum esse rettulere».

ALKABITIUS, *op. cit.*, diff. III, p. 302, 3–4: «Et omnis planeta ex quo tegitur a radiis Solis, donec appareat de sub radiis, vocatur combustus; et dum incipit intrare radios, dicitur incoepisse comburi; et dum absconditur sub radiis et fuerit prope Solem per XII gradus dicitur oppressus ... Et ex quo apparent tres altiores de sub radiis et incipiunt oriri, id est, apparere mane ante Solem, hoc est cum fuerint propinquiores circulo hemisphaerii orientalis, donec veniant ad oppositionem, vocantur orientales dextri; et ex quo transierint oppositionem donec coniungantur iterum Soli, vocantur occidentales sinistri».

ALKABITIUS, *op. cit.*, diff. III, p. 301, 1: «De esse autem planetarum ab invicem tractemus, scilicet quid accidat quinque planetis erga luminaria; et dicemus ex hoc quod dixit Ptolemeus de almugea, hoc est de visione invicem faciei ad faciem, hoc est, cum fuerit inter planetam et Solem dum fuerit planeta occidentalis, id est, dum sequitur Solem tantum, quantum est inter domum illius planetae et domum Solis de signis, aut cum fuerit inter ipsum planetam et Lunam, cum fuerit orientalis a Luna, id est, dum succedit Luna tantum, quantum est inter domum planetae et domum Lunae ex signis, id est, cum fuerit planeta tantum distans a Sole post, quantum distat domus eius a domo Solis. Similiter de Luna dicitur ...». PTOLEMAEUS, *op. cit.*, Cap. 23 «In almugea idest in visione ad invicem facie ad faciem et in alchinara, id est splendore et in his similibus – ... Item quod planetae proprie habeant almugea similiter ostenditur. Cum eorum unusquisque aliquam in figura cum Sole vel Luna societatem habuerit, modo qui dicitur almugea, hoc est ut sit inter unumquodque et Solem ac Lunam ex longitudine, quantum est inter ipsius domum et domum Solis, aut Lunae, ex longitudine; veluti si Venus esset in sextili aspectu alterius luminaris, ita quod a Sole occidentalis vel a Luna foret orientalis, esset tunc in almugea. Hic est ergo modus almugea uniuscuiusque planetarum cum luminaribus ...».

ALKABITIUS, *op. cit.*, diff. III, p. 315–316, 1: «De his suae accidunt planetis ad se invicem. – Esse autem illorum erga se invicem; hoc est, idem quod coniunctio, id est, cum fuerint duo planetae in duobus signis aspicientibus se, et fuerit levior in signo suo minus gradibus quam fuerit ponderosior in signo suo; fuerintque inter eos VI gradus vel infra; tunc dicitur quod levior eat ad coniunctionem ponderosioris; et cum gradus eorum fuerint aequales, perficitur coniunctio eorum; et cum transierit eum, erit ab eo separatus: coniunctio haec dicitur coniunctio longitudinis. Coniunctio vero latitudinis est, ut duo planetae iungantur per latitudinem. Et si fuerit applicatio coniunctionis, oportet ut sit latitudo eorum aequalis in una parte».

ALKABITIUS, *op. cit.*, diff. III, p. 318–319, 14: «Sequitur alfaziom, id est frustratio. Haec quoque fit cum aliquis planeta petit coniunctionem alterius planetae, sed antequam perveniat ad eum, mutatur iste in aliud signum et erit aliquis planeta in paucis gradibus aspiciens ipsum signum, et erunt radii eius in initio signi. Cumque exierit sequens planeta de primo signo, iungitur isti aspicienti, et annullatur coniunctio quam habebat cum illo, scilicet cum primo». *Ibidem*, p. 318, 12: «Inde sequitur almenen, id est, refrenatio, quae fit quando planeta vult coniungi alteri; sed antequam iungatur accidit retrogradatio, et sic destruitur eius coniunctio». *Ibidem*, p. 317, 8: «Sequitur prohibitio, et fit duobus modis. Uno scilicet ex coniunctione, hoc est, cum fuerint tres planetae in uno signo sed in diversis gradibus; et fuerit ponderosior plus gradibus; tunc ille qui est medius, prohibet priorem, illum scilicet qui est minus gradibus ne iungatur ponderosiori donec pertranseat eum. Secundo modo, ut duo planete sint in uno signo, et levior iungatur ponderosiori, alter quoque iungatur eidem ponderosiori per aspectum: ille ergo qui est cum eo in uno signo, aspicientem prohibet a ponderosioris coniunctione; si fuerint tamen gradus illius qui iungitur, et ipsius qui aspicit, aequales, id est, unius numeri. Si vero ille qui aspicit fuerit propior gradui ponderosioris, erit coniunctio aspicientis».

ALKABITIUS, *op. cit.*, diff. III, p. 319, 15–16: «Hinc sequitur abscissio luminis, hoc est quando aliquis planeta petit coniunctionem alterius, et fuerit in secundo signo a signo illius cui iungitur alter planeta; sed antequam iungatur ei, prius fit ille qui est in secundo signo retrogradus, coniungiturque ei et abscindit lumen suum a planeta qui volebat coniungi ei. Similiter si fuerit planeta iens ad coniunctionem alterius planetae, et ipse alter planeta cui vult iungi, petit coniunctionem alterius planetae se ponderosioris; sed antequam perveniat levis ad gradus ponderiosioris, iungitur ipse ponderosus alteri seipso ponderosiori, et abscindit lumen illius a planeta primo leviori». *Ibidem*, p. 316, 2: «Et cum separatur unus planeta ab alio et nulli planetarum iungitur, quamdiu in eodem signo fuerit, dicitur cursu vacuus».

ALKABITIUS, *op. cit.*, diff. III, p. 317, 9: «Et si coniungitur planeta domino illius signi in quo fuerit, vel domino exaltationis seu domino caeterarum dignitatum in quibus fuerit, dicitur pulsare, id est, mittere naturam illius planetae domini, scilicet eiusdem dignitatis ad eum».

ALKABITIUS, *op. cit.*, diff. III, p. 316, 3: «Et cum fuerit planeta in aliquo signo, et aliquis planeta non aspexerit hoc signum, talis planeta quamdiu in eodem fuerit, dicitur feralis vel agrestis».

ALKABITIUS, *op. cit.*, diff. I, p. 29–30, 1: «De domibus planetarum. Habent quoque planetae in his signis potestates, quasdam per naturam quasdam per accidens. Quae sunt per naturam sunt hae: Domus, Exaltatio, Terminus, Triplicitas, Facies. De illis autem quae per accidens sunt, loco convenienti tractabimus».

ALBUMASAR, *op. cit.*, I, 1: «*In tractatu vero quinto* sunt XXXII differentie. In dignitatibus seu potestatibus planetarum in signis, ut domus, exaltationes, termini ac cetera».

ALBUMASAR, *op. cit.*, V, 1–22: «*Differentia i*, in dignitatatibus planetarum in signis. – *Differentia ii*, in causa domorum planetarum secundum quod putaverunt astrologi. – *Differentia iii*, in causa domorum planetarum secundum quod convenit dictis Ptholemei. – *Differentia iiii*, in causa domorum circuli secundum quod convenit dictis Hermetis Abaidimon. – *Differentia v*, in causa ⟨exaltationis⟩ planetarum secundum quod putaverunt quidam astrologorum. – *Differentia vi*, in causa exaltationum planetarum secundum quod putavit Ptolemeus. – *Differentia vii*, in causa exaltationum planetarum secundum quod convenit Hermeti. – *Differentia viii*, in diffinitione diversitatis terminorum planetarum et esse eorum. – *Differentia viiii*, in terminis Egiptiorum. – *Differentia x*, in terminis Ptholemei. – ... *Differentia xiiii*, in dominis triplicitatum. – *Differentia xv*, in faciebus et eorum dominis secundum quod convenit sapientibus Feriz et Behil i.e. Babilonie, et Mizor i.e. Egipti. – *Differentia xvi*, in faciebus et eorum dominis secundum quod dixerunt Indi, et nominatur aldurugen. – *Differentia xvii*, in naubaharat signorum, que sunt novene, secundum quod convenit dictis Indorum. – *Differentia xviii*, in duodenariis signorum et dominis uniuscuiusque gradus omnis signi. – *Differentia xviiii*, in gradibus masculinins et femininis. – *Differentia xx*, in gradibus lucidis et tenebrosis fuscis as vacuis. – *Differentia xxi*, in puteis planetarum in signis. – *Differentia xxii*, in gradibus augentibus fortunam».

ALKABITIUS, *op. cit.*, diff. I, p. 30, 1: «Domus sunt hae: Aries et Scorpius, domus Martis; Taurus et Libra, domus Veneris; Gemini et Virgo, domus Mercurii; Cancer, domus Lunae; Leo, domus Solis; Sagittarius et Pisces, domus Iovis; Capricornus et Aquarius, domus Saturni». *Ibidem*, p. 37–38, 1: «De exaltationibus planetarum – Hae sunt exaltationes planetarum. Sol exaltatur in Ariete, hoc est, in XIX gradu eius. Luna in III gradu Tauri. Saturnus in XXI gradu Librae, Iuppiter in XV gradu Cancri (..) In septimo autem signo ab exaltatione uniuscuiusque planetae, in simili gradu erit eius descensio. Verbi gratia, sicut Sol exaltatur in XIX gradu Arietis, ita in XIX gradu Librae cadit, et sic de caeteris». *Ibidem*, p. 57–58 s.n.: «De terminis planetarum – Sunt quoque planetarum in signis termini vel fines. Quia in unoquoque signo habent quinque planetae terminos, per diversos gradus dispositos: nam ab initio Arietis usque ad sextum gradum eiusdem Arietis, est terminus Iovis; et a sexto usque ad XII terminus Veneris, et a XII usque ad XX terminus Mercurii ...». *Ibidem*, p. 39–40, 1: «De triplicitatibus – Triplicitates vero sic distinguimus. Omnia tria signa, quae in una natura videntur concordare, faciunt triplicitatem, et eodem nomine vocantur, hoc est, Triplicitas. Aries ergo, Leo et Sagittarius faciunt triplicitatem primam, quia unumquodque istorum signorum est igneum, masculinum, diurnum, calidum, scilicet et siccum, cholericum, sapore amarum. Est quoque haec triplicitas orientalis. Cuius domini sunt in die Sol, et in nocte Iuppiter, et particeps in die ac nocte est Saturnus. Triplicitas secunda est ex Tauro, Virgine et Capricorno ...». *Ibidem*, p. 64, 1: «De faciebus signorum, et cui planetarum attribuuntur. Facies autem signorum sunt hae. Unumquodque signum dividitur in tres partes aequales. Quaelibet pars conflatur ex X gradibus, et vocatur facies, quarum initium est a primo gradù Arietis. Prima ergo facies est a primo gradu Arietis usque in X et datur Marti. Secunda usque ad XX gradum, et datur Soli, qui succedit ei in ordine circulorum. Tertia usque ad finem praedicti signi, et est Veneris. Similiter prima facies Tauri est Mercurii, qui Veneri succedit, et ita usque in finem signorum».

ALKABITIUS, *op. cit.*, diff. I, p. 193: «De potestatibus accidentalibus planetarum – Ex potestatibus quoque planetarum accidentalibus est alhaiz idest similitudo. Et hoc cum fuerit planeta diurnus in die supra terram, et in nocte sub terra; et planeta nocturnus in nocte super terram, et in die sub terra, et cum hoc si fuerit masculinus planeta in signo masculino, et planeta foemininus in signo foeminino, dicitur esse in sua similitudine, id est in suo alhaiz. Et erit fortitudo eius ut viri fortitudo in loco eius profectus idest acquisitionis atque fortunae». *Ibidem*, p. 97 s.n.: «De gradibus augmentantibus fortunam – Et in hoc circulo sunt quidam gradus, qui dicuntur augentes fortunam, qui in ista tabula descripti sunt».

ALKABITIUS, *op. cit.*, diff. I, p. 95 s.n.: «De gradibus putealibus – Et in signis sunt quidam gradus, qui vocantur putei. Cum fuerit planeta in aliquo eorum dicitur esse in puteo ut est sextus gradus Arietis et caetera, ut in hac tabula sequenti ostendetur».

ALKABITIUS, *op. cit.*, diff. I, pp. 184–185 s.n.: «De gaudiis planetarum in domibus – Et unusquisque planeta habet in unaquaque istarum domorum quandam potestatem, ex potestatibus scilicet accidentalibus, quae dicitur gaudium. Quia Mercurius gaudet in ascendete, Luna in domo tertia, Venus quoque in quinta, et Mars in sexta, Sol in nona, Iuppiter in undecima, Saturnus in duodecima».

CAPUT VII

PS. – PTOLEMAEUS, *Liber centum verborum cum expositione Haly*, v. 50: «Non oblivisceris esse centum viginti coniunctiones que sunt in stellis erraticis: in illis nam est maior scientia eorum quae fiunt in hoc mundo suscipienti incrementum et decrementum»

HALY, *Glosa super* v. 50: «120 coniunctiones continentur in omni ad quod perveniunt coniunctiones septem planetarum. Et sunt hae: binaria, ternaria, quaternaria, quinaria, senaria, septenaria ...»

ALBUMASAR, *Flores*, I, 1–7 (*Speculum*'s text here simply lists the titles of this section of Albumasar's work). «De gravitate et levitate annone; de pluviis; de bellis et guerris; de pestilentia; de terre motu; de stellis fixis quid operantur in revolutionibus ac nativitatibus; de domino anni quomodo infortunatur ex stellis; de latitudine, ortu et occasu planetarum in signis.»

CAPUT VIII

ALBOHALI, *De iudiciis nativitatum*, cap. I: «De nutritione». «Primum omnium necessarium est prescire scientiam nutritionis, in qua debes considerare dominos triplicitatis ascendentis et dominos triplicitatis solis in die signi coniunctionis vel preventionis que fiunt ante nativitatem ... Incipies autem cum dominis triplicitatis ascendentis scilicet primo et secundo domino. Ac si fuerit unus ex illis in ascendente vel medio celi vel in undecimo aut quarto loco liber ab impedimentis et a maleficis stellis, significat nutritionem bonam si deus voluerit. Verum si fuerint ambo cadentes et impediti a malis aspice dominos triplicitatis solis si fuerit nativitas diurna, vel domino triplicitatis lune, si fuerit nativitas nocturna. Qui si fuerint in bonis locis, liberi ab impedimentis et ab infortuniis significant nutritionem».

ALKABITIUS, *Enarratio elementorum astrologiae*, p. 376: «Hylech id est locus vite in nativitatibus».

ALBOHALI, *op. cit.*, cap. II: «Quando nutritionem inveneris nato et volueris scire quantitatem vitae eius, quere hylech, in nativitatibus quidem diurnis a sole incipiendo, qui si fuerit in angulis, vel succedentibus angulorum in signo masculino vel in quarta masculina et aspexerit eum dominus domus eius, vel dominus termini eius vel dominus exaltationis, triplicitatisve eiusdem vel faciei, tunc poterit esse hylech. Et non potest sol esse hylech neque alius planeta nisi respiciat eum aliquis de dominis quinque dignitatum essentialium idque iudicium perpetuo est observandum in omnibus hylegiis. Quod si modo predicto sol non fuerit hylech aspice lunam que si fuerit in angulis, vel succedentibus angulorum vel in signo feminino vel in quarta feminina, et aspexerit eam aliquis de dominis quinque dignitatum predictorum erit hylech. Sin diverso modo luna sese habuerit, non erit hylech. Tunc enim si fuerit nativitas coniunctionalis, quere hylech ascendenti sicut quaesivisti soli et lune. Porro si neque hunc hilegium decerni potest, quere hylech parti fortunae. Quod si neque hinc evenerit, similiter quaere hylech ab aliquo illorum, qui dignior fuerit in gradu coniunctionis vel oppositionis luminarium que fuerit ante nativitatem ... Quando vero sol fuerit hylech et non habuerit alchocoden aspicientem, quere hylech gradui ascendentis. Qui si fuerit hylech et non habuerit alchochoden, quere hylech gradui coniunctionis vel preventionis que fuerit ante nativitatem. Tandem si nullus istorum hylech habuerit alchocoden, erit natus imbecillis vitae et parvi temporis»

ALKABITIUS, *op. cit.*, p. 379: «Alcochoden qui est significator vite idest dominus annorum vel dans annos».

ALBOHALI, *op. cit.*, cap. III: «De alcochoden et quid significet de vita»: «Quando sciveris hylech et volueris scire alchocoden, aspice dominum termini hylech, et dominum domus et dominum exaltationis eius, triplicitatisve aut faciei et si aliquis illorum aspexerit hylech, ipse est alchochoden et si respexerint eum duo ex illis vel tres, vel omnes, ille qui habuerit plures dignitates at fuerit ei proprior in gradibus alcochoden erit».

Ibidem cap. IV: «Quantum addant vel subtrahant stelle annis alcochoden»: «Quando igitur cognoveris quantum significavit alcochoden de annis mensibus et diebus et volueris scire quid addiderint ei vel minuerint planetae ab eodem sic facito: considera diligenter si fuerit cum eo fortuna iuncta vel aspexerit eum trino vel sextili aspectu, addet ei annos suos minores et si fuerit mediocris in fortitudine, tot menses et si fuerit imbecillior dies vel horas ...».

Ibidem: «... Sed quando sciveris quot annos dederit alcochoden dirige hylech cum gradibus ascensionis, quousque perveniat ad corpora vel radios malorum. Quando enim illuc pervenerit, significat destructionem nati, salva tamen dei omnipotentia».

Ibidem, cap. VI: «De testimoniis, quae significant super nativitates regum»: Aspice in primis gradum ascendentis. Si enim fuerit in ipso aliqua de stellis fixis luminosis primae magnitudinis vel secundae de natura bonorum planetarum vel fuerit iuncta gradui medii caeli vel alicui luminarium et maxime soli in nativitatibus diurnis, ac lunae in nativitatibus nocturnis aut si iunctae fuerint duobus locis ex istis, vel tribus, natusque ex genere regum fuerit vel mereatur habere regnum significat regnum altissimum ...».

Ibidem, cap. VII: «De prosperitate et adversitate nati»: «Quod si pars fortunae et dominus eius liberi a malis et orientales ex angulis aspexerint ascendens significant fortunam nati durabilem et magnitudinem precii existimationisque eius ... Quando dominus triplicitatis eius luminaris, quod habet dignitatem, fuerit impeditus, aspice partem fortunae que si fuerit in angulis et fortunae ac mali pariter eam aspexerint significat mediooritatem fortunae ac prosperitatis ...».

Ibidem, cap. XVI: «De fortuna parentuum ac de significatis quartae domus»: «Pro fortuna patrum considera in diurna nativitate locum Solis, in nocturna Saturni. Tam in nocturna quam in diurna vero genesi, aspicies partem patruum et dominum eius et pro ambobus parentibus signum quartum ab ascendente, cum suo domino. Porro in re matris iudicium facies in die a Venere, in nocte a Luna in die ac nocte pariter, a parte matris et domino eius, quod si inveneris significatores patris et matris, aut plures ex illis in multitudine testimoniorum ac dignitatum in angulis vel succedentibus angulorum et fortuna fuerint in domo patrum et domini triplicitatis eius luminaris quod obtinet dominium temporis in locis circuli fortibus feliciter positi fuerint, significant prosperitatem parentum et bonum statum ac multitudinem gaudii eorum». Cfr. cap. XVII «De spatio vitae patris»; cfr. cap. XVIII «De spatio vitae matris»: Dixit Ptolomaeus: In vita matris aspice, cum nativitas fuerit diurna, Venerem, quae si aspexerit ascendens, dirige eam ad corpus et radios malorum, per gradus ascensionum et unicuique gradui ascensionum da unum annum. Si vero hoc non aspexerit ascendens Luna tamen mittente eo radios dirige eam sicut diximus de Venere. Porro si neque Luna aspexerit ascendens, dirige gradum medii celi. Verum in nativitatibus nocturnis incipe a Luna, postea a Venere, deinde a gradu medii celi, postea considera quot anni provenerint ex gradibus ascensionum. Si enim totidem anni fuerint quot significabat planeta, qui habet maiorem dignitatem in die in loco Veneris, in nocte vero in loco Lunae, et domino domus eius in die autem et in nocte pariter in gradu medii celi et dominus eius et in parte matris et domino eius vel si prope eum numerum annorum aequaverint gradus ascensionum per directionem inventi, significatur mors matris in eodem anno». Cfr. cap. XIX «De inveniendo hylech in vita parentum» «Aspice in nativitatibus diurnis solem qui si fuerit in signo angulorum vel succedenti et aspexerit eum aliquis de dominis quinque dignitatum, sol erit hylech quem diriges patri et planeta dominus dignitatis aspiciens solem erit alcochoden».

CAPUT IX

MESSALLACH, *De receptione planetarum*, Prologus: «Invenit quidam vir ex sapientibus librum ex libris secretorum astrorum de illis quorum thesaurizaverant reges exposuitque eum et patefecit eius intentionem in omnibus quibus indigent homines in rebus suis de interrogationibus fuitque ex eo qui posuit et patefecit in rebus interrogationum: utrum sit res an non et quando erit si fieri debet, et quando apparebit quod non sit, et si fieri non debet et quid prohibet eam quod non sit et per quem fit et unde fieri debet; et scientia hius rei et eius expositio est in planetis septem et in domibus eorum duodecim et in exaltationibus planetarum septem, et in eorum descensionibus et in coniunctionibus eorum in separationibus quoque et in receptionibus eorum ad invicem; et in reditu receptionis et pulsatione eorum dispositionis ab invicem. Eritque ad quem pervenerit dispositio significator iussu dei, qui si fuerit in esse effectus rei significabit eius effectum et si fuerit in esse prohibitionis significabit eius prohibitionem iussu dei».

ZAHEL, *De interrogationibus*: «Quaestio de re aliqua ad duodecim signa pertinente si fiet vel non»: «Et si interrogatus fueris de aliqua re ex rebus que sunt in duodecim signis da ascendens et dominum eius et lunam significatores illius viri qui te interrogat signum rei quesite et dominum unius rei quesite; post hoc aspice dominum ascendentis et lunam et fortiorem eorum, illum scilicet qui in angulo fuerit et qui aspexerit ascendens et ab eo incipe. Quod si aliquis eorum iunctus fuerit domino rei perficietur illa res petitione interrogantis et si inveniris dominum rei iunctum cum domino ascendentis perficietur ipsa res cum levitate ac studio interrogantis sine petitione et sine aliqua infortunitate. Et si inveneris dominum ascendentis aut lunam in loco rei aut inveneris dominum rei in ascendente perficietur res nisi sit ascendens impeditus et dominus eius in descensione sua aut combustus in eo tunc non erit. Et si inveneris dominum ascendentis aut lunam iungi alicui planete in loco cause aut dominum cause inveneris iunctum alicui planete in ascendente et habuerit ipse planeta testimonium in eo de domo aut exaltatione vel triplicitate et cetera perficietur, et si non fuerit aliquid de omnibus que diximus aspice tunc in translatione luminis ad lunam vel ad aliquem planetarum levium; quem si inveneris separatum a domino ascendentis et iunctum domino rei aut separatum a domino rei et iunctum domino ascendentis perficitur res per manus legatorum et eorum qui discurrunt inter utrosque. Et si non inveneris inter eos planetam qui defert lumen unius eorum ad socium suum, tunc aspice in collectione luminis quod si inveneris dominum rei et dominum ascendentis utrosque scilicet iunctos uni planete se ponderosiori et ipse planeta aspexerit locum rei aut fuerit in ascendente aut in medio celi, perficitur res per manus iudicis aut viri qui mittuntur. Ab his ergo tribus modis sit effectus omnium rerum: primo a coniunctione domini ascendentis et lune ac domini rei; secundo ut aliquis planeta deferat inter eos lumen idest separetur ab uno eorum et coniungatur alteri tunc fit res per manus legatorum; tertio a collectione luminis idest ut sint iuncti ambo alicui planete se ponderosiori qui coniungat lumen eorum accipiens fortitudinem utrorumque; eritque acceptabile inter eos eius iudicium aut per virum qui auxiliabitur in eadem re.

Ex his ergo tribus primis capitulis fit effectus rerum. Post hec aspice sicut predixi tibi ad receptorem dispositionis ab aliquo eorum qui est planeta ponderosior, sive dominus ascendentis fuerit aut dominus rei et planeta, qui colligit lumen, si fuerit a malis liber in angulis vel in sequentibus angulorum et non fuerit retrogradus neque combustus, nec cadens ab angulis, perficietur ipsa res post adeptionem eius, et si fuerit receptor retrogradus solvetur postquam putaverit se eam adeptum fuisse».

ALKABITIUS, *op. cit.*, p. 360: «Annimodar quod est investigatio gradus ascendentis alicuius nativitatis», p. 381: «Almutam est qui preest nativitati».

CAPUT X

ZAHEL, *De electionibus*, §1: «Omnes concordati sunt quod electiones sint debiles nisi in regibus habentur. Isti licet debilitentur eorum electiones radices idest nativitates eorum que confortant omnem planetam debilem in itinere, mulieribus vero et mercatoribus et his qui sequuntur non eligas aliquod nisi supra nativitates eorum et revolutiones annorum illorum et super nativitates eorum filiorum, quorum autem ista ignorantur, accipiantur eis interrogationes et sciatur effectus rei eorum ex eis, postea eligatur eis secundum hoc quia, dum interrogaverit quis de semetipso, iam pervenit ex nativitate sua ad bonum vel malum quare ipse te interrogavit et iudica ei secundum interrogationem suam».

HALY, *De electionibus horarum*, ms. Firenze, Biblioteca Nazionale Centrale, J X 20 (S. Marci 163), f. 26^{r-v}: «Quidam astrologorum cum eligere vellent alicui incipiebant ab eo interrogatione de re qua volebant incipere ... Item iudicia interrogationum non sunt ita rata sicut iudicia nativitatum. Nativitates enim res naturales, interrogationes vero sunt similes naturalibus».

CAPUT XI

THEBIT, *De Imaginibus*, cap. V, 36, p. 186: «Cum hoc volueris facere, incipies primum accipere eius interrogationem certissimam cum intentione radicali ...».

Ibidem, cap. II, 10, p. 182: «Secundum hoc exemplar facies cum volueris destructionem alicuius civitatis vel regionis: facies imaginem sub ascendente eiusdem civitatis et infortunabis dominum domus vite eius id est ascendentis, et infortunabis dominum domus mortis eius, et infortunabis dominum ascendentis et lunam et dominum domus lune ac dominum domus domini ascendentis, impediesque decimum et dominum eius si quiveris». cap. I, 6 p. 181 (see above).

Ibidem, cap. I, 5–9, p. 181: «Et iam tradidi tibi in hoc libro meo quasdam regulas quas constitui exemplaria ex quibus est opus imaginum ad effugandos scorpiones. Quod cum volueris exercere, incipies operari ascendente Scorpione, faciesque imaginem scorpionis ex ere vel stagno aut plumbo vel argento vel auro .... Et sculpes in ea nomen ascendentis et domini eius et dominum hore diei in qua fueris et nomen lune; et sit luna in Scorpione. 6) Et infortunabis ascendens pro posse tuo, infortunabis quoque dominum ascendentis in domo mortis prout quiveris, et infortunabis etiam dominum ascendentis prout poteris, aut iungatur malo in quarto vel septimo. 7) Cumque haec feceris, sepelies eam versam id est capite deorsum, et dices dum sepelies eam: Hec est sepultura illius et illius speciei ut non intret in illum locum. 8) Et sepelies eam in medio loci a quo volueris fugare ipsam speciem, vel in domo eius habitationis aut in loco eius collectionis. Et si feceris quatuor imagines secundum hanc dispositionem, et sepelieris unamquamque earum in unaquaque quarta ipsius loci a quo volueris fugare ipsam speciem, erit validius et melius».

CAPUT XII

Cfr. ARISTOTELES, *Ethica Nicomachea*, I, 4 (1096 a 11–16); in the transl. by Guilelmus de Moerbeke, ed. in Thomas Aquinas's *In Ethicorum*, L. I, VI, cap. IV: «Etsi ardua tali quaestione facta propter amicos veros introducere ideas. Videbitur autem utique melius esse forsitan et oportere et pro salute veritatis et familiaria destruere, specialiterque et philosophos existentes. Ambobus enim existentibus amicis, sanctam praehonorare veritatem». See also *Aristoteles latinus*, XXVI, 1–3, III–IV: *Ethica Nicomachea*, transl. Roberti Grosseteste Linconiensis: A. Recensio pura; B. Recensio recognita, ed. R. A. Gauthier, Leiden, Brill-Bruxelles, Desclée de Brouwer, 1972–1973, p. 379/22-25.

PTOLEMAEUS, *Almagestum*, fol. aii: «Nos autem laboravimus ut in amore scientie sempiternorum manentium usque ad terminum quem eorum conditor eis imposuit: in sequentibus hius nostri libri addamus etc.».

ALBUMASAR, *Introductorium Maius*, transl. Ioannes Hispalensis, Tract, I, diff. ii (ms. Laur. Plut. 29.12, fol. 6^{r}): «Ex hoc quoque intelligitur dignitas magisterii astrorum quia magisterium astrorum altius est et loca eius sunt planetae quae non corrumpuntur nec recipiunt effectum neque decrementum usque ad tempus quod deus voluerit». Cfr. *Ibidem*, fol. 14^{r}: «Dixerunt universi antiqui philosophi quod efficitur ex motibus planetarum naturalibus ac durabilibus effectus naturalis ac durabilis qui fit usque ad tempus quod deus voluerit».

ALBUMASAR, *op. cit.*, I, diff. V (ms. Laur. Plut. 29.12, f. 16^{v}): «Et iam dixit Philosophus quod planetae sint animati. Et per motus eorum naturales significant concordiam animae rationalis et vitalis cum corpore».

ALBUMASAR, *op. cit.*, I, diff. V, cap. «De tertia secta» (ms. Laur. Plut. 29.12, fol. 17^{v}): «Planetis autem etsi sunt eis animae rationabiles non eligunt tamen nec indigent electione propter longitudinem eorum ab impedimentis».

ALBUMASAR, *op. cit.*, I, diff. III (ms. Laur. Plut. 29.12, fol. 11^{r}): «...et non secundum electionem eiusdem Solis fuerit introitus eius in hanc partem neque effectus ipsarum rerum et earum corruptio, sed propter adventum eius per motum naturalem in ipsam quartam.».

ALBUMASAR, *op. cit.*, I, diff.III (ms. Laur. Plut. 29.12, fol. 11^{r}): «Vidi qualiter pertinximus creatorem moventem res ex rebus apparentibus et notis quae pertingunt sensibus, qui sit scilicet sempiternus habens virtutem absque essentia finis immobilis et incorruptibilis altissimi, sit nomen eius benedictum et exaltatum exaltatione maxima».

ALBUMASAR, *op. cit.*, VI, diff. XXVI (ms. Laur. Plut. 29.12, fol. 79^{v}): «Domus quoque nona vocata est domus peregrinationis et motionis fidei quoque atque operum bonorum propter reversionem eius ad Iovem ...».

Ibidem (ms. Laur. Plut. 29.12, fol. 79^{v}-80^{r}): «etiam quia Iuppiter et Venus sunt fortunae. Fortunarum autem sunt duae species quarum una est fortuna huius mundi, altera est fortuna futuri saeculi. Et fortuna alterius saeculi est dignior fortuna huius mundi et hoc quaeritur per fidem. Et quia Iuppiter plus est fortuna quam Venus ideo facta est ei significatio super fidem per quam quaeritur fortuna futuri saeculi quae est dignior. Et facta est Veneri significatio super fortunas huius mundi ex ludis et gaudio atque laetitia».

ALBUMASAR, *op. cit.*, VI, diff. I (ms. Laur. Plut. 29.12, f. 69^{v}): «Virgo est duum corporum suntque ei tres species. Et ascendit in prima facie illius puella quam vocamus Celchuis Darosthal, et est virgo pulchra atque honesta et munda prolixi capilli et pulchra facie, habens in manu sua duas spicas. Et ipsa sedet supra sedem stratam et nutrit puerum dans ei ad comedendum ius in loco qui vocatur Abrie, et vocant ipsum puerum quaedam gentium Ihesum cuius interpretatio arabice est Eiceh. Et ascendit cum ea vir sedens super ipsam sedem».

CAPUT XIII

PS. – PTOLEMAEUS, *Liber centum verborum cum expositione Haly*, v. 5, f. 47v: « Optimus astrologus multum malum prohibere poterit quod secundum stellas venturum est, cum earum naturam praesciverit: sic enim praemuniet eum cui malum futurum est ut possit illud pati ».

HALY, *Glosa super v.* 5 (PTOLEMAEUS, *op. cit.*): « Videmus quod idem opus non est aequale suscipientibus, et receptorem ad maius vel minus suscipiendum vertere possumus. Ideo peritus astrologus, cum timuerit ne malum eveniat, convertet ipsum in futuri mali contrarium, ut cum ipsum malum evenerit non tantum ei applicetur quantum si ex improviso contingeret, verbi gratia, si aliquis temperatus esset bene, cuius nativitatem sciremus, tunc si aliquam infirmitatem ex Martis natura sibi venturam videremus, cuius complexionem ad frigiditatem verteremus ut infirmitas adveniens eam in temperantiam verteret. Similiter operaberis in caeteris planetis cum praesciverit quod ex eorum natura venturum sit ».

HALY, *Glosa super v.* 8 (PTOLEMAEUS, *op. cit.*): « Sapiens est illa anima quae scit illud quod diximus de fortitudinibus caeli: et eius adiutorium est quando aliquod bonum alicui eventurum cognoverit ei res sic aptare praecipiat, ut illud bonum venturum maius ac melius eveniat quam eveniret nisi sic cum praemuniret: ut iam locuti sumus de hoc sufficienter in quinto capitulo ».

PTOLEMAEUS, *op. cit.*, v. 8, f. 48r: « Anima sapiens ita adiuvabit opus stellarum quemadmodum seminator fortitudines naturales ».

CAPUT XIV

Cfr. ALBUMASAR, *Introductorium maius*, transl. Ioannes Hispalensis, tr. 1, diff. v, cap. *De tertia secta*, (ms. Laur. Pl. 29.12, f. 14v). « Tertia secta est quorundam disputantium qui contradixerunt scientiae astrorum et dixerunt quod planetis non sit significatio supra res de his quae fiunt in hoc mundo. Et hac ratione in actione usi sunt ut dicerent quod stellae non significarent id quod possibile est scilicet, sed tantum necessarium et impossibile. Sed nunc narremus ratiocinationem quorundam antiquorum qui repulerunt possibile, demum confirmemus possibile.

Postea ostendamus quod planetae /(15r) significent possibile. Quia quidam qui repulerunt iudicia astrorum circa possibile hac ratiocinatione usi sunt ut dicerent quod Philosophus dixit quod esse rerum in mundo esset trinum: necessarium ut ignis calidus, et impossibile ut homo volans, possibile vero ut homo scribens. Et stellae significant duo principia, necessarium scilicet et impossibile. Possibile autem non significant: ergo magisterium astrorum falsum est. Quidamque astrologorum et multi antiquorum philosophorum qui erint affirmantes significationes planetarum supra res quae efficiuntur in hoc mundo confirmatione necessaria et unde provenisset ad eos haec sententia inextricabilis perversa pigritarentur ac deficerent in responsione eius repulerunt possibile, et dixerunt quod essent principia duo, necessarium scilicet et impossibile tantum quia scimus res per affirmationem et negationem ita vel non et eorum interpretatio est inventio et privatio naturalis significat inventionem, ut vero privationem; et inventio est principium necessarium, privatio vero est principium impossibile (dialetici latini modos vocant). Inventio autem et privatio sunt quae vocantur propositiones contradictiones vel contradictoriae, quia cum vera fuerit una pars irritietur altera. Et impossibile est ut vera sit utraque in una re in eadem hora. Ut duo viri quorum unus dixerit: erit cras pluvia, et alter dixerit: non erit, necesse est unus eorum sit verax quod est necessarium, et alter mentiatur quod est impossibile. Similiter si quis diceret hodie: cras eveniet aliqua res, si venerit ipsa res in crastino ita eveniet quia eventus eius fuit necessarius. Et si dixerit: non eveniet, et non evenerit, ideo non eveniet quia eventus eius fuit impossibilis; et ita si unus eorum verax fuerit alter mentietur. Similiter si dixerit ille: ambulabis, et ambulaverit, ideo ambulavit quia necessarium fuit ut ambularet, et e converso. Et dixerunt: homines iudicant secundum opus suum in rebus, et hoc est quia est necessarium. Quod si non fuerint operati, ideo prohibiti sunt ex opere eius quia impossibile est ut operetur illud. Omnis ergo res quae fit necesse est ut fiat... non significant stellae nisi ista, et quod omnino non sit impossibile. Quorum dicta Philosophus dissoluit et affirmavit possibile per multas rationes. Postea necessarium quod possibile revertitur ad necessarium vel ad impossibile. Ratiocinatio autem prima eius est in confirmatione possibilis; quia dixi quod necessarium et impossibile sint nota in tribus temporibus: necessarium scilicet per inventionem suam naturaliter et impossibile per naturam suam. Opera autem sunt his contratria quae sunt possibilia; ut scientia nostra in sole qui fuerit lucidus in tempore praeterito et nec est lucidus eritque lucidus in futuro. Et sicut scientia nostra in igne qui fuerit calidus et nunc est calidus/(15^{v}) et erit calidus in futuro. Et similiter si dixerimus quod ignis et aer et aqua et terra quae admodum nunc sunt ita erant et erunt iam novimus quod sit verum in temporibus tribus: et hoc est principium necessarium. Et principium impossibile est ut si diceris: homo volavit aut homo volat nunc, et possibile est ut volet in futuro. Et similiter si diceris quod ignis fuerit frigidus et est frigidus eritque frigidus, erit hoc impossibile in tribus temporibus, et est mendacium. Iamque efficitur necessarium cum possibile notum in tribus temporibus per inventionem et impossibilitatem per naturam suam. Opera autem non sunt ita quia si dixerit homo in transacto eram operans bonum et ego nunc operor bonum, non poterit tamen dicere: ego in futuro operor bonum absque dubio, quia ignorat utrum possibile ei sit hoc vel non. Igitur nescit homo quid velit operari antequam operet eum certissime et absque dubio. Non ergo sunt hec necessaria sed possibilia. Et sua est. Invenitur ergo possibile. Secunda ratiocinatio est quia dixit quod necessarium et impossibile unumquodque eorum est in universo tempore equaliter, possibile autem non est ita aqualiter ob hoc quod vita sit inventa in universis hominibus equaliter. Et calor equaliter est in omni igne, non recipit augmentum aliquid ex eo vel detrimentum. Et similiter universa impossibilia eorum longitudo est equalis ab universa specie quia de universis hominibus praedicatur aequaliter qui non volent. Et quod ignis sit non frigidus. Opera autem non sunt ita quia ex specie hominum quidam operantur bonum, quidam vero malum. Et quidam eorum operantur bonum vel malum plus aliis. Si igitur universa necessaria sunt omni speciei aequaliter et impossibilia longe sunt ab omni specie aequaliter et ipsa non corrumpuntur, opera autem eorum non sunt aequalia in omni specie sed corrumpuntur et permutantur a bono in malum et malo in bonum, et in tempore post tempus et ex paucitate in multitudinem et ex multitudine in paucitatem, et ipsa recipiunt augmentum et diminutionem. Ergo possibile invenitur. Ratiocinatio quoque tertia est inventione possibilis quod homo cogitat et consulit in hoc quod est possibile. Unde homo cum voluerit aedificare domum cogitat in ea et in qualitate eius quod vult aedificare. Et consulit de eo. Et cum verificatum fuerit eius propositum super aedificium cogitabit et eliget in quali die incipit; et cum voluerit peregrinari consulit utrum peregrinetur vel non, et utrum sit ei melius peregrinari in litore quam in mari. Cumque verificatum fuerit eius propositum super peregrinationem cogitabit atque consulet in quali die peregrinaverit in qua die voluerit. /(16^{r}) Cum voluerit quoque seminare cogitabit et consulet de eo quod voluerit ex seminibus. Postea eliget quod voluerit de eo quod cogitat et super quod consultus fuerit. Similiter cum voluerit societatem alicuius hominis cogitat et consulet quis hominum erit melior, demum eliget quem voluerit et similiter operatur in rebus particularibus, ut homo qui cogitat et dicit: quid comedam hodie, aut quid potabo, vel quali vestimento induar hodie, vel quali platulo sedebo. Et cum fuerit sanus in membris et sensibus dicit: aspiciam illum vel non, et alloquar illum vel non. In his et horum similibus possibilis est electio. Horum omnium et his similium initia surgunt in cogitatione. Fines vero convertuntur ad opus vel ad dimissiones. Porro necessarium et impossibile inveniuntur in rebus inventione naturali, in cogitatione tamen, quia homo per cogitationem scit quod vita necessaria est homini viventi, et quod impossibile est ut volet. Et si essent res aut necessaria vel impossibilia tamen et necessarium ac impossibile ex necessitate non indigerent homines cogitatione nec consultu. Et si esset cogitatio et consultus in electione alicuius rei ⟨alicuius rei⟩ ex re alia frustra quia impossibile est homini ut cogitet cogitationem incertis aut consulet aliquem de igne utrum comburat vel non; quia est comburens necessario. Non enim cogitat in caeteris dicens utrum volet homo vel non, quia impossibile est ut volet. Quarta vero ratiocinatio est quod in rebus necessariis et impossibilibus sit, una fortitudo utrum sint vel non sint omnino. Et videmus plurimis rebus duas fortitudines videlicet utrum sit quemadmodum est, aut non sit ut pannus integer quod si dimittitur superiorem suum ante abscissionem remanebit donec veterescat per dies. Si vero abscissus fuerit recipit ascissionem. Et sicut ferrum aut plumbum vero et caetera quae fiunt ex eis quae si dimissa fuerint remanebit unumquodque eorum... /(16^{v}) Cum autem cessaret Philosophus ex confirmatione possibilis, dixit quod revertatur ad necessarium et impossibile: ut si quis diceret ambulem cras vel non, quia res et eius contrarium possibilia sunt ei. Cum autem ambulaverit efficitur eius ambulatio possibile quia antequam ambularet erat ambulatio possibile ei. Cum vero ambulavit ablata est ab eo possibilitas et effecta est in diffinitionem necessarii. Quod si non ambulaverit in crastino, efficitur in diffinitionem impossibilis quia ambulatio non fuit ei apta. Quia igitur patefactum est nobis esse possibilis, patet nunc quod sit planetis significatio super tria principia quae sunt necessarium, possibile, [et impossibile]. Dicimus quoque quod omne individuum in hoc mundo ex individuis animalium et sementum et metallorum sit compositum ex quatuor elementis, ex igne scilicet, et aere et aqua et terra, quia sunt inventa in omni individuo. Et unumquodque elementorum recipit augmentum et diminutionem et conversionem in invicem quia fit calor infra calorem et aer humidior aere et aqua quoque frigidior aqua et terra gravior terra. Et convertuntur in invicem. Si igitur in omni elemento est singulariter fortitudo per quam recipit corruptionem, et individua sunt composita ex his quatuor elementis, est ergo individuis fortitudo recipiendi augmentum et diminutionem, convertendi in invicem. Et per motum signorum vel planetarum super ea fit motus eorum et receptio corruptionis et compositionis. Signa ergo et planetae sunt significantia esse quatuor elementorum, et corruptionis eorum atque compositionis in individuis nutu Dei. Alius namque qui est homo est compositus ex anima vitali et rationali et ex quatuor elementis. Et iam dixit Philosophus quod planetae sint animati. Et per motus eorum naturales significant concordiam animae rationalis et vitalis cum corpore nutu Dei, quemadmodum praecessit in praecedentibus. Animae igitur rationali est fortitudo cogitationis et electionis. Corpori vero /(17^r) fortitudo receptionis possibilium. Cum autem significaverint planetae concordiam animae vitalis et rationalis et corporis iam significant necessarium et impossibilia ac possibilia, quia homini animato est vita quae est necessaria, et impossibile ex volatu, et possibile quia recipit infirmitatem et sanitatem et calorem, ac frigus, humiditatem quoque et siccitatem. Et est in eo ut cogitet ex in rebus multis, et eligit unam earum. Et fortitudo per quam eligit rem ex alia re per cogitationem in ea est homini super omnia animalia. Possibilitas vero receptionis rei est ex suo contrario corporibus. Opera autem nostra fiunt per praecedentiam cogitationis nostrae in re quam volumus operari. Cumque fuerit praecedens in cogitatione quod opus rei et eius contrarium sit possibile, operabimus unum eorum et consulemus de eo. Astrologus autem aspicit res illas in quibus est fortitudo possibilitatis ad receptionem rei et eius contrarium ad quod revertitur res. Et non aspicit proprietatem eorum, quia non aspicit astrologus in magisterio astrorum, utrum sit ignis comburens vel non, quia scit quod comburat, nec aspicit in significatione planetarum utrum sit frigida nix aut non, quia scit quod frigida est; sed aspicit utrum comburat ignis cras corpus ex corporibus recipientibus combustionem aut non, et utrum refrigeret cras nix aliquam rem ex rebus recipientibus frigus vel non; et utrum sit cras pluvia vel non ... an illum ambulet cras vel non; aspicit enim in his rebus utrum possibile est ut sit vel non sit. Cumque significaverint planetae per motus suos naturales aliquam rem ex rebus quae non fiet, erit impossibile ut fiat. Cum ergo significaverint effectus rei ex rebus in hora significationis, erit eius effectus ex necessario. Et si significaverint quod fiet in tempore futuro, significatio eius super effectum eiusdem rei fit per fortitudinem usque ad tempus quo fiet. Cumque fuerit res efficitur eius effectus ex necessario ... Hunc igitur patefactum est nobis, quod planetae sint significatores super possibile aut super [electionem]. Et hoc fit in duobus: uno scilicet et in compositione, ut possibile quoniam est individuo hominis ad receptionem rei et eius contrarium, et electionem rei ad semetipsum. Et secundo in rebus quae significant effectum in tempore futuro sicut diximus, et sicut significatio possibile atque electionem quae est hominis, similiter significant quod homo /(17ᵛ) non eligit nisi quod significaverint planetae, quia electio eius ad rem et eius contrarium fit per animam rationabilem quae complectitur animae in individuo per significationes planetarum; vero homo in cuius corpore est receptio possibilium ad motum et eius dimissionem ut surgat vel non surgat; et in fortitudine animae est electio unius istorum aut dimissio eius. Et iam pervenit ad diffinitionem necessarii vel impossibilis, quia possibile et electio pervenit ad unum eorum absque dubio nisi quod non eligit nisi quod significaverint planetae ex necessario vel impossibili».

CAPUT XV

HALY, *De eleccionibus horarum*; I, 1, ed. p. 329: «Opera iudiciorum astrorum certa esse a Ptholomeo rege patenti ratione probatum est ... Ex hoc enim opere constat, ut cum sciveris horam impregnationis cuiuslibet mulieris vel animalis sciemus per eam quid fiat de hoc semine donec inspiretur, et quid usque quo exeat ex utero, et quid fiat de eo usque ad diem obitus, sicut in opera Astrologiae dicitur».

Ibidem: «Astrologi non iudicaverunt per nativitates nisi quod hora impregnationis vix potest certificare, Tholomeus autem inquit horam nativitatis significare ... initium» (ms. FN J X 20, f. 19ᵛ): «horam nativitatis secundum esse inicium».

Ibidem: «Quod ita fit cum elegerimus alicui horam conceptioni prosequendo iudicia librorum nativitatum, accidunt et bona nato, quoque secundum libros nativitatum astrologus futura praenuntiavit, et eodem modo dicemus in plantatione arborum, et in serendis seminibus, et in edificatione quoque urbium, necnon in omni inceptione».

Cfr. Ps. – PTOLEMAEUS, *Liber centum verborum*, V. 19: « Si quis purgatorium acceperit Luna cum Iove existente abreviabitur eius opus et effectus ipsius minuetur».

Cfr. *Canon pro minutionibus et purgationibus*, «Tangere etiam membria ferro Luna esistente in signo ipsum membrum respiciente periculosum est ut dicit idem Ptolemaeus ... Secundo Campanus assignat causam dicens: Tangere cum ferro membrum illud vulnerando est causativum doloris et dolor causat fleuma (CLM: reuma) propter quod inquit: in cirurgia cavendum est ab incisione in membro Luna existente in signo significationem habentem supra illud membrum» (BT, p. 23).

Ibidem: «Item narrat Campanus se vidisse hominem imperitum in astris qui in periculo squinantie minuerat sibi de brachio luna existente in geminis quod signum dominatur super brachia et absque ulla manifesta egritudine excepta modica brachii inflacione die septimo mortuus est. Novit etiam quendam ut asserit patientem fistulam in capite membri virilis et ipsum fuisse incisum luna existente in scorpione quod signum dominatur super partem illam corporis et eadem hora incisionis in manibus tenentium obiit nulla (alia *add.* CLM) causa concurrente» (BT, p. 24).

Ps. – Ptolemaeus, *Liber centum verborum*, v. 22: «Nova vestimenta facere vel exercere et luna in Leone timendum, maxime si fuerit impedita.

Haly, *Glosa*: «Signa vero fixa vitanda sunt in omnibus quae intendimus alterare, et precipue signum Leonis».

CAPUT XVI

Ps. – Ptolemaeus, *Liber centum verborum*, v. 9, fol 48^v: «Vultus huius seculi sunt subiecti vultibus celestibus: et ideo sapientes qui imagines faciebant stellarum introitum in celestes vultus inspiciebant, et tunc operabantur quod debebant».

Thebit, *De imaginibus*, ii, 17: «Hec est imago destructionis illius regionis... ».

Thebit, *op. cit.*, 69–70: «Cumque feceris hoc sculpes nomina odii in dorso imaginis. Si autem fuerit ad dilectionem, scribes nomina dilectionis in medio imaginis».

Thebit, *op. cit.*, 63, 69: «Et nominabis imaginem nomine sui famoso (publico)». «Et sculpes in imagine nomen eius famosum».

THEBIT, *op. cit.*, 5: «Et sculpes in ea nomen ascendentis et domini eius et dominum horae diei in qua fueris et nomen lune».

THEBIT, *op. cit.*, 7: «Cumque hec feceris, sepelies eam versam id est capite deorsum, et dices dum sepelies eam: "Hec est sepultura illius et illius speciei et non intret in illum locum". Et sepelies eam in medio loci a quo volueris fugare ipsam speciem, vel in domo eius habitationis, aut in loco eius collectionis».

CAPUT XVII

Cfr. ISIDORUS HISPALENSIS, *Ethymologiarum libri xx*, Oxford 1950: De l. VIII, cap. IX (De Magis): «Varro dicit divinationis quatuor esse genera, terram, aquam, aerem et ignem. Hinc geomantiam, hydromantiam, aeromantiam, pyromantiam dicta».

Cfr. *ibidem*, l. IX, cap. ii: «Garamantes populi Africae prope Cyrenas inhabitantes, a Garamante rege Apollinis filio nominati, qui ibi ex suo nomine Garama oppidum condidit. Sunt autem proximi gentibus Aethiopum. De quibus Vergilius (Ecl. 8, 44): 'Extremi Garamantes'. Extremi autem quia saevi et a consortio humanitatis remoti».

BIBLIOGRAPHY

1. PRINTED PRIMARY SOURCES

Albumasar, *De magnis coniuctionibus, annorum revolutionibus ac eorum perfectionibus octo continens tractatus*, Augsburg, E.Ratdolt, 1489.

Albumasar, *De revolutionibus nativitatum*, greek transl. ed. by D. Pingree, Leipzig, Teubner, 1968.

Albumasar, *Flores*, Venezia, E. Ratdolt, 1488.

Albumasar, *Introductorium in astronomiam*, Augsburg, E.Ratdolt, 1489.

Aegidius Romanus, *Errores philosophorum*, Critical Text with Notes and Introduction by J.Koch; English Translation by J.O.Riedl, Milwaukee, Marquette U.P., 1944.

Aegidius Romanus, *Errores philosophorum*. First Ed.in P. Mandonnet, *Siger de Brabant*, Louvain 1911, II, pp.1-25.

Albertus Magnus, *Opera omnia*, ed. P.Jammy, Paris 1651, 21 vols.

Albertus Magnus, *Opera omnia*, ed. E. Borgnet, Paris 1890-1899, 38 vols.

Albertus Magnus, *Opera omnia, ad fidem codicum manuscriptorum edenda... curavit Institutum Alberti Magni Coloniense*, Münster, Aschendorff, 1960 (in progress).

Albertus Magnus, *Book of Minerals*, transl. by D. Wyckoff, Oxford 1967.

Albertus Magnus, *Commentaar op Euclides' Element der Geometrie*, Inleidende studie, analyse en uitgave van Boek I v.P.M.J.E. Tummers, Nijmegen 1984, 2 vols.

Albertus Magnus, *De animalibus*, ed. H. Stadler, Münster 1916-21 (= Beiträge z. Gesch. d Philos. u. Theol. d. Mittel., 15-16).

Albertus Magnus, *De caelo*, ed. P. Hossfeld, *Opera omnia*, V, I, Münster, Aschendorff, 1971.

Albertus Magnus, *De causis proprietatum elementorum*, ed. P. Hossfeld, in *Opera Omnia*, V, Münster, Aschendorff, 1971.

Albertus Magnus, *De fato*, ed. P. Simon, in *Opera omnia*, XVII, 1, Münster, Aschendorff, 1975.

Albertus Magnus, *De vegetabilibus*, ed. H. Stadler-C. Jessen, Berlin 1867.

Albertus Magnus, *Drei ungedruckte Teile der 'Summa de creaturis'*, hg. v. M. Grabmann, Leipzig 1919 (= Quellen und Forschungen zur Geschichte der Dominikaner in Deutschland, H.13).

Albertus Magnus, *In Dionysium de divinis nominibus*, ed. P. Simon, in *Opera omnia*, XXXVII, Münster, Aschendorff, 1971.

Albertus Magnus, In *II Sententiarum*, in *Opera omnia*, ed. Jammy, XV, Paris 1651.

Albertus Magnus, *Metaphysica*, ed. B. Geyer, in *Opera omnia*, XVII, 1-2, Münster, Aschendorff, 1960-1964.

Albertus Magnus, *Summa theologiae sive de mirabili scientia Dei libri I pars I, quaestiones 1-50A*, ed. W. Kübel et E.G. Vogels, Münster, Aschendorff, 1978

Albertus Magnus, *Super Matthaeum*, in Opera*omnia*, XXI/1-2, ed. B. Schmidt, Münster 1987.

Albertus Magnus, *Summa theologiae*, in *Opera omnia*, ed.Jammy, XVII-XVIII, Paris 1651.

Ps. Albertus Magnus, "The *De occultis naturae* Attributed to Albertus Magnus", ed. by P. Kibre, *Osiris*, XIII, 1958, pp. 157-183.

Ps. Albertus Magnus, *Libellus de alchimia*, Transl., Intr. and Notes by V. Heines, Berkeley-Los Angeles, University of California Press, 1959.

Ps. Albertus Magnus, *The Book of Secrets[...]also a Book of the Marvels of the World*, ed. by M. R. Best and F.H.Brightman, Oxford, at the Clarendon Press, 1973.

Al-Bitruji, *On the Principles of Astronomy*, ed. B. R. Goldstein, New HavenLondon, Yale U.P. 1971, 2 vols. (= Yale Studies in the History of Science and Medicine, 7)

Aristoteles, *Ethica Nicomachea: Translatio antiquissima ...Transl. Roberti Grosseteste Lincolniensis*, ed. R. Gauthier, Leiden, Brill, 1972-1974 (= Aristoteles latinus, XXVI).

Aristoteles, *Metaphysica, Libri I-X,XII-XIV, Translatio anonyma sive 'Media'*, ed. G. Vuillemin-Diem, Leiden, Brill, 1976 (= Aristoteles Latinus, XXV,2).

Averroes, *Tahafut al-Tahafut. The incoherence of the Incoherence*, Transl. from the Arabic with Introd. and Notes by S. van den Bergh, London, Luzac, 1954, 2 vols. (= Unesco Collection of great works.Arabic series–E.J.W.Gibb Memorial, New Series, 19).

Bernardus de Trilia, *Quaestiones disputatae de cognitione animae separatae*, ed. P. Künzle, Bern 1969 (= Corpus Philosophorum Medii Aevi).

Bernardus de Trilia, *Quaestiones de cognitione animae separatae*, ed. S. Martin, Toronto, Pontifical Institute of Mediaeval Studies, 1965 (= Studies and Texts, XI).

Bernardus de Virduno, *Tractatus super totam astrologiam*, ed. P.P. Hartmann, Werl 1961 (= Franziskanische Forschungen, 15).

Bonaventura, *Collationes de decem praeceptis*, in *Opera omnia*, V, Quaracchi 1891.

Bonaventura, *Collationes de donis Spiritus Sancti*, in *Opera Omnia*, V, Quaracchi 1891.

Bonaventura, *Collationes in Hexaëmeron*, in *Opera Omnia*, V, Quaracchi 1891.

Bonaventura, *Collationes in Hexaëmeron*, ed. F. Delorme, Quaracchi 1934.

Bonaventura, *Commentaria in IV libros Sententiarum*, in Opera omnia, II, Quaracchi 1882-1902.

Denifle H. and E. Chatelain eds., *Chartularium Universitatis Parisiensis*, I, Paris 1899.

Campanus of Novara and Medieval Planetary Theory. 'Theorica Planetarum', ed. with Intr., Engl. Translation and Commentary by F. S. Benjamin and G. J. Toomer, Madison, University of Winsconsin Press, 1971.

Dietrich von Freiberg, *Opera omnia*, ed. K. Flasch et aliï, Hamburg, F. Meiner, 1977-1985.

Dondaine H.-F. ed., 'Le *"De 43 questionibus"* de Robert Kilwardby', *Archivum Fratrum Praedicatorum*, XLVII, 1977, pp.5-50.

Gerson Jean, *Opera omnia*, ed. L.-E. Dupin, The Hague 1728 (2nd ed.).

Gerson Jean, *Oeuvres complètes*, ed. P. Glorieux, II, Paris 1960.

Godefroid de Fontaines, *Quodlibet XII*, in *XIV Quodlibeta*, ed.M.De Wulf-A.Pelzer-J.Hoffmans, Louvain 1904-1935.

Guilelmus de Sancto Amore, *Opera*, Constantiae 1632.

Hermannus de Carinthia, *De essentiis*, ed. Ch. Burnett, Leiden, Brill, 1982.

Johannes Hispalensis, *Epitome totius astrologiae*, Nürnberg 1548.

Katalog der Werke des hl. Albertus Magnus in einer Handschrift der Lütticher Universitätsbibliothek, ed. P. Simon, in *Zur Geschichte und Kunst im Erzbistum Köln. Festschrift für W. Neuss*, Düsseldorf 1960, pp. 79-88.

Katalog der Dominikanerschriftsteller, cf. 'Secondary Literature' under Auer P. ed., *Ein neuaufgefundener Katalog*.

Klopsch P. ed., *Pseudo-Ovidius de vetula*. Untersuchungen und Text, Leiden und Köln, Brill, 1967.

Laurenti Pignon catalogi et chronica. Accedunt catalogi Stamensis et Uppsalensis Scriptorum O.P., ed. G. G. Meersseman, Roma 1936, (= Monumenta Ordinis Praedicatorum Historica, XVIII)

Malevicz M. H. ed., "Libellus de efficacia artis astrologiae", *Mediaevalia philosophica polonorum*, XX, 1974, pp. 3-95.

Nemesius of Emesa, *De natura hominis*. Transl. by Burgundio Pisanus, Critical Ed. G. Verbeke-J. Moncho, Paris-Louvain 1975 (= Corpus latinum Commentariorum in Aristotelem graecorum).

Petrus Abelardus, *Expositio in Hexaëmeron*, Paris 1885 (= Migne, Patrologia Latina, vol. 178), cols. 730-784.

Petrus de Abano, *Il 'Lucidator dubitabilium astronomiae'. Opere scientifiche inedite*, ed. G. Federici Vescovini, Padova, Programma e 1 + 1 Editori, 1988.

Petrus de Abano, *Conciliator*, Venetiis 1565, Reprint ed. by E. Riondato and L. Olivieri, Padova, Antenore, 1985.

Petrus de Prussia, *Alberti Magni Vita* [1486], in Albertus Magnus, *De adherendo Deo*, Antwerp 1621.

Picatrix. The latin Version, ed. by D. Pingree, London, The Warburg Institute, 1986 (= Studies of the Warburg Inst., XXXIX).

Pico della Mirandola Giovanni, *Disputationes adversus astrologiam divinatricem*, ed. E. Garin, Firenze 1946-1952.

Pico della Mirandola Giovanfrancesco, *De rerum praenotione*, in his *Opera omnia*, Basileae 1572.

Pierre d'Ailly, *De ymagine mundi*[...]*Vigintiloquium de concordantia astronomicae veritatis cum theologia.De concordia astronomiae cum theologia. Elucidarius astronomicae concordiae cum theologica et historica narratione*, Augsburg [1480 ca.].

Raymundus Lullus, *Declaratio per modum dialogi edita*, hrsg. v. O. Keicher, Münster 1909 (= Beiträge zur Geschichte der Philosophie u. Theologie des Mittelalters, VII, 4-5).

Richard of Wallingford, *An edition of his Writings with introductions*, English translations and commentary by *John D. North, Oxford, at the Clarendon Press, 1976.*

Robertus Grosseteste, *De cometis*, in *Die philosophischen Schriften*, ed. L.Baur, Münster 1912 (= Beiträge zur Geschichte der Philosophie und Theologie des Mittelalters, 9).

Robert Kilwardby, *De ortu Scientiarum*, ed. by A.G. Judy, Leiden, Brill, 1976 (= Auctores Britannici Medii Aevi, IV).

Robotham D.M. ed., *The pseudo-Ovidian 'De vetula'*, Text with Introd. and Notes, Amsterdam 1968.

Rodulphus a Noviomago, *Legenda de Alberto Magno*, ed. H.C. Scheeben, Köln 1928 (2nd Ed.).

Roger Bacon, *De retardatione accidentium senectutis*, in *Opera hactenus inedita*, ed. A. G. Little and E. Withington, Oxford 1928.

Roger Bacon, *Opus tertium, Opus minus, Compendium philosophiae*, in *Opera quaedam hactenus inedita*, ed.J.S.Brewer, London 1859.

Roger Bacon, *Opus maius*, ed. J.H. Bridges, Oxford 1897-1900.

Roger Bacon, *Part of the Opus tertium*, ed. A.G. Little, Aberdeen 1912.

Roger Bacon's Philosophy of nature. A critical edition, with English Translation, and Notes of *De multiplicatione specierum* and *De speculis comburentibus*, by D.C. Lindberg, Oxford. Clarendon Press, 1983.

Roger Bacon, *Un fragment inédit de l'Opus tertium de R.Bacon*, précédé d'une étude sur ce fragment par P.Duhem, Quaracchi 1909.

Roger Bacon, *Secretum secretorum*, in *Opera hactenus inedita*, ed. R. Steele, V, Oxford 1920.

Siger de Brabant, *Quaestiones in tertium de anima*, ed. B. Bazan, Louvain-Paris 1972.

Siger de Brabant, *De anima...De aeternitate mundi*, ed. Barsotti, Münster, Aschendorff, 1972 (Texta et documenta)

Siger de Brabant, *De aeternitate mundi*, ed. B. Bazàn, Louvain, Publications universitaires, 1972 (= Philosophes médiévaux, 13).

Siger de Brabant, *De aeternitate mundi*, First Ed. in Mandonnet, Siger*de Brabant et l'Averroisme latin au XIIIe siècle*, II, Louvain 1911(2nd Ed.).

Simon P. ed., 'Ein Katalog der Werke des hl. Albertus Magnus in einer Handschrift der Lütticher Universitätsbibliothek', in *Zur Geschichte und Kunst im Erzbistum Köln. Festschrift für W. Neuss*, Düsseldorf 1960, pp.79-88.

Thabit ben Qurra, *Four Astrological Works in Latin*, ed. by F. Carmody, Berkeley 1941.

Thomas Aquinas, *De substantiis separatis*, ed. by H.-F.Dondaine, in *Opera omnia iussu Leonis XIII P.M. edita*, T.XL [= Opuscula I], Romae, ad S.Sabinam, 1969, pp.D1-D80.

Thomas Aquinas, *Responsio de 108 articulis. Responsio de 43 articulis. Responsio de 36 articulis. Responsio de 6 articulis... [De secreto]*, ed. by H.-F.Dondaine, in *Opera omnia iussu Leonis XIII P.M. edita*, T.XLII [= Opuscula III], Roma, Editori di S.Tommaso, 1979, pp.259-371.

Thomas Aquinas, *De operationibus occultis naturae. De iudiciis astrorum. De sortibus*, in *Opera omnia iussu Leonis XIII P.M. edita*, T.XLIII [= Opuscula IV], Roma, Editori di S. Tommaso, 1976, pp.159-141.

Thomas Aquinas, *De iudiciis astrorum[...] De sortibus [...] Responsio ad Ioannen Vercellensem generalem magistrum O.P. de articulis 42*, ed. R.Verardo, in Thomas' *Opuscula Theologica I*, Torino-Roma 1954.

Thomas Aquinas, *Sententia libri de anima*, ed. R.-A. Gauthier, in *Opera omnia iussu Leonis XIII P.M. edita*, T.XLIV Roma-Paris, Editori di S. Tommaso-Vrin, 1984.

Thomas Aquinas, *The Letter of St. Thomas Aquinas 'De occultis operibus naturae'*, ed. by J. B. Mc Allister, Washington D.C., 1939.

Thomas Aquinas, *Treatise on Separate Substances*, A Latin-English Ed. of a newly-established Text by F.J. Lescoe, West Hartford, Conn., St.Joseph College, 1963.

Thomas de Cantimpré, *De naturis rerum: Prologue, Bk.III and Bk.XIX*, ed.J. B. Friedmann, Montréal, Bellarmin, 1974 (= La science de la nature: théorie et pratique. Cahiers d'études médievales.II)

Witelo, *Perspectivae liber primus*, crit. Ed. trad. inglese di and english Translation by S. Unguru, (= Studia copernicana, XV), Wrocław 1977

Witelo, *De causa primaria poenitentiae in hominibus et de natura daemonum*, crit. Ed. by G. Burchardt, in his volume *List Witelona do Ludwika we Lwówku slaskim*, Wrocław 1979 (= Studia copernicana, XIX), pp. 153-208.

2. SECONDARY LITERATURE*

Agrimi J. and Crisciani C., 'Albumazar nell'astrologia di Ruggero Bacone', *ACME. Annali della Facoltà di Lettere e Filosofia ... Milano*, XXV, 1972, pp. 315-338.

Albert der Grosse, seine Zeit, sein Werk, seine Wirkung, hg. v. A. Zimmermann, Berlin-New York, W. de Gruyter, 1981 (= Miscellanea mediaevalia, 14).

Albert der Grosse und die deutsche Dominikanerschule. Philosophische Perspektive (Tagung, 15-18 Oktober 1984), Special Issue of the *Freiburger Zeitschrift für Philosophie und Theologie*, XXXII/1-2, 1985.

Albert le Grand (Le bienheureux), Special Issue of the *Revue thomiste*,XXXVI, 1932.

Albert the Great, Special Issue of the *Southwestern Journal for Philosophy*, X, 3, 1979.

Albert the Great. Commemorative Essays, Ed. and with an Introd. by F. J. Kovach-R. W. Shanan, Norman, University of Oklahoma Press, 1980.

Alberto Magno. Atti della settimana albertina, Rome [1932]

Albertus Magnus and the Sciences. Commemorative Essays, ed. Weisheipl J.A., Toronto, Pontifical Institute of Mediaeval Studies, 1980.

Albertus Magnus. Ausstellung zum 700. Todestag, Köln, Historisches Archiv der Stadt Köln, 1980.

Albertus Magnus, Doctor Universalis. 1280-1980, hg. v. G. Meyer and A. Zimmermann, Mainz, M. Grünewald, 1980.

Albertus-Magnus-Festschrift, hg. v. M. Manser und G.M. Häfele, Special Issue of *Divus Thomas* (Freiburg), X/2-3, 1932.

Alessio F., 'Un secolo di studi su Ruggero Bacone (1848-1957)', *Rivista critica di storia della filosofia*, XIV, 1959, pp. 81-102.

Allard G.H., 'Réactions de trois penseurs du XIIIe siècle [Thomas, Albert et Roger Bacon] vis-à-vis de l'alchimie', in *La science de la nature: théories et pratiques*, Montréal, Bellarmin, 1974 (= Cahiers d'études médiévales, 2) pp. 97-106.

Alonso M., 'Juan Sevillano. Sus obras proprias y sus traducciones', *Al Andalus*, XVIII,1953, pp.18-41.

Alonso M., 'Notas sobre los traductores toledanos Domingo Gundisalvo y Juan Hispano', *Al Andalus*, VIII, 1943, pp. 155-188.

*I list here all the books and articles quoted in my footnotes as well as many others I found relevant to the special problems of astrology in Albert, the authenticity of the *Speculum astronomiae* or other natural works attributed to him or to Roger Bacon, the autographic or manuscript tradition of Albertus' writings concerning natural and astrological topics (even if for sake of space I decided not to use them in footnotes). For what concerns other sides of Albertus's thought I refer the reader to the various lists of research on Albertus (from M. Weiss's *Primordia*, 1898 and 1905, to the bibliographies in *Revue thomiste* (Laurent and Congar); *Bulletin Thomiste; Maître Albert*, all of them in 1931; Käppeli, *Bibliographia albertina*, Roma 1931; *Revue des sciences philosophique et théologiques*, 1932 (Congar); *Angelicum*, 1944 (Walz and Pelzer); *The Modern Schoolman*, 1959-1960 (Catania); *Revista da Universidade catolica de São Paulo*, 1961 (Schoyans); *The Tomist*, 1980 (O' Meara) , and, last but not least, to the lengthy one printed in *Albertus Magnus and the Sciences*, ed. J.A.Weisheipl, Toronto, The Pontifical Institute of Medieval Studies, 1980, pp.585-617.

Altner H., 'Albertus Magnus: ein Wegweiter', in *Naturwissenschaftliche Forschung in Regensburger Geschichte*, ed. J.Barthel, Regensburg, Mittelbayerische Druckerei, 1980, pp.9-28.

Anawati G. C., 'Albert le Grand et l'Alchemie', in *Albert der Grosse, seine Zeit, sein Werk, seine Wirkung*, hg. v. A. Zimmermann, Berlin-New York, W. de Gruyter, 1981, pp. 126-133.

Ashley B. M., 'St.Albert and the Nature of Natural Science', in *Albertus Magnus and the Sciences*, ed. J.A. Weisheipl, Toronto 1980, pp. 73-101.

Auer P., *Ein neuaufgefundener Katalog der Dominikanerschriftsteller*, Paris 1933 (= S. Sabinae Dissertationes Historicae, II).

Avena A., 'Guglielmo da Pastrengo e gli inizi dell'umanesimo a Verona', in *Atti dell'Accademia di agricoltura, scienze, lettere, arte e commercio di Verona*, LXXXII (S. IV, vii), Verona 1907, pp. 229-85.

Balss H., *Albertus Magnus als Biologe*, Stuttgart 1947.

Balss H., *Albertus Magnus als Zoologe*, München 1928 (= Beiträge z. Geschichte u. Literatur der Naturwissenschaften u. Medizin, 11-12).

Barker Price B., *The Astronomy of Albertus Magnus*, Unpublished Ph. D. Thesis, University of Toronto 1983.

Barker Price B., 'The Physical Astronomy and Astrology of Albertus Magnus', in *Albertus Magnus and the Sciences*, ed. J.A. Weisheipl, Toronto 1980, pp. 155-186.

Bataillon L. J., 'Status quaestionis sur les instruments et techniques de travail de St. Thomas et St. Bonaventure', in *1274. Année charnière. Mutations et continuités*, Paris 1977, pp. 647-657.

Bauer U., *Der 'Liber Introductorius' des Michael Scotus in der Abschrift CLM 10268 der Bayerischen Staatsbibliothek München*, München, tuduv Studie, 1983.

Bayle P., *s.v.* 'Albert le Grand', *Dictionnaire historique et critique*, I, Paris 1820, pp. 358-364.

Bazan B.C., Franzen G., Wippel J.W. and Jacquart D., *Les questions disputées et les quéstions quodlibétiques dans les Facultés de Théologie, de Droit et de Médicine*, Turnhout, Brepols, 1985.

Bedoret H., 'L'auteur et le traducteur du *Liber de causis*', *Revue néoscolastique de philosophie*, XLI, 1938, pp. 519-533.

Benjamin F.S. and Toomer G.J., cf. Primary literature.

Bennett R.F., 'P. Mandonnet O.P. and Dominican Studies', *History* (London), XXXIV/4, 1939 December, pp.193-205.

Bertola E., 'Alano di Lilla, Filippo il Cancelliere ed una inedita quaestio sull'immortalità dell'anima umana', *Rivista di filosofia neoscolastica*, LXII, 1970, pp. 245-271.

Bérubé C., *De la philosophie de la sagesse chez Saint Bonaventure et Roger Bacon*, Roma, Biblioteca Serafico-cappuccina, 1976.

Bezold F. von, 'Astrologische Geschichtskonstruktion in Mittelalter' , in his *Aus Mittelater und Renaissance*, München-Berlin 1918, pp. 164-195, 399-411.

Bianchi L., *Il vescovo e i filosofi. La condanna parigina del 1277 e l'evoluzione dell'aristotelismo scolastico*, Bergamo, Lubrina, 1990.

Bigalli D., *I tartari e l'Apocalisse*, Firenze, La Nuova Italia, 1971.

Bjornbö A.A. – Vogl S., *Alkindi, Tideus and Pseudo-Euklid*, Leipzig 1912 (= Abhandlungen zur Geschichte der mathematischen Wissenschaften mit Einschluss ihrer Anwendungen, 26,3)

Birkenmajer A., *Études d'histoire des sciences et de la philosophie du Moyen Age*, Wrocław-Warszawa-Krakòw, Ossolineum, 1970.

Birkenmajer A., 'La bibliothèque de Richard de Fournival, poète et érudit français du XIII siècle' [1922], in his *Études d'histoire des sciences et de la philosophie du Moyen Age*, Wroclaw-Warszawa-Kraków, Ossolineum, 1970.

Birkenmajer A., 'Zur Bibliographie Alberts des Grossen', *Philosophisches Jahrbuch des Görresgesellschaft*, XXXVI, 1924, pp. 270-272.

Boll F., Bezold C. and Gundel W., *Sternglaube und Sterndeutung. Die Geschichte und das Wesen der Astrologie*, Darmstadt, Wissenschaftliche Buchgesellschaft, 1965.

Bonney F., 'Autour de Jean Gerson. Opinions de théologiens sur les superstitions et la sorcellerie au début du XVe siècle', *Le Moyen Age*, LXXVII, 1971, pp. 85-98.

Bormans J.H., 'Thomas de Cantimpré indiqué comme une des sources oû Albert le Grand et surtout Maerlant ont puisé les matériaux de leurs écrits sur l'histoire naturelle', *Bulletin de l'Academie Royal des Sciences de Belgique*, XXIX/1, Bruxelles 1852, pp.132-159.

Bouché-Leclercq A., *L'astrologie grecque*, Paris, 1899.

Bougerol, G., *Introduction à l'étude de Saint Bonaventure*, Paris, Vrin, 1990.

Bouyges M., 'Roger Bacon a-t-il lu des textes arabes?', *Archives d'histoire doctrinale et littéraire du Moyen Age*, V, 1930, pp. 311-315.

Brady I., 'Two sources of the *Summa de homine* of Saint Albert the Great', *Recherches de théologie ancienne et médievale*, XX, 1953, pp. 222-271.

Brady I.,'Background of the Condemnation of 1277: master William of Baglione', *Franciscan Studies*, VIII, 1970, pp. 5-48.

Brehm E., 'R. Bacon's Place on the History of Alchemy', *Ambix*, XXIII, 1976, pp. 53-58.

Brusadelli M. [= Giovanni Semeria], 'Lo *Speculum astronomiae* di Ruggero Bacone', *Rivista di filosofia neoscolastica*, VI, 1914, pp. 572-579.

Burnett C.F.S., 'A Group of Arabic-latin Translations', *Journal of the Royal Asiatic Society*, 1977, pp. 62-118.

Burnett C.F.S., 'An Apocryphal Letter from the Arabic Philosopher al-Kindi to Theodore, Frederick II's Astrologer', *Viator*, XV, 1984, pp. 151-166.

Bürke B., *Das neunte Buch des lateinischen grossen Metaphysik-Kommentars von Averroes. Textedition und Vergleich mit Albert dem Grossen und Thomas von Aquin*, Bern 1969.

Callus D.A., 'S. Tommaso d'Aquino e S. Alberto Magno', *Angelicum*, XXXVI, 1960, pp. 133-161.

Callus D.A., 'Une oeuvre récemment découverte de Saint Albert le Grand: *De XLIII problematibus ad magistrum ordinis (1271)*', *Revue des sciences philosophiques et théologiques*, XLIV, 1960, pp. 243-261.

Carmody F., *Arabic Astronomical and Astrological Science in Latin Translation*, Berkeley and Los Angeles 1956.

Caroti S., 'Alberto Magno "Doctor universalis"', *Cultura e scuola*, 90, 1984, pp. 110-115.

Caroti S., 'Alberto Magno e la scienza: bilancio di un centenario', *Annali dell'Istituto e Museo di Storia della Scienza di Firenze*, VI, 1981, pp. 17-44.

Caroti S., *L'astrologia in Italia*, Roma , Newton Compton, 1983.

Caroti S., *La critica contro l'astrologia di Nicole Oresme*, Roma 1979 (= Accademia Nazionale dei Lincei. Memorie. Classe di Scienze morali storiche e filologiche, S. VIII, xxiii, 6).

Caroti S., 'Problèmes textuels et lexicographiques dans l'oeuvre scientifique d'Albert le Grand', *Annali dell'Istituto e Museo di Storia della Scienza di Firenze*, VI, 1981, pp. 187-202.

Caroti S. and Zamponi S., 'Note', *Annali dell'Istituto e Museo di Storia della Scienza di Firenze*, V/2, 1980, pp. 111-117.

Carozzi C., 'Le monde laic suivant Humbert de Romans'; 'Humbert de Romans et l'union avec les Grecs'; 'Humbert de Romans et l'histoire', in *1274. Année charnière. Mutations et continuités*, Paris 1977, pp. 233-241, 491-494, 849-861.

Carton R., *L'expérience physique chez Roger Bacon*, Paris, 1924.

Carton R., *L'expérience mystique de l'illumination interieure chez Roger Bacon*, Paris 1924.

Carton R., *La synthèse doctrinale de Roger Bacon*, Paris 1924.

Charles E., *Roger Bacon, sa vie, ses oeuvres, ses doctrines*, Paris 1861.

Charmasson T., *Recherches sur une technique divinatoire: la géomancie dans l'occident médiéval*, Genève-Paris, Droz, 1980.

Chatillon J., 'L'exercice du pouvoir doctrinal dans la chrétienté du XIIIe siècle: le cas d'Etienne Tempier', in *Le pouvoir* (Colloque de l'Institut Catholique), Paris, Beauchesne, 1978, pp. 13-45.

Chenu M.-D., Le *De spiritu imaginativo* de Robert Kilwardby, *Revue des sciences philosophiques et théologiques*, XV, 1926, pp. 507-517.

Chenu M.-D., 'Les réponses de s.Thomas et de Kilwardby à la consultation de Jean de Verceil', in *Mélanges Mandonnet*, Paris 1930, I, pp. 191-222.

Chenu M.-D., 'Aux origines de la "science moderne"', *Revue des sciences philosophiques et théologiques*, XXIX, 1940, pp. 206-217.

Choulant L., 'Albertus Magnus in seiner Bedeutung für die Naturwissenschaften', *Janus. Zeitschrift für Geschichte und Literatur der Medizin*, I, 1846, pp. 138-139.

Cipolla C., 'Attorno a Giovanni Mansionario e a Guglielmo da Pastrengo', in *Miscellanea Ceriani*, Milano 1910, pp.743-788.

Collins J., *The Thomistic Philosophy of the Angels*, Ph. D. Dissertation, Washington D.C., 1947 (= The Catholic University of America, Philosophical Studies, vol. 89).

Congar Y.M.J., 'Aspects écclésiologiques de la querelle entre mendiants et séculiers', *Archives d'histoire doctrinale et littéraire du Moyen Age*, XXVIII, 1961, pp. 35-151.

Congar Y., '"In dulcedine societatis quaerere veritatem". Notes sur le travail en équipe chez St. Albert et chez les Prêcheurs au XVIIIe siècle', in *Albertus Magnus Doctor Universalis. 1280-1980*, hrgs. v. G. Meyer u. A. Zimmermann, Mainz, M. Grünewald, 1980, pp.47-58.

Cortabarria Beitia A., 'Las obras y la filosofia de Alfarabi y Al Kindi en los escritos de S. Alberto Magno', *Estudios filosoficos*, I, 1951-52, pp. 191-209.

Cortabarria Beitia A., 'Al Kindi vu par Albert le Grand', *Mélanges de l'Institut d'études Orientales* [MIDEO], XIII , 1977, pp. 117-146.

Cortabarria Beitia A., *De Alpharabii et Alkindi operibus in scriptis Alberti Magni*, Las Caldas de Besaya 1953.

Cortabarria Beitia A., 'Deux sources de S. Albert le Grand, al Bitruji et al Battani', *Mélanges de l'Institut d'études orientales* [MIDEO], XV, 1982, pp.31-52.

Cortabarria Beitia A., 'El astronomo Alpetragio en las obras de S.Alberto Magno', *La Ciudad de Dios*, CXCIII, 1974, pp.503-533.

Cortabarria Beitia A., 'Fuentes arabes de San Alberto. Albumsasar', *Estudios filosoficos*, XXX/84, 1981, pp.283-299.

Cortabarria Beitia A.,'La eternidad del mundo a la luz de las doctrinas de san Alberto Magno', *Estudios filosoficos*, X, 1961, pp.5-39.

Cranz F.E., 'The Publishing History of the Aristotle Commentary of Thomas Aquinas', *Traditio*, XXXIV, 1978, pp. 157-192.

Creytens R., 'Hugues de Castello astronome dominicain du XIVe siècle', *Archivum fratrum praedicatorum*, XI, 1941, pp. 95-108.

Crisciani C. and Gagnon C., *Alchimie et philosophie au Moyen Age. Perspectives et problèmes*, Montréal, L'aurore-Univers, 1980.

Crowley T., *Roger Bacon. The Problem of the Soul in his Philosophical Commentaries*, Louvain-Dublin 1950.

Dales R.C., 'Mediaeval Deanimation of the Heavens', *Journal of the History of Ideas*, XLI, 1980, pp.531-550.

Dales R.C., 'Robert Grosseteste's Place in Mediaeval Discussion of the Eternity of the World', *Speculum*, LXI, 1986, pp.543-563.

d'Alverny M.-Th., 'Abélard et l'astrologie', in *Pierre Abélard, Pierre le Vénérable* (Colloque intern. du CNRS, N.546), Paris, CNRS, 1975 , pp. 613-630.

d'Alverny M.-Th., 'Avendauth?', in *Homenaje a Millâs Vallicrosa*, I, Barcelona 1954, pp.34-35.

d'Alverny M.-Th., 'Avicenna latinus. II', *Archives d'histoire doctrinale et littéraire du Moyen Age*, XXXVII, 1962, pp.217-233.

d'Alverny M.-Th., 'Les nouveaux apports dans les domains de la science et de la pensée au temps de Philippe Auguste: la philosophie', in *La France de Philippe Auguste. Le temps de mutations. Colloque intern. du CNRS*, Paris, CNRS, pp. 864-880.

d'Alverny M.-T.,'Translations and Translators', in *Renaissance and Renewal in the Twelfth Century*, L. Benson and G. Constable eds., Cambridge Mass., C.D.Lanham, 1982.

d'Alverny M.-Th., 'Un témoin muet des luttes doctrinales du XIIIe siècle', *Archives d'histoire doctrinale et littéraire du Moyen Age*, XXIV, 1949, pp.223-248.

d'Alverny M.-Th. and Hudry F., eds., 'Al-Kindi De radiis', *Archives d'histoire doctrinale et littéraire du Moyen Age*, XLI, 1974, pp. 139-260.

Daniel E. R., 'Roger Bacon and the *De seminibus Scripturarum*', *Mediaeval Studies*, XXXIV, 1972, pp.462-467.

De astronomia Alphonsi Regis. Proceedings of the Symposium on Alphonsine Astronomy held at Berkeley (August 1985) together with other papers on the same subject, ed. by M. Comes, R. Puig Aguilar, J. Samsò, Barcelona, Instituto "Millàs Vallicrosa" de Historia de la Ciencia Arabe–Universidad de Barcelona, 1987.

De Libera, A., *Albert le Grand et la philosophie*, Paris, Vrin, 1990.

De Libera A., 'Ulrich de Strasbourg, lecteur d'Albert le Grand', *Freiburger Zeitschrift für Philosophie und Theologie*, XXXII, 1985 (=*Albert der Grosse und die deutsche Dominikanerschule*, pp. 105-136.)

Delisle L., *Le cabinet des manuscrits de la Bibliothèque Nationale*, II, Paris 1874, pp. 514-536.

Delorme A., 'La morphogenèse d'Albert le Grand dans l'embryologie scholastique', *Revue thomiste*, XXXVI, 1931, pp. 352-360.

Denifle H., 'Quellen zur Gelehrtengeschichte des Predigerordens im 13. und 14. Jahrhundert', *Archiv für Literatur- und Kirchengeschichte des Mittelalters*, II, 1886, pp. 165-252.

Denomy A.J., 'The *De Amore* of Andreas Capellanus and the Condemnation of 1277', *Mediaeval Studies*, VIII, 1946, pp. 107-149.

Destrez J., 'La lettre de Saint Thomas d'Aquin dite au lecteur de Venise d'après la tradition manuscrite', *Mélanges Mandonnet*, I, Paris 1930, pp. 103-189.

Dezani S., 'S. Alberto Magno: l'osservazione e l'esperimento', *Angelicum*, XXI, 1944 (= *Serta albertina*),pp. 45-47.

Die Kölner Universität im Mittelalter. Geistige Wurzeln und soziale Wirklichkeit, hg. v. A. Zimmermann [und] G. Vuillemin-Diem, Berlin-New York, W. de Gruyter, 1989 (= Miscellanea mediaevalia, 20).

Donati S., 'La dottrina di Egidio Romano sulla materia dei corpi celesti', *Medioevo*, XII, 1986, pp. 229-280

Dondaine A., 'Un catalogue de dissensions doctrinales entre les maîtres parisiens de la fin du XIIIe siècle', *Recherches de théologie ancienne et medievale*, X, 1938, pp. 374-394.

Dondaine A., 'S. Thomas et les traductions latines des *Métaphysiques* d'Aristote', *Bulletin Thomiste*, III (1931-1933), pp. 199*-213*.

Dondaine A., *Secrétaires de Saint Thomas*, Roma 1956.

Dondaine A. and Peters J., 'Jacques de Tonengo and Giffredus d'Anagni auditeurs de Saint Thomas', *Archivum Fratrum Praedicatorum*, XXIX, 1959, pp. 52-72.

Dondaine H.F., 'Le '*De 42 quaestionibus*' de Robert Kilwardby', *Archivum fratrum Praedicatorum*, XLVII, 1977, pp. 5-50.

Druart Th.-A., 'Astronomie et astrologie selon Farabi', *Bulletin de Philosophie médievale*, XX, 1978, pp. 43-47.

Dufeil M.M., *Guillaume de Saint-Amour et la polémique universitaire parisienne 1250-59*, Paris 1972.

Duhem P., *Le système du monde*, Paris 1958, 10 vols.

Duin J., *La Doctrine de la providence dans les écrits de Siger de Brabant*, Louvain 1954 (= Philosophes Médievaux, 3).

Easton S.C., *Roger Bacon and His Search for a Universal Science*, New York 1952.

Eckert W.P., 'Albert der Grosse als Naturwissenschaftler', *Angelicum*, LVII, 1980, pp. 477-495.

Edwards G.M., 'The two redactions of Michael Scot's *Liber introductorius*', *Traditio*, XLI, 1985, pp. 329-340.

Emmerson R.K. and R. B. Herzman, 'Antichrist, Simon Magus and *Inferno* XIX', *Traditio*, XXXVI, 1980, pp. 373-398.

Evans G.R., '"Inopes verborum sunt latini". Technical Language and Technical Terms in the Writings of St. Anselm and Commentators of the Mid-twelfth Century', *Archives d'histoire doctrinale et litteraire du Moyen Age*, XLIII, 1976, pp. 113-134.

Faes de Mottoni B., 'La dottrina dell' "anima mundi" in Guglielmo d'Alvernia, nella *Summa halensis*, in Alberto Magno', *Studi medievali*, XXII, 1981, pp. 283-297.

Fauser W., *Die Werke des Albertus Magnus in ihrer handschriftlichen Überlieferung. I: Die echten Werke. Codices manuscripti operum Alberti*, Münster, Aschendorff, 1982 (= Alberti Magni Opera. Tomus subsidiarius I, pars prima).

Federici Vescovini G., 'Albumasar in Sadan e Pietro d'Abano', in *La diffusione delle scienze islamiche nel Medio Evo europeo*, Roma, Accademia Nazionale dei Lincei, 1987, pp. 29-55.

Federici Vescovini G., *'Arti' e filosofia nel sec. XIV. Studi sulla tradizione aristotelica e i moderni*, Firenze, Nuove Edizioni E. Vallecchi, 1983.

Federici Vescovini G., *Astrologia e scienza. La crisi dell'aristotelismo sul cadere del Trecento e Biagio Pelacani da Parma*, Firenze, Nuove Edizioni E. Vallecchi, 1979.

Federici Vescovini G., 'La teoria delle immagini di Pietro d'Abano', in *Die Kunst und das Studium der Natur von 14. bis 16. Jahrhundert*, hg. v. W. Prinz und A. Beyer, Weinheim,

VHC Chemie Verlag, 1987 (= Acta humaniora, 1987), pp. 213-236.

Federici Vescovini G., 'Pietro d'Abano e l'astrologia-astronomia', *Centro Italiano di Storia dello Spazio e del Tempo. Bollettino*, n.5, no date, pp. 9-28.

Federici Vescovini G., 'Pietro d'Abano e le fonti astronomiche greco-arabo-latine (a proposito del *Lucidator dubitabilium astronomiae*)', *Medioevo*, XI, 1985, pp. 65-95.

Federici Vescovini G., 'Peter of Abano and Astrology', in *Astrology, Science and Society. Historical Essays*, ed. by P. Curry, Woodbridge/Suffolk, Boydell, 1987, pp.19-39.

Federici Vescovini G., 'Su alcune testimonianze dell'influenza di Alberto Magno come 'metafisico', 'scienziato' e 'astrologo' nella filosofia padovana del cadere del secolo XIV. Angelo di Fossombrone e Biagio Pelacani da Parma', in *Albert der Grosse. Seine Zeit, sein Werk, seine Wirkung*, hg. v. A. Zimmermann, Berlin-New York , W. de Gruyter, 1981, pp. 155-176.

Federici Vescovini G., 'Un trattato di misura dei moti celesti, il *De motu octavae spherae* di Pietro d'Abano', in *Mensura. Mass, Zahl, Zahlensymbolik*, Berlin-New York, W. de Gruyter, 1984 (= Miscellanea mediaevalia, 16/2), pp. 277-293.

Ferckel C., 'Die *Secreta mulierum* und ihr Verfasser', *Sudhoffs Archiv*, XXXVIII, 1954, pp. 267-274.

Field J. V., 'Astrology in Kepler's Cosmology', *Astrology, Science and Society. Historical Essays*, ed. by P. Curry, Woodbridge/Suffolk, Boydell, 1987, pp. 143-170.

Filthaut E., 'Um die *Quaestiones de animalibus* Alberts des Grossen', in *Studia albertina*, hg. v. H. Ostlender, Münster 1952, pp. 112-127.

Fioravanti G., 'La "scientia sompnialis" di Boezio di Dacia', in Accademia delle Scienze di Torino, *Atti della Classe di Scienze Morali*, CI, 1966-67, pp. 329-369.

Fisher N.W. and Unguru S., 'Experimental Science and Mathematics in Roger Bacon's Thought', *Traditio*, XXVII, 1971, pp. 353-378.

Flasch K., *Aufklärung im Mittelalter? Die Verurteilung von 1277*, Das Dokument des Bischofs von Paris übersetzt und erklärt, Mainz , DVB, 1989.

Flasch K., 'Von Dietrich zu Albert', *Freiburger Zeitschrift für Philosophie und Theologie*, XXXII, 1985, pp. 7-26.

Fleming, 'R. Bacone e la Scolastica', *Rivista di filosofia neoscolastica*, VI, 191, p. 541.

Frank I.W., 'Albertus Magnus, der Wissenschaftler und Dominikaner', in Entrich M. ed., *Albertus Magnus. Sein Leben und seine Bedeutung*, Graz–Wien–Köln n.d.[1980], pp. 61-81.

Frank I.W., 'Zum Albertus-Autograph in der Österreichischen Nationalbibliothek und zum 'Albertinismus' der Wiener Dominikaner im Spätmittelalter', *Albertus Magnus, Doctor Universalis. 1280-1980*, hg. v. G. Meyer und A. Zimmermann, Mainz 1980, pp.89-117.

Frankowska-Terlecka M., 'Scientia as conceived by Roger Bacon', *Organon* (Warszawa), VI, 1969, pp. 213-231.

Fries A., 'Hat Albertus Magnus in Paris studiert?', *Philosophie und Theologie*, LIX, 1984, pp. 415-429.

Fries A. and Illing K., *s.v.* 'Albertus Magnus', in *Deutsches Literatur des Mittelalters. Verfasserlexicon*, I, Berlin, 1977, cols. 124-139.

Garin E., *L'età nuova*, Napoli, Morano, 1969.

Garin E., *Medioevo e Rinascimento*, Bari, Laterza, 1954.

Garin E., *Lo zodiaco della vita*, Bari, Laterza, 1966.

R. A. Gauthier, 'Arnoul de Provence et la doctrine de la *phronesis'*, *Revue du Moyen Age latin*, XIX, 1963, pp. 129-170.

George N.F., 'Albertus Magnus and Chemical Technology in a Time of Transition', in *Albertus Magnus and the Sciences*, ed. J.A. Weisheipl, Toronto 1980, pp. 235-261.

Geyer B., 'Albertus Magnus und die Entwicklung der scholastischen Metaphysik', in *Die Metaphysik im Mittelalter*, Berlin. W. de Gruyter, 1963 (= Miscellanea Mediaevalia, II), pp. 3-13.

Geyer B., 'Das Speculum astronomiae kein Werk des Albertus Magnus', *Münchener Theologische Zeitschrift*, IV, 1953, pp. 95-101 (see the same in *Studien zur historischen Theologie. Festgabe für F. X. Seppelt*, hg. v. W. Düring u. B. Panzram, München 1953, pp. 95-101).

Geyer B., 'Der alte Katalog der Werke des hl. Albertus Magnus', in *Miscellanea G. Mercati*, Città del Vaticano 1946, II, pp. 398-413.

Geyer B., 'Die handschriftliche Verbreitung der Werke Alberts des Grossen als Masstab seines Einflusses', in *Studia mediaevalia in hon. R .J. Martin*, Bruges 1949, pp. 221-228.

Geyer B., 'Die mathematischen Schriften des Albertus Magnus', *Angelicum*, XXXV, 1958, pp. 159-175.

Geyer B., 'Die Übersetzungen der Aristotelischen Metaphysik bei Albertus Magnus und Thomas von Aquin', *Philosophisches Jahrbuch*, XXX, 1917, pp. 392-402.

Geyer B., 'Die ursprüngliche Form der Schrift Albert des Grossen *De animalibus* nach dem kölner Autograph', in *Aus dem Geisteswelt des Mittelalters. Studien und Texte M. Grabmann gewidmet*, hg. v. A. Lang, J. Lechner und M. Schmaus, Münster 1935, (Beiträge zur Geschichte der Philosophie und Theologie des Mittelaters, Supplementband, III/2), pp. 578-591.

Geyer B., 'Zur Datierung des Aristotelesparaphrases des hl. Alberts des Grossen', *Zeitschrift für katholische Theologie*, LVI, 1932, pp. 423-436.

Geyer B., 'Die von Albertus Magnus in *De anima* benutzte Aristotelesübersetzung und die Datierung dieser Schrift', *Recherches de théologie ancienne et médiévale*, XXII, 1955, pp. 322-326.

Gilbert A.H., 'Notes on the Influence of the *Secretum Secretorum*', *Speculum*, III, 1928, pp. 84-98.

Gilson E., 'Albert le Grand à l'Université de Paris', *La vie interieure*, 1933 (Mai), pp. 9-28.

Gilson E., 'L'âme raisonnable chez Albert le Grand', *Archives d'histoire doctrinale et littéraire du Moyen Age*, XVIII, 1943-1945, pp. 1-72.

Gilson E., *La philosophie au Moyen Age*, Paris, Payot, 1952 (2nd ed.).

Gilson, E., *La philosophie de Saint Bonaventure*, Paris, Vrin, 1953.

Glorieux P., 'Bibliothèques des maîtres parisiens. Gérard d'Abbeville', *Recherches de théologie ancienne et médiévale*, XXXVI, 1969, pp. 148-183.

Glorieux P., 'Etude sur la *Biblionomia* de Richard de Fournival', *Recherches de théologie ancienne et médiévale*, XXX, 1963, pp. 205-231.

P. Glorieux, *La Faculté des Arts et ses maîtres au XIIIe siècle*, Paris 1917.

P. Glorieux,*Littérature quodlibétique*, Paris 1935.

Glorieux P., 'Les polemiques "contra geraldinos"', *Recherches de théologie ancienne et médievale*, VI, 1934, pp. 5-41; VII, 1935, pp. 129-155; IX, 1937, pp. 61-65.

Glorieux P., *Répertoire des maîtres en théologie au XIIIe siècle*, Paris 1933.

Goergen J., *Des hl. Albertus Magnus Lehre von der göttlichen Vorsehung und dem Fatum*, Vechta i. Oldenburg 1932.

Goergen J., 'Untersuchungen und Erlauterungen zu den *Quaestions de fato, de divinatione, de sortibus* des Mag. Alexander', *Franziskanische Studien*, 19, 1932, pp. 13-38.

Goldstein B.R.- Pingree, D., 'Levi ben Gerson's Prognostication for the Conjunction of 1345', *Transactions of the American Philosophical Society*, 80/6, 1990, pp. 1-60.

Golubovich P.G., *Biblioteca bio-bibliografica della Terra Santa e dell'Oriente francescano*, I, Quaracchi, 1906.

Gomez E., O.P., 'San Alberto Magno y sus obras en la Universidad de Oxford', *Divus Thomas* (Piacenza) XXXV, S. III, IX, 1932, pp. 633-643.

Gorce M.M., 'La lutte "contra Gentiles" à Paris au XIIIe siècle', *Mélanges Mandonnet*, Paris, 1930, I, pp. 223-243.

Grabmann M., 'Aegidius von Lessines', in his *Mittelalterliches Geistesleben*, München II, 1936, pp. 512-530.

Grabmann M., 'Albertus Magnus Theolog, Philosoph und Naturforscher', *Philosophisches Jahrbuch des Görresgesellschaft*, LXI, 1951, pp. 473-480.

Grabmann M., 'Andreas Capellanus und Bischof Stephan Tempier', *Speculum*, VII, 1932, pp. 75-79.

Grabmann M., 'Aristoteles im Werturteil des Mittelalters', in his *Mittelalterliches Geistesleben*, II, München 1936, pp. 63-102.

Grabmann M., 'Bernard von Trilia und seine 'Quaestiones de cognitione animae coniunctae [...] separatae", *Divus Thomas* (Freiburg), XIII, 1935, pp. 385-399.

Grabmann M., 'Das Werk *De amore* des Andreas Cappellanus und das Verurteilungsdekret des Bischofs Stephan Tempier von Paris von 7. März 1277', *Speculum*, VII, 1932, pp. 75-79.

Grabmann M., 'Der Einfluss Alberts des Grossen auf das mittelalterliche Geistesleben', in his *Mittelalterliches Geistesleben*, II, München, 1936, pp. 324-412.

Grabmann M., 'Die *Summa de astris* des Gerards von Feltre', in his *Mittelaterliches Geisetsleben*, II, München 1937, pp. 254-279.

Grabmann M., *Die Werke des hl. Thomas von Aquin*, Münster 1920; 1931 2nd ed.; 1949 3rd ed. (= Beiträge zur Geschichte der Philosophie und Theologie des Mittelalters, 22, 1-2).

Grabmann M., 'Einzelgestalten aus der Dominikaner- und Thomistenschule' [I. Aegidius de Lessines], in his *Mittelalterliches Geistesleben*, II, München 1937, pp.512-529.

Grabmann M., *Forschungen über die lateinischen Aristotelesübersetzungen des XIII. Jahrhunderts*, Münster 1916 (= Beiträge zur Geschichte der Philosophie und Theologie des Mittelalters, 17/5-6).

Grabmann M., *Gesammelte Akademieabhandlungen*, I-II, Paderborn, Schöning, 1979 (= Veröffentlichungen des Grabmann-Instituts der Universität-München, N.F., 25).

Grabmann M., *Guglielmo di Moerbeke O.P., il traduttore delle opere di Aristotele*, Roma, ad S. Sabinae, 1946.

Grabmann M., 'Zur philosophischen und naturwissenschaftlichen Methode in den Aristoteleskommentaren Alberts des Grossen', *Angelicum*, XXI, 1944 (= *Serta albertina*),pp. 50-64.

Grant E., 'The condemnation of 1277, God's absolute power, and physical thought in the late Middle Ages', *Viator*, 10, 1979. pp. 211-244.

Grant E., 'The condemnation of 1277', *Cambridge History of Later Mediaeval Philosophy*, Cambridge 1982, pp. 537-539.

Grant E., 'Mediaeval and Renaissance scholastic Conceptions of the Influence of the Celestial Region on the Terrestrial', *Journal of Mediaeval and Renaissance Studies*, XVII, 1987, pp. 1-23.

Gregory T., 'Discussioni sulla "doppia verità"', *Cultura e scuola*, I, 1962, pp. 99-106.

Gregory T., 'Filosofia e teologia nella crisi del XIII secolo', *Belfagor*, XIX, 1964, pp. 1-16.

Gregory T., 'Forme di conoscenza e ideali di sapere nella cultura medievale', *Giornale critico della filosofia italiana*, LXVII (LXIX), 1988, pp. 1-62.

Gregory T., 'La Filosofia medievale: i secoli XIII-XIV', in *Storia della filosofia*, ed. by M. Dal Pra, Milano 1976, VI, pp. 1-232.

Gregory T., 'I sogni e gli astri', in *I sogni nel Medioevo. Seminario del Lessico Intellettuale europeo*, Roma, Edizioni dell'Ateneo, 1985, pp. 111-148.

Gregory T., 'La nouvelle idée de nature et de savoir scientifique au XIIe siècle', in*The Cultural Context of Medieval Learning*, eds. J. E. Murdoch and E. D. Sylla, Dordrecht, Reidel, 1975, pp. 193-218.

Gregory T., 'L'idea di natura nella filosofia medievale prima dell'ingresso della fisica di Aristotele', in *La filosofia della natura nel Medioevo. Atti del III Congresso Internazionale di Filosofia Medievale* [1964], Milano 1966, pp. 27-65.

Gregory T., 'Temps astrologique et temps chrétien', in *Le temps chrétien de la fin de l'Antiquité au Moyen Age*, Paris 1984 (= Colloque international du CNRS, 604), pp. 557-573.

Gregory T., 'The Neoplatonic Inheritance', in *A History of Twelfth-Century Western Philosophy*, ed. P. Dronke, Cambridge, Cambridge U.P., 1988, pp. 54-80.

Grignaschi M., 'La diffusion du *Secretum secretorum (Sirr-al 'Asrar)* dans l'Europe occidentale', *Archives d'histoire doctrinal et littéraire du Moyen Age*, LV, 1980 (but 1981), pp. 7-69.

Guillaume de Moerbeke. Recueil d'études à l'occasion du 700ème anniversaire de sa morte (1286), ed. par Brams J. et Vanhamel W., Leuven, University Press, 1989.

Hadrianus a Krizovlian, 'Controversia doctrinalia inter magistros franciscanos et Sugerium', *Collectanae franciscana*, XXVII, 1957, pp. 127-131.

Hackett J.M.G., 'The Attitude of Roger Bacon to the Scientia of Albertus Magnus', in *Albertus Magnus and the Sciences*, ed. J. A. Weisheipl, Toronto, 1980, pp. 53-72.

Hakkett J.M.G., *The Meaning of Experimental Science ("Scientia experimentalis") in the Philosophy of Roger Bacon*, unpublished Ph. D. Thesis, Toronto 1983.

Harmening D., *Superstitio. Überlieferungs- und theoriegeschichtliche Untersuchungen zur kirchlich-theologischen Aberglaubenliteratur des Mittelaters*, Berlin, E.Schmidt, 1979.

Haskins Ch. H., *Studies in the History of Mediaeval Science*, Cambridge, Mass. 1924.

Haubst R., 'Die Fortleben Alberts des Grossen bei Heymerich von Kamp und Nikolaus von Kues', in *Studia albertina*, ed. H. Ostlender, Münster 1952, pp. 420-447.

Heintke F., *Humbert von Romans*, Berlin 1933 (= Historische Studien, H. 222); 2nd ed. Berlin 1965 .

Hertling G. von, *Albertus Magnus. Beiträge zur seiner Würdigung*, Münster 1914; 2nd ed. (= Beiträge zur Geschichte der Philosophie und Theologie des Mittelalters, 14, 5-6).

Hillgarth J. N., *Ramon Lull and Lullism in fourteenth century France*, Oxford, At the Clarendon Press, 1971.

Hissette R., 'Albert le Grand et l'expression *diluvium ignis*', *Bulletin de philosophie médiévale*, XXII,1980, pp.78-81.

Hissette R., 'Albert le Grand et Thomas d'Aquin dans la censure parisienne du 7 Mars 1277', *Studien zur mittelaterlichen Geistesgeschichte und ihre Quellen*, Berlin-New York, W. de Gruyter, 1982 (= Miscellanea mediaevalia, XV), pp. 226-246.

Hissette R., *Enquête sur les 219 articles condamnés à Paris le 7 mars 1277*, Louvain, Publ. Universitaires-Paris, Vander Oyez, 1977.

Hissette R., 'Etienne Tempier et ses condamnations', *Recherches de théologie ancienne et médievale*, 48, 1980, pp. 231-270.

Hissette P.,'Notes sur la réaction antimoderniste d'Etienne Tempier', *Bulletin de philosophie médievale*, XXII, 1980, pp. 88-97.

Hissette G., review of *Speculum Astronomiae*, eds. S. Caroti, M. Pereira, S. Zamponi, P. Zambelli, Pisa 1977, *Bulletin de théologie ancienne et médievale*, XII, 1979, pp.484-485.

Histoire Littéraire de la France, XXI, Paris 1842.

Hofmann J.E., 'Über eine Euklid-Bearbeitung die dem Albertus Magnus zugeschrieben wird', *Proceedings of the International Congress of Mathematicians* [1958], ed. J.A. Todd, Cambridge, Cambridge U.P., 1960, pp. 554-566.

Hossfeld P., *Albertus Magnus als Naturphilosoph und Naturwissenschaftler*, Bonn, Albertus Magnus Institut, 1983.

Hossfeld P., 'Albertus Magnus über die Ewigkeit aus philosophischer Sicht', *Archivum Fratrum Praedicatorum*, LVI, 1986, pp. 31-48.

Hossfeld P.,'Albertus Magnus über die Natur des geographischen Orts', *Zeitschrift für Religions- und Geistesgeschichte*, XXX, 1978, pp. 107-115.

Hossfeld P., '"Allgemeine und umfassende Natur" nach Albertus Magnus', *Philosophia naturalis*, XVIII, 1980, pp. 479-492.

Hossfeld P., 'Der Gebrauch der Aristotelischen Übersetzung in den *Meteora* des Albertus Magnus', *Medieval Studies*, 42, 1980, pp. 395-406.

Hossfeld P., 'Die eigenen Beobachtungen des Albertus Magnus', *Archivum Fratrum Praedicatorum*, LII, 1983, pp. 147-174.

Hossfeld P., 'Die Lehre des Albertus Magnus von den Kometen', *Angelicum*, LVII, 1980, pp. 533-541.

Hossfeld P., 'Die Lehre des Albertus Magnus über die Milchstrasse', *Philosophia naturalis*, XX, 1983, pp. 108-111.

Hossfeld P., 'Die naturwissenschaftlich-naturphilosophische Himmelslehre Alberts des Grossen (nach seinem Werk *De caelo et mundo*)', *Philosophia naturalis*, XI, 1969, pp. 318-359.

Hossfeld P., 'Die Physik des Albertus Magnus (Teil I, die Bücher 1-4)', *Archivum Fratrum Praedicatorum*, LV, 1985, pp. 49-65.

Hossfeld P., 'Die Ursachen der Eigentümlichkeiten der Elemente nach Albertus Magnus', *Philosophia naturalis*, XIV, 1973, pp. 197-209.

Hossfeld P., '"Erste Materie" oder "Materie im allgemeinen" in den Werken des Albertus Magnus', in *Albertus Magnus, Doctor Universalis. 1280-1980*, hg. v. G. Meyer und A. Zimmermann, Mainz 1980, pp. 205-234.

Hossfeld P., 'Grundgedanken in Alberts des Grossen Schrift 'Ueber Entstehung und Vergehen"', *Philosophia naturalis*, XVI, 1977, pp. 191-204.

Hossfeld P., 'Über die Bewegung- und Veränderungsarten bei Albertus Magnus', in *Die Kölner Universität im Mittelalter*, hg. v. A. Zimmermann, Berlin-New York, W. de Gruyter, 1989 (= Miscellanea mediaevalia, XX), pp. 128-143.

Hossfeld P., 'Zum Euklidkommentar des Albertus Magnus', *Archivum Fratrum Praedicatorum*, LII, 1982, pp. 115-133.

Huber M., 'Bibliographie zu Roger Bacon', *Franziskanische Studien*, LXV, 1983, pp. 98-102.

Huber Legnani M., *Roger Bacon Lehre der Anschaulichkeit*, Freiburg i.B., Hochschuleverlag, 1984.

Hugonnard-Roche H., *L'oeuvre astronomique de Themon Juif*, Genève, Droz, 1973.

Janssens E., 'Les premiers historiens de la vie de Saint Thomas', *Revue néoscolastique de philosophie*, XXVI, 1924, pp.201-214, 452–476.

Jehl R., *Melancholie und Acedia. Ein Beitrag zu Anthropologie und Ethik Bonaventuras*, Paderborn, F.Schöning, 1984.

"Imago mundi":la conoscenza scientifica nel pensiero bassomedievale, Todi, Accademia tudertina, 1983 (= Convegni del Centro di Studi sulla spiritualità medievale. XXI)

Jourdain Ch., 'N. Oresme et les astrologues', *Revue des questions historiques*, XVIII, 1875, pp. 137-159; reprinted in his *Excursions historiques et philosophiques à travers le Moyen Age*, Paris 1888, pp. 559-586.

Käppeli Th., 'Fra Baxiano von Lodi Adressat der Responsio ad Lectorem Venetum des hl. Thomas', *Archivum Fratrum Praedicatorum*, XIII, 1943, pp. 181-182.

Käppeli Th., *Scriptores Ordinis Preadicatorum Medii Aevi*, I-III, Roma, ad S.Sabinae, 1970-1980.

Kaiser R., 'Die Benutzung proklischer Schriften durch Albert den Grossen', *Archiv für Geschichte der Philosophie*, XLV, 1963, pp.1-22.

Kaiser R., 'Versuch einer Datierung der Schrift Alberts des Grossen *De causis et processu universitatis, Archiv für Geschichte der Philosophie*', XLV, 1963, pp. 125-136.

Kaiser R., 'Zur Frage der eigenen Anschauung Alberts des Grosses in seinen philosophischen Kommentaren. Ein grundsätzliche Betrachtung', *Freiburger Zeitschrift für Philosophie und Theologie*, IX, 1962, pp. 53-62.

Kennedy E.S. and Pingree D., *The Astrological History of Masha'allah*, Cambridge Mass., Harvard U.P.,1971.

Kibre P., 'Albertus Magnus on Alchemy', in *Albertus Magnus and the Sciences*, ed. J.A. Weisheipl, Toronto 1980, pp. 187-201.

Kibre P., 'The *Alkimia minor* ascribed to Albertus Magnus', *Isis*, XXXI, 1940, pp. 267-300; XXXIX, 1949, pp. 267-306.

Kibre P., 'Alchemical Writings Attributed to Albertus Magnus', *Speculum*, XVII, 1942, pp. 499-519.

Kibre P., 'Further Manuscripts containing Alchemical Tracts Attributed to Albertus Magnus', *Speculum*, XXXIV, 1959, pp. 238-247.

Kibre P., 'An alchemical Tract ascribed to Albertus Magnus', *Speculum*, XXXV, 1944, pp. 303-316.

Kibre P., 'The *De occultis naturae* attributed to Albertus Magnus', *Osiris*, XI, 1954, pp. 23-39.

Kibre P., 'Albertus Magnus and Alchemy', in *Albertus Magnus and the Sciences*, ed. J. A. Weisheipl, Toronto, 1980, pp.187-202.

Kieckhefer R., *Magic in the Middle Ages*, Cambridge U. P. 1989 (= Cambridge Medieval Textbooks).

Klauck K., 'Albertus Magnus und die Erdkunde', in *Studia albertina*, hg. v. H. Ostlender, Münster 1952, pp. 234-248.

Klibansky R., *The Continuity of the Platonic Tradition during the Middle Ages*, London 1950.

Koch J., 'Philosophische und theologische Irrtumslisten von 1270-1329', in his *Kleine Schriften*, II, Roma, 1973, pp. 423-450.

Kovach F.J., 'The enduring Question of Action at a Distance in Saint Albert', *Albert The Great. Commemorative Essays*, F.J. Kovach and R. W. Shanan eds., Norman, Okl., The University of Oklaoma Press, 1980, pp. 169-235.

Kristeller P.O., 'The School of Salerno: Its Development and Its Contribution to the History of Learning', in his *Studies in Renaissance Thought and Letters*, Roma, Edizioni di Storia e Letteratura, 1956, I, pp. 495-551.

Kübel W., 'Die Übersetzungen der Aristotelischen Metaphysik in den Frühwerken Alberts des Grossen', *Divus Thomas* (Freiburg), XI, 1933, pp. 241-268.

Kübel W., *s.v.* 'Albertus Magnus', in *Lexicon für Theologie und Kirche*, I, Freiburg i.B., Herder, 1975, 2nd ed., cols. 285-287.

P. Künzle, *s. v.* 'Bernardo de Trilia', *Enciclopedia filosofica*, I, Firenze 1967, 2nd ed., cols. 873-874.

P. Künzle , 'Notes sur les questions disputées *De spiritualibus creaturis* et *De potentia Dei* de Bernard de Trilia,O.P.', *Bulletin de philosophie médiévale*, VI, 1964, pp. 87-90.

Kunitzsch P., *Der Almagest. Die Syntaxis Mathematica des Claudius Ptolemaus in arabisch-lateinischer Übersetzung*, Wiesbaden, O.Harrasowitz, 1974.

Kusche B., 'Zur *Secreta mulierum* Forschung', *Janus*, LXII, 1975, pp. 103-123.

Lafleur C., *Quatre introductions à la philosophie au XIIIe siècle. Textes critiques et étude historique.*, Montréal Institut d'études mèdiévales-Paris, Vrin, 1988.

Langlois Ch. V., 'Siger de Brabant', in *Revue de Paris*, VII, 1900, pp. 60-96.

Lemay R., *Abu ma'shar and Latin Aristotelianism in the Twelfth Century. The Recovery of Aristotle's Natural Philosophy through Arabic Astrology*, Beirut, The Catholic Press, 1962 (= American University. Publications of the Faculty of Arts and Sciences. Oriental Series N. 38).

Lemay R., 'Dans l'Espagne du XIIe siècle: les traductions de l'arabe au latin', *Annales E.S.C.*, XVIII, 1963, pp. 639-665.

Lemay R., 'De la scolastique à l'histoire par le truchenment de la philologie', in *La diffusione delle scienze islamiche nel Medio Evo Europeo*, Roma, Accademia Nazionale dei Lincei, 1987, pp. 399-535.

Lemay R., 'Le Nemrod de l'*Enfer* de Dante', *Studi danteschi*, XL, 1963, pp. 57-128.

Lemay R., '*Libri naturales* et sciences de la nature dans la scolastique latine du XIIIe siècle', *XIVth Intern. Congress of the History of Science*, Tokyo, 1974, Off-print from the unpublished *Proceedings*.

Lemay R., 'Mythologie païenne éclairant la mythologie chrétienne chez Dante: le cas des Géants', in *Revue des études italiennes*, XI,1965 (= Special Issue under the title *Dante et les mythes*), pp.237-279.

Lemay R., 'The Teaching of Astronomy in mediaeval Universities, principally at Paris in the 14th Century', *Manuscripta*, XX, 1976, pp. 197-217.

Liebermann M., 'Chronologia gersoniana', *Romania*, LXX, 1948, pp. 51-67; LXXIII, 1952, pp. 480-498; LXXIV, 1953, p. 289-337; LXXVI, 1955, pp. 289-333.

Lindberg D.C., 'On the Applicability of Mathematics to Nature: R. Bacon and his Predecessors', *British Journal for the History of Science*, XV, 1982, pp. 3-25.

Lindberg D.C., 'Roger Bacon's Theory of the Rainbow: Progress or Regress?', *Isis*, LVII, 1966, pp. 235-242.

Livesey S.J. and Rouse R.R., 'Nimrod the Astronomer', *Traditio*, 37 (1981), pp. 203-266.

Litt Th., *Les corps célestes dans l'univers de saint Thomas d'Aquin*, Louvain-Paris, Nauwelaerts, 1963.

Löffler K., *Kölnische Bibliotheksgeschichte im Umriss*, Köln 1923.

Lottin O., 'L'influence littéraire du chancellier Philippe sur les théologiens préthomistes', *Recherches de théologie ancienne et médievale*, II, 1930, pp. 311-326.
Lottin O., 'Problèmes concernantes la *Summa de creaturis* et le Commentaire de Sentences de St. Albert le Grand', *Recherches de théologie ancienne et médiévale*, XVII, 1950, pp. 319-328.
Mabille M., 'Pierre de Limoges copiste de manuscripts', *Scriptorium*, XXIV, 1970, pp.45-47.
Mabille M., 'Pierre de Limoges et ses méthodes de travail', in *Hommage à A. Boutemy*, ed. G.Cambier, Bruxelles, Latomus, 1976, pp. 244-251.
Mc Vaugh M. and Behrends F., 'Fulbert of Chartres' Notes on Arabic Astronomy', *Manuscripta*, XV, 1971, pp. 172-177.
Madkour I., 'Astrologie en terre d'Islam', in *Arts libéraux et philosophie. IVe Congrès international de Philosophie médievale*, Montréal-Paris, 1969.
Mahoney E.P., 'Albert the Great and the 'Studio Patavino' in the Late Fifteenth and Early Sixteenth Centuries', in *Albertus Magnus and the Sciences*, ed. J.A. Weisheipl, Toronto 1980, pp. 537-563.
Mandonnet P., 'Polémique averroiste de Siger de Brabant', *Revue thomiste*, V, 1897, pp. 95-110.
Mandonnet P., 'Roger Bacon et la composition des trois *Opus*', *Revue néoscolastique de philosophie*, XX, 1913, pp. 52-68, 164-180.
Mandonnet P., 'Roger Bacon et le *Speculum astronomiae* (1277)', *Revue néoscolastique de philosophie*, XVII, 1910, pp. 313-335.
Mandonnet P., *Siger de Brabant et l'Averroisme latin au XIIIe siècle*, I-II, Louvain 1911 (2nd ed.).
Mandonnet P., 'Gilles de Lessines et son tractatus *De crepuscolis*', *Revue néoscolastique de philosophie*, 22, 1920. pp. 190-194.
Mandonnet P., review of Destrez, 'La lettre' and Chenu, 'Les Réponses de S. Thomas et de Kilwardby', *Bulletin thomiste*, III, 1930, pp. 129-139.
Mandonnet P.,*s.v.* 'Albert-le-Grand', in *Dictionnaire de théologie catholique*, I, Paris 1930, pp. 666-674.
Mansion A., 'Sur le texte de la version latine médievale de la *Métaphysique* et de la *Physique* d'Aristote dans les éditions des commentaires de s. Thomas', *Revue néoscolastique*, XXXIV, 1932, pp. 65-69.
Meersseman G., 'In libris Gentilium non studeant', *Italia medievale e umanistica*, I, 1958, pp. 1-13.
Meersseman G., *Introductio in Opera omnia Alberti Magni*, Bruges 1931.
Meylan H., *Les 'Quaestiones' de Philippe le chancellier*, Paris 1927.
Meyer P., 'Traités en vers provençaux sur l'astrologie et la géomancie', *Romania*, XXVI, 1897, pp. 225-275.
Miethke J., 'Papst, Ortsbischof und Universität in der Pariser Thologenprozess', in *Die Auseinandersetzung an der Pariser Universität im XIII Jahrhundert*, Berlin-New York, W. de Gruyter, 1976 (= Miscellanea Mediaevalia,10), pp. 52-94.
1274. Année charnière. Mutations et continuité, Paris, SEVPEN, 1977 (= Colloque international du CNRS, n. 558).
Millás Vallicrosa J.M., *Estudios sobre Azarquiel*, Madrid-Granada 1950.
Millás Vallicrosa J.M., *Estudios sobre historia de la ciencia española*, Barcelona, Consejo Superior de Investigaciones Cientificas, 1949.

Millás Vallicrosa J.M., 'Las mas antiguas traducciones arabes hechas en España', in *Oriente ed Occidente nel Medioevo.* (Convegno dell'Accademia Nazionale dei Lincei), Roma, Acc. Naz. d. Lincei, 1971, pp. 383-389.

Millás Vallicrosa J.M., *Las traducciones orientales en los manuscritos de la Biblioteca catedral de Toledo,* Madrid, Consejo Superior de Investigaciones Cientificas, 1942.

Millás Vallicrosa J.M., 'Una obra astronomica disconocida de Johannes Avendaut Hispanus', *Osiris,* I, 1936, pp. 451-476.

Minio-Paluello L., 'Note sull'Aristotele latino medievale: I. La *Metaphysica vetustissima* comprendeva tutta la metafisica?', *Rivista di filosofia neoscolastica,* XLII, 1950, pp. 222-237.

Minio Paluello L., *s.v.* 'William of Moerbeke', *Dictionnary of Scientific Biography,* IX, New York 1974, pp. 434-440.

Minio Paluello L., *Opuscula. The Latin Aristotle,* Amsterdam, Hakkert,1972,

Mojsisch B., 'Grundlinien der Philosophie Alberts des Grossen', *Freiburger Zeitschrift für Philosophie und Theologie,* XXXII, 1985 (= *Albert der Grosse und die deutsche Dominikanerschule*), pp. 28-44.

Molland A. G., 'Mathematics in the Thought of Albertus Magnus', in *Albertus Magnus and the Sciences,* ed. J.A. Weisheipl, Toronto 1980, pp. 463-478.

Molland A. G., 'Mediaeval Ideas on scientific Progress', *Journal of the History of Ideas,* XXIX, 1978, pp.561-577.

Molland A. G., 'Roger Bacon as a Magician', *Traditio,* XXX, 1974, pp. 445-460.

Morpurgo P., *Il 'Liber Introductorius' di Michele Scoto,* in *Accademia Nazionale dei Lincei. Rendiconti della Classe di Scienze Morali-storiche e filologiche,* VII, 34, Roma 1979, pp. 149-161.

Mothon J.P., *Vita del b. Giovanni da Vercelli, sesto Maestro Generale dell'O.P.,* Vercelli 1903.

Nallino C.A., 'Astrologia e astronomia presso i Musulmani', 'Storia dell'astronomia presso gli Arabi nel Medio Evo', in his *Raccolta di scritti editi e inediti,* ed. by M. Nallino, V, Roma, Istituto per l'Oriente, 1944, pp.1-86, 87-329.

Nardi B., *Nel mondo di Dante,* Roma, 1944,

Nardi B., *Saggi sull'aristotelismo padovano dal secolo XIV al XVI,* Firenze, Sansoni, 1958.

Nardi B., *Studi di filosofia medievale,* Roma, Edizioni di Storia e Letteratura, 1960.

Nardi B., 'Discussioni dantesche: II. intorno al Nemrot dantesco e ad alcune opinioni di R. Lemay', *L'Alighieri. Rassegna di bibliografia dantesca,* VI, 1965, pp. 42-55.

Neufeld H., Zum Problem des Verhaltnisses der Theologischen Summe Alberts des Grossen zur Theologischen Summe Alexanders von Hales, *Franziskanische Studien,* 27, 1940, pp. 22-56, 65-87.

North J. D., 'Astrology and the Fortunes of Churches', *Centaurus,* XXIV, 1980, pp. 181-211.

North J.D., *Horoscopes and History,* London, The Warburg Institute, 1986.

North J.D., 'Celestial Influence. the Major Premiss of Astrology', in *'Astrologi hallucinati'. Stars and the End of the World in Luther's Time,* ed. P. Zambelli, Berlin-New York, W.de Gruyter, 1986, pp. 45-100.

North J.D., 'Mediaeval Conceptions of celestial Influence', in Curry P., *Astrology Science and Society,* Woodbridge/ Suffolk, Boydell, 1987, pp.5-18.

Nowotny K.A., 'Einleitung', in H.C. Agrippa, *De occulta philosophia,* hg. u. erlautert v. K. A. Nowotny, Graz 1967, pp. 387-466.

Nykl A.R., 'Dante, *Inferno* XXXI/67', in *Estudios dedicatos a Menéndez Pidal,* Madrid 1952, III, pp. 321-24.

Olivieri L., *Pietro d'Abano e il pensiero neolatino. Filosofia, scienza e ricerca dell'Aristotele greco tra i secoli XIII e XIV*, Padova, Antenore,1988.

Omez R., 'St Thomas d'Aquin et l'astrologie', *La Tour St Jacques*, 1956, nr. 4, pp. 36-38.

Orlando T.A., 'Roger Bacon and the 'Testimonia gentilium de secta christiana'', *Recherches de théologie ancienne et médievale*, XLIII, 1976, pp. 202-218.

Ostlender H., 'Die Autographe Alberts des Grossen', in *Studia albertina*, hg. v. H. Ostlender, Münster 1952, pp. 3-21.

Ostlender H., 'Das Kölner Autograph des Matthaeus Kommentars Alberts des Grossen', *Jahrbuch des Kölnischen Geschichtesvereins*, XVII, Köln 1935, pp. 129-142.

Pack R. A. 'Pseudoaristotelian Chyromancy', *Archives d'histoire doctrinale et littéraire du Moyen Age*, XXXVI, 1969, pp. 189-241.

Pack R. A., 'Pseudo-Aristoteles: Chiromantia', *Archives d'histoire doctrinale et littéraire du Moyen Age*, XXXIX, 1972, pp. 289-320.

Palitzsch F., *Roger Bacons zweite Schrift über die kritischen Tage* (Med. Dissertation 1918), Borna-Leipzig 1919.

Pangerl F., 'Studien über Albert den Grossen', *Zeitschrift für katholische Theologie*, XXXVI, 1912, pp. 304-346, 512-549, 784-800.

Paravicini Bagliani A., 'Campano da Novara e il mondo scientifico romano duecentesco', *Novarien*, N.14, 1984, pp. 99-110.

Paravicini Bagliani A., *Cardinali di Curia e 'familiae' cardinalizie*, Padova, Antenore, 1972 (= Italia sacra,18).

Paravicini Bagliani A., 'La scienza araba nella Roma del Duecento', in *La diffusione delle scienze islamiche nel Medio Evo europeo*, Roma, Accademia Nazionale dei Lincei, 1987, pp. 103-166.

Paravicini Bagliani A., 'Medicina e scienze della natura alla corte di Bonifacio VIII: uomini e libri', in *Roma anno 1330. Atti del Congresso Intern. di Storia dell'arte medievale* (Maggio 1980), Roma, L'erma di Breitschneider, 1983, pp. 787-789.

Paravicini Bagliani A., 'Nuovi documenti su Guglielmo da Moerbeke', *Archivum Fratrum Praedicatorum*, VII, 1982.pp. 141-143.

Paravicini Bagliani A., 'Un matematico nella corte papale del secolo XIII: Campano da Novara', *Rivista di storia della chiesa in Italia*, XXVIII, 1973, pp.107-108.

Paré G., *Les idées et les lettres au XIIIe siècle*, Montréal 1947.

Partington J.R., 'Albertus Magnus on Alchemy', *Ambix*, I, 1937, pp. 3-20.

Paschetto E., *Demoni e prodigi. Note su alcuni scritti di Witelo e di Oresme*, Torino, Giappichelli, 1978.

Paschetto E., 'Il *De natura daemonum* di Witelo', *Atti dell'Accademia delle Scienze di Torino*, 109, 1974-1975, pp. 231-271.

Pattin A., 'Un recueil alchimique', *Bulletin de philosophie médiévale*, XIV, 1972, pp. 89-107.

Pedersen O., 'A Fifteenth-Century Glossary of Astronomical Terms', *Classica et Mediaevalia. Dissertationes. IX. F. Blatt dedicata*, Copenhagen 1983, pp.584-594.

Pedersen O., 'The "Corpus astronomicum" and the Traditions of mediaeval latin astronomy', in *Studia copernicana*, XIII, Warszawa 1975, pp.57-97.

Pedersen O., 'The Origins of the *Theorica planetarum*', *Journal of the History of Astronomy*, XII, 1981, pp. 113-123.

Pedersen O., 'The *Theorica Planetarum* Literature of the Middle Ages', *Classica et Mediaevalia*, XXIII, 1962, pp. 225-232.

Pelster F., 'Alberts des Grossen neuaufgefundene Quaestionen zu den aristotelischen Schrift *De animalibus'*, *Zeitschrift für katholische Theologie*, 1922, pp. 332-334.

Pelster F., 'Alberts des Grossen Jugendaufenthalt in Italien', *Historisches Jahrbuch*, XLII, 1922, pp. 102-105.

Pelster F., 'Beiträge zur Aristotelesbenutzung Alberts des Grossen', *Philosophisches Jahrbuch*, XLVI, 1933, pp. 450-463; XLVII, 1934, pp. 55-64.

Pelster F., 'De traditione manuscripta operum S. Alberti Magni', in *Alberto Magno. Atti della settimana albertina*, Roma 1932, pp. 107-126.

Pelster F., 'Die beiden ersten Kapiteln der Erklärung Alberts des Grossen *De animalibus* in ihrer ursprünglichen Fassung', *Scholastik*, X, 1935, pp. 229-240.

Pelster F., 'Die griechisch-lateinische Metaphysikübersetzungen des Mittelalters', in *Festgabe Clemens Beaumker zur 70. Geburtstag*, Münster 1923 (= Beiträge zur Geschichte der Philosophie und Theologie des Mittelalters. Supplementband II), pp.89-118.

Pelster F., 'Die Übersetzungen der aristotelischen Metaphysik bei Albertus Magnus und Thomas von Aquin', *Gregorianum*, XVI, 1935, pp. 325-348, 531-561; XVII, 1936, pp.377-406.

Pelster F., *Kritische Studien zum Leben und zu den Schriften Alberts des Grossen*, München 1918 (cf. 2nd. Ed. Freiburg i.B. 1920; Stimmen der Zeit. Erganzungshefte: II.4).

Pelster F., 'Neue philosophische Schriften Alberts des Grossen', *Philosophisches Jahrbuch der Görresgesellschaft*, XXXVI, 1923, pp. 150-174.

Pelster F., 'Neuere Forschungen über den Aristoteles-Übersetzungen des 12. und 13. Jahrhundertes: eine kritische Übersicht', *Gregorianum*, XXX, 1949, pp. 52-77.

Pelster F., *s.v.* 'Albertus Magnus', in *Lexicon für Theologie und Kirche*, I, Freiburg 1930, cols. 214-217.

Pelster F., 'Um die Datierung von Alberts des Grossen Aristotelesparaphrase', *Philosophisches Jahrbuch des Görresgesellschaft*, XLVIII, 1935, pp. 443-461.

Pelster F.,'Zur Datierung der Aristotelesparaphrase des hl. Alberts des Grossen', *Zeitschrift für katholische Theologie*, LVI, 1932, pp. 423-436.

Pelster F., 'Zur Datierung einiger Schriften Alberts des Grossen', *Zeitschrift für katholische Theologie*, XLVII, 1923, pp. 475-482.

Pereira M., 'Campano da Novara autore dell'*Almagestum Parvum'*, *Studi medievali*, 19 (1978), pp. 769-776.

Pingree D., *s.v.* 'Abu Mashar', *Dictionary of Scientific Biography*, I, New York 1970, pp. 32-39.

Pingree D., 'Astronomy and Astrology in India and Iran', *Isis*, 54, 1963, pp. 229-246.

Pingree D., 'Historical Horoscopes', *Journal of American Oriental Society*, LXXXII, 1962, pp.487-502.

Pingree D. see under Kennedy E.S.

Planzer D., 'Albertus-Magnus-Handschriften in mittelalterlichen Bibliothekskatalogen des deutschen Sprachgebietes', *Divus Thomas* (Freiberg), X, 1932, pp. 246-276.

Pluta O., 'Albert von Köln und Peter von Ailly', *Freiburger Zeitschrift für Philosophie und Theologie*, 32, 1985, pp. 261-271.

Pouillon H., 'Le premier traité des proprietès transcendantes . La *Summa de bono* du Chancellier Philippe', *Revue néoscolastique de philosophie*, XLII, 1939, pp. 40-77.

Ratzinger J., *Die Geschichtstheologie des hl. Bonaventura*, München und Zürich 1959.

Raymond O.N.C., 'Docteurs franciscains et doctrines franciscaines', *Études franciscaines*, XXXI/1, 1914, pp. 94-95.

Rencontres de culture dans la philosophie médiévale. Traductions et traducteurs de l'antiquité tardive au XIVe siècle, J. Hamesse and M. Fattori eds., Louvain-la-Neuve-Cassino 1990 (= Publ. de l'Institut d'Études médiévales. Rencontres de Philosophie médiévales, 1): especially pertinent the papers by Ch. Burnett, J.S. Gil, J. Brams.

Riddle J.M. and Mulholland J. W., 'Albertus Magnus on Stones and Minerals', in *Albertus Magnus and the Sciences*, ed. J.A. Weisheipl, Toronto 1980, pp. 203-204.

Robert P., 'St. Bonaventure, Defender of Christian Wisdom', *Franciscan Studies*, III, 1943, pp. 159-179.

Robinson P., 'The Seventh Centenary of Roger Bacon', *The Catholic University Bulletin*, 1914, fasc. I.

Roensch F.J., *Early Thomistic School*, Dubuque/Iowa 1964.

Roger Bacon Essays, ed. by A. G. Little, Oxford 1914.

Rouse R.H., 'Manuscripts Belonging to Richard de Fournival', *Revue d'histoire des textes*, III, 1973, pp. 253-269.

Ruggiero F., 'Intorno all'influsso di Averroè su S. Alberto Magno', *Laurentianum*, IV, 1963, pp. 27-58.

Russell J.L., 'St Thomas and the Heavenly Bodies' (review of Litt, *Les corps célestes*), *Heytrop Journal*, VIII, 1967, pp. 27-39.

Sabbadini R., *Le scoperte dei codici latini e greci ne' secoli XIV e XV*, Firenze 1967 (2nd ed.).

Saffrey H.D., 'S. Thomas et ses secrétaires. A propos du livre du R.P. A. Dondaine', *Revue des sciences philosophiques et théologiques*, XLI, 1957, pp. 49-74.

Salman D., 'Albert le Grand et l'averroisme latin', *Revue des sciences philosophiques et théologiques*, XXIV, 1935, pp. 38-64.

Salman D., 'Saint Thomas et les traductions latines des *Metaphysiques*', *Archives d'Histoire doctrinale et littéraire du Moyen Age*, VII, 1932 (but 1933), pp.77-122.

Salembier L., *Petrus ab Alliaco*, Lille 1886.

Saxl F., *Verzeichnis astrologischer und mythologischer illustrierten Handschriften der National-Bibliothek in Wien*, Hamburg 1927.

Sbaralea J. H., *Supplementum et castigatio ad Scriptores trium Ordinum S. Francisci*, Roma 1806.

Scheeben H. C., 'Albert der Grosse und Thomas von Aquin in Köln', *Divus Thomas* (Freiburg), IX, 1931, pp. 28-34.

Scheeben H. C., *Albert der Grosse. Zur Chronologie seines Lebens*, Vechta, Albertus Magnus Verlag, 1931 (= Quellen und Forschungen zur Geschichte des Dominikanerordens in Deutschland, gegr. v. P. von Loë, Bd. 27).

Scheeben H. C., *Albertus Magnus*, Köln, Bachem, 1955 (2nd ed.): cfr. pp.197-224: 'Appendix. Albertus in der Legende. Albertus als Magier'.

Scheeben H. C., *De Alberti Magni discipulis*, Roma, 1932.

Scheeben H. C., 'Les écrits d'Albert le Grand d'après les catalogues' *Revue Thomiste*, XXXVI, 1931, pp.260-292.

Scheeben H. C., *'Zur Chronologie des Lebens Alberts des Grossen'*, *Divus Thomas* (Freiburg), X, 1932, pp. 231-245.

Schmitt C. and Knox D., *Pseudo-Aristoteles Latinus: A Guide to Latin Works falsely attributed to Aristoteles before 1500*, London, The Warburg Institute, 1985.

Schneider J. M., 'Aus Astronomie und Geologie des hl. Alberts des Grossen', *Divus Thomas* (Freiburg), X, 1932, pp. 52-60.

Schneider T., *Die Einheit des Menschen*, Münster 1973 (= Beiträge zur Geschichte der Philosophie und Theologie des Mittelalters, N. F., 8).

Schneyer J. B., 'Predigten Alberts des Grossen in der Handschrift Leipzig Universitätsbibliothek 683', *Archivum Fratrum Praedicatorum*, XXXIV, 1964, pp. 45-106.

Schooyans M., 'La distinction entre philosophie et théologie d'après les commentaires aristotéliciens de s. Albert le Grand', *Rivista da Universidade Católica de Sao Paulo*, XVIII, 1959, pp. 255-279.

Scritti pubblicati in occasione del VII centenario della nascita di R. Bacone, A. Gemelli ed., Special Issue of the *Rivista di filosofia neoscolastica*, VI, 1914 (publ. also under the title of *Il VII centenario di R. Bacone. Studi e commenti*, Firenze, 1914).

Seidler E., 'Die Medizin in der *Biblionomia* des Richard de Fournival', *Südhoff's Archiv*, 51, 1967, pp.44-54.

Serta albertina, Special Issue of *Angelicum*, XXI, 1944.

Sezgin F., *Geschichte des arabischen Schriftums*, V, Leiden, 1974.

Sherwood Taylor F., *St. Albert Patron of Scientists*, Oxford 1950 (= Aquinas Pages, n. 14), pp. 3-14.

Sighart J., *Albertus Magnus*, Regensburg 1857.

Siraisi N., 'The Medical Learning of Albertus Magnus', in *Albertus Magnus and the Sciences*, ed. J.A. Weisheipl, Toronto 1980, pp. 379-403.

Siraisi N.G., *Arts and Sciences at Padua: The 'Studium' of Padua before 1350*, Toronto, Pontifical Institute for Medieval Studies, 1973.

Stabile G., *s.v.* 'Biondo, Michelangelo', in *Dizionario biografico degli italiani*, X, Roma 1968, pp. 560-563.

Steele R., 'R. Bacon and the State of Science in the XIII Century', in *Studies in the History and Method of Science*, ed. C. Singer, Oxford 1921, II, pp. 121-150.

Stein G., 'S. Alberto Magno e l'astronomia', *Angelicum*, XXI, 1944, pp. 182-191.

Steinschneider M., 'Die europäischen Übersetzungen aus dem Arabischen', in *Sitzungsberichte der K. Akademie der Wissenschaften in Wien*, Ph.-Hist. Kl., Bd. 151, 1906.

Steinschneider M., 'Zum *Speculum astronomiae* des Albertus Magnus, über die darin angeführten Schriftstellern und Schriften', *Zeitschrift für Mathematik und Physik*, XVI, 1871, pp. 357-396.

Steinschneider M., 'Zur Geschichte der Übersetzungen aus dem Indischen ins Arabische', *Zeitschrift der Deutschen morgenländischen Gesellschaft*, XXV, 1871, pp. 378-428.

Steneck N.H., 'Albert the Great on the Classification and Localization of the Internal Senses', *Isis*, LXV, 1974, pp. 193-211.

Strunz F., *Albertus Magnus. Weisheit und Naturforschung im Mittelalter*, Wien 1926.

Strunz F., 'Albertus Magnus und die Naturforschung des Mittelalters', in *Das Mittelalter in Einzeldarstellung*, hg. v. H. Leitmeier, Wien 1930.

Strunz F., *Astrologie, Alchimie, Mystik: ein Beitrag zur Geschichte der Naturwissenschaften*, München 1928.

Studia albertina. Festschrift für B.Geyer zum 70. Geburtstag, hg. v. H. Ostlender, Münster 1952 (= Beiträge zur Geschichte der Philosophie und Theologie des Mittelalters, Supplementband IV).

Sturlese L., review of Albertus Magnus, *Speculum astronomiae*, in *Annali della Scuola Normale Superiore di Pisa*, Cl. di Lett., s. III, vii, 1977, p. 1616.

Sturlese L., 'Note su Bertoldo di Moosburg O.P. (III.'La mano di Bertoldo e gli autografi di Alberto')', *Freiburger Zeitschrift für Philosophie und Theologie*, XXXII, 1985, pp. 257-259
Sturlese L., 'Il *De animatione caeli* di Teodorico di Freiberg', in *Xenia Medii Aevi Historiam illustrantia oblata T.Kaeppeli O.P.*, Roma 1978, pp. 175-247.
Sturlese L., *s.v.* 'Dietrich von Freiberg', in *Deutsche Literatur des Mittelalters. Verfasserlexicon*, II, Berlin 1979, pp. 127-137.
Sturlese L., 'Saints et magiciens: Albert le Grand en face d'Hermès Trismegiste', *Archives de philosophie*, XLIII, 1980, pp. 615-634.
Tannery P., *Mémoires scientifiques*, IV, Paris 1920.
Teetaert A., 'Deux Questions inédites de Gérard d'Abbeville en faveur du clergé séculier', in *Mélanges Auguste Pelzer*, Louvain 1947, pp. 347-388.
Théry G., 'Note sur l'aventure bélénienne de Roger Bacon', *Archives d'histoire doctrinale et littéraire du Moyen Age*, XVIII, 1950-51, pp. 129-147.
Thomassen B., *Metaphysik als Lebensform. Untersuchungen zur Grundlegung der Metaphysik im Metaphysikkommentar Alberts des Grossen*, Münster, Aschendorff, 1985 (= Beiträge zur Geschichte der Philosophie und Theologie des Mittelalters, N.F. 27).
Thorndike L., 'A Bibliography composed around 1300 A. D. of Works in Latin on Alchemy, Geometry, Perspective, Astronomy and Necromancy', *Zentralblatt für Bibliothekswesen*, LV, 1938, pp. 357-360.
Thorndike L., *A History of Magic and of Experimental Science*, New York, Columbia U.P., 1923-1958, 8 vols.
Thorndike L., 'Aegidius of Lessines on Comets', in *Studies and Essays in the History of Science and Learning offered to G. Sarton*, New York 1946, pp. 403-414.
Thorndike L., 'Albumasar in Sadan', *Isis*, XLV, 1954, pp. 22-32.
Thorndike L., 'Further consideration of the *Experimenta, Speculum Astronomiae* and *De secretis mulierum* ascribed to Albertus Magnus', *Speculum*, XXX, 1955, pp. 413-443.
Thorndike L., 'John of Seville', *Speculum*, XXXIV, 1959, pp. 20-38.
Thorndike L., ed., *Latin Treatises on Comets between 1238 and 1368 A. D.*, Chicago 1945.
Thorndike L.,'Notes on some astronomical, astrological and mathematical Mss. of the Bibliothèque Nationale, Paris', *Journal of the Warburg and Courtauld Institutes*, XX, 1957, pp. 142-152.
Thorndike L., 'Notes upon some mediaeval Manuscripts', *Isis*, L, 1959, pp. 45-46.
Thorndike L., 'Notes upon some Mediaeval Latin Astronomical, Astrological and Mathematical Manuscripts at the Vatican Library', *Isis*, XLIX, 1958, pp. 34-49.
Thorndike L., 'Pierre de Limoges on the Comet of 1299', *Isis*, XXXVI, 1945-1946, pp. 3-7.
Thorndike L., 'Roger Bacon and the Experimental Method in the Middle Ages', *Philosophical Review*, XXIII, 1914, pp. 271-298.
Thorndike L., 'Some little known astronomical and mathematical manuscripts', *Osiris*, VIII, 1948, pp. 41-72.
Thorndike L., 'The latin Translation of astrological Works by Messahala', *Osiris*, XII, 1956, pp. 49-72.
Thorndike L., 'Traditional Medieval Tracts concerning engraved astrological Images', in *Mélanges Auguste Pelzer*, Louvain 1947, pp. 217-274.
Thorndike L., *Michael Scot*, London 1965.
Toomer G.J., *s.v.* 'Campanus', in *Dictionary of Scientific Biography*, III, New York 1971, pp. 23-29.

Tinivella F., 'Il metodo scientifico in S. Alberto Magno e Ruggero Bacone', *Angelicum*, XXI, 1944 (= *Serta albertina*), pp.65-83.

Tschackert P., *Peter von Ailly*, Gotha 1877.

Tummers P.M., 'Albertus Magnus' View on the Angle with Special Emphasis on His Geometry and Metaphysics', *Vivarium*, XXII/1 (1984), pp. 35-36.

Tummers P.M., 'The Commentary of Albert on Euclid's Elements of Geometry', in *Albertus Magnus and the Sciences*, ed. J.A.Weisheipl, Toronto, 1980, pp.479-500.

Ueberweg F. and Geyer B., *Die Geschichte der patristischen und scholastischen Philosophie*, Berlin 1928.

Ullmann B.L., 'The Library of the Sorbonne in the XIVth Century' in *The Septicentennial Celebration of the Funding of the Sorbonne*, Chapel Hill 1963, pp. 33-47.

Ullmann B.L., 'The Sorbonne Library and the italian Renaissance', in his *Studies in the italian Renaissance*, Roma 1955, pp. 41-53.

Van de Vyver A., 'Les plus anciennes traductions latines medievales (Xe-Xie siècles)des traités d'astronomie et d'astrologie', *Osiris*, I, 1936, pp. 657-691.

Vandewalle B., 'Roger Bacon dans l'histoire de la philologie, Roger Bacon et le *Speculum astronomiae*', *La france franciscaine*, XII, 1929 , pp. 178-196.

Vanni Rovighi S., 'Alberto Magno e l'unità della forma sostanziale nell'uomo', in *Medioevo e Rinascimento. Studi in onore di B. Nardi*, Firenze 1955, pp. 753-778.

Van Steenberghen F., 'Deux monographies sur la synthèse philosophique de Saint Thomas', *Revue philosophique de Louvain*, LXI, 1963, pp. 82-91.

Van Steenberghen F., 'La filosofia di Alberto Magno', *Sapienza*, XVIII (1965), pp. 381-393.

Van Steenberghen F., *La Philosophie au XIIIe siècle*, Louvain, Publ. Universitaires-Paris, B. Nauwelaerts, 1966.

Van Steenberghen F, 'Le *De quindecim problematibus* d'Albert le Grand', in *Mélanges Auguste Pelzer*, Louvain 1947, pp. 415-440.

Vansteenkiste C., 'Autori arabi e giudei nell'opera di S. Tommaso', *Angelicum*, XXXVII, 1960, pp. 336-401.

Vansteenkiste C., 'Il quinto volume del nuovo Alberto Magno', *Angelicum*, XXXIX, 1962, pp. 205-220.

Vansteenkiste C., *s.v.* 'Giles of Lessines', *New Catholic Encyclopaedia*, VI, New York 1967, p. 484.

Verbeke G., *Het wetenschappelijk Profiel van Willem van Moerbeke*, Amsterdam–London, B.N. Noord-hollandsche Uitgevers Maatschappij, 1975 (= Mededelingen der K. Nederlanse Akademie van Wetenschappen. AFD. Letterkunde. Nieuwe Reeks, Deel 38–N.4).

Vernet J., *La cultura hispano-arabe en Oriente y Occidente*, Barcelona-Caracas-Mexico 1978.

von Loë P., 'Albert der Grosse auf dem Konzil von Lyon', *Literarische Beilage der kölnischen Volkszeitung*, LV/29, 1914, pp. 225-226.

von Loë P., 'De vita et scriptis B. Alberti Magni', *Analecta Bollandiana*, XIX, 1900, pp. 257-284; XX, 1901, pp. 273-316; XXI, 1902, pp. 361-371.

Vuillemin-Diem G., 'Die Metaphysica media. Uebersetzungsmethode und Textverständnis', *Archives d'histoire doctrinale et littéraire du Moyen Age*, 42, 1975 [but 1976], pp. 7-69.

Vuillemin-Diem G., 'Jacob von Venedig und die Übersetzer der *Physica Vaticana* und *Metaphysica Media*', *Archives d'histoire doctrinale et litteraire du Moyen Age*, XLI, 1974 (but 1975), pp. 7-25.

Vuillemin-Diem G., 'Les traductions greco-latines de la *Métaphysique* au Moyen-Age', *Archiv für die Geschichte der Philosophie*, XLIX, 1967, pp. 7-71.

Vuillemin-Diem G., 'Recensio Palatina und Recensio Vulgata. Wilhelm von Moerbeke doppelte Redaktion der Metaphysikübersetzung', *Aristotelische Erbe im Arabisch-lateinischen Mittelater*, hg. v. A. Zimmermann, Berlin-New York, W. de Gruyter, 1986, pp.289-366.

Wagner C., 'Alberts Naturphilosophie im Licht der neueren Forschung (1979-1983)', in *Freiburger Zeitschrift für Philosophie und Theologie*, XXXII, 1985 (= *Albert der Grosse und die deutsche Dominikanerschule*), pp. 65-104.

Wallace W.A., s. v. 'Albert' in *Dictionary of Scientific Biography*, I, New York 1970, pp.99-101.

Wallace W.A., *The scientific methodology of Theodoric of Freiberg*, Fribourg Schwz. 1959.

Walz A., 'L'opera scientifica di Alberto Magno secondo le indagini recenti', *Sapienza*, V, 1952, pp.442-452.

Walz A., *Saint Thomas d'Aquin*, Louvain-Paris 1962.

Wedel Th. O., *The Medieval Attitude Towards Astrology Particularly in England*, New Haven 1920 (= Yale Studies in English, LX).

Weisheipl J. A., 'Albertus Magnus and the Oxford Platonists', *Proceedings of the American Catholic Association*, XXXII, 1958, pp. 124-139.

Weisheipl J. A., *s. v.* 'Albert' in *New Catholic Encyclopaedia*, I, New York 1967, pp. 254-258.

Weisheipl J. A., 'The Axiom "Opus naturae est opus intelligentiae"', *Albertus Magnus, Doctor Universalis. 1280-1980*, hg. v. G. Meyer und A. Zimmermann, Mainz , M. Grünewald, 1981, pp. 441-464.

Weisheipl J. A., 'The Celestial Movers in Medieval Physics', in *The Dignity of Science. Studies in the Philosophy of Science Presented to W.H. Kane*, ed by J.A. Weisheipl, Washington 1961, pp. 150-190.

Weisheipl J. A., 'The Celestial Movers in Medieval Physics', *The Thomist*, XXIV, 1961, pp. 286-326.

Weisheipl J. A., 'The Commentary of St Thomas on the *De Caelo* of Aristotle', *Sapientia*, XXIX, 1974, pp. 11-34.

Weisheipl J. A., 'The *Problemata determinata XLIII* ascribed to Albertus Magnus', *Medieval Studies*, XXII, 1960, pp. 323-327.

Weisheipl J. A., 'The Life and Works of St. Albert the Great'; 'Albert's Works on Natural Sciences (*libri naturales*) in Probable Chronological Or', in *Albertus Magnus and the Sciences*, ed. by J. A. Weisheipl, Toronto 1980, pp. 13-53, 565-577.

Weisheipl J. A., *Thomas d'Aquino and Albert his Teacher*, Toronto, Pontifical Institute of Mediaeval Studies, 1980 (= The E.Gilson Series, 2).

Werminghoff W., 'Die Bibliothek eines Konstanzer Officials [Johann von Kreuzlingen, J.U.D.] aus dem Jahre 1506', *Zentralblatt für Bibliothekswesen*, XIV, 1897, pp. 290-298.

Wicki N., *s.v.* 'Philip der Kanzler und die Pariser Bischofswahl von 1227-1228', *Freiburger Zeitschrift für Philosophie und Theologie*, V, 1958, pp.318-326.

Wicki N., *s.v.* 'Philip der Kanzler', *Lexicon für Theologie und Kirche*, VIII, Freiburg, Herder, 1963, pp. 452-453.

White L., 'Medical Astrologers and late Mediaeval Technology', *Viator*, VI, 1975, pp. 297-308.

Wielockx R., 'Le ms. Paris. lat. 16096 et la condemnation du 7 mars 1277', *Recherches de théologie ancienne et médievale*, XLVIII, 1981, pp.227-237.

Wielockx R., 'La censure de Gilles de Rome', *Bulletin de philosophie médievale*, XXII, 1980, pp. 87-88.

Wippel J. F., 'The Condemnations of 1270 and 1277 at Paris', *The Journal of Mediaeval and Renaissance Studies*, VII, 1977, pp. 169-202.

Wippel J.F., *The Metaphysical Thought of Geoffrey of Fontanes. A Study in the late 13th Century Philosophy*, Washington 1981.

Witzel T., *s.v.* 'Roger Bacon', *The Catholic Encyclopedia*, XIII, New York 1913, pp. 111-116.

Wolfson, H.A., 'The Plurality of Immovable Movers in Aristotle and Averroes', *Harvard Studies in Classical Philology*,LXIII, 1958, pp.233-253.

Wyckoff D., 'Albertus Magnus on Ore Deposits'. *Isis*, 49, 1958, pp. 109-122.

Zambelli P., 'Albert le Grand et l'astrologie', *Recherches de théologie ancienne et médiévale*, XLIX, 1982, pp. 141-158.

Zambelli P., 'Da Aristotele a Abu Ma'shar, da Richard de Fournival a Guglielmo da Pastrengo, un'opera controversa di Alberto Magno', *Physis*, 1974, pp. 26. (see the same paper also printed under the title 'Per lo studio dello *Speculum astronomiae*', in *Actas del 5° Congreso internacional de Filosofia Medieval* [1972], Madrid 1979, II, pp. 1377-1391.)

Zambelli P., 'Il problema della magia naturale nel Rinascimento', *Rivista critica di storia della filosofia*, XXV, 1973, pp. 271-296; now reprinted in her *L'ambigua notura della magia*, Milano, Il Saggiatore, 1991, pp. 121-152.

Zambelli P., 'Mediaeval and Renaissance Hermetists versus the Problem of Witchcraft', in *Hermeticism and the Renaissance*, eds. A. G. Debus and I. Merkel, Washington, Associated University Presses, 1988, pp. 125-151

ADDENDUM 1992

Faes de Mottoni B., 'Bonaventura e la caduta degli angeli', *Doctor seraphicus*, XXXVIII, 1991, pp. 97-113.

Gregory T., 'Théologie et astrologie dans la culture médiévale: un subtil face-à-face', *Bulletin de la Société française de Philosophie*, LXXXIV, 1990 (but 1991), pp. 101-130.

INDEX OF NAMES

* As usual authors who lived until the 15th Century are listed under their first name uniformed under the modern english spelling. Names of the supposed authors of pseudoepigraphic writings are also listed. Documents extensively quoted are also indexed: but it proved impossible to register every peculiar spelling of the names there quoted.

INDEX OF MANUSCRIPTS

MIX
Papier aus verantwortungsvollen Quellen
Paper from responsible sources
FSC® C105338

If you have any concerns about our products,
you can contact us on
ProductSafety@springernature.com

In case Publisher is established outside the EU,
the EU authorized representative is:
Springer Nature Customer Service Center GmbH
Europaplatz 3, 69115 Heidelberg, Germany

Printed by Libri Plureos GmbH
in Hamburg, Germany